Outstanding Contributions to Archaeology

Series Editor

Harold Mytum, University of Liverpool, Archaeology, Classics and Egyptology, Liverpool, UK

Outstanding Contributions to Archaeology enables a focus on the work of an individual or research group in contributing important advances in archaeological research. Each volume is devoted to a major contribution by an eminent archaeologist. The series will cover contributions to archaeology with a thematic, period or regional focus, and can emphasise their theoretical, methodological or interpretive contribution.

A typical volume of Outstanding Contributions to Archaeology contains:

- A short autobiography by the archaeologist to whom the volume is devoted (or if they are deceased, an equivalent biography), including personal, wider societal and professional aspects.
- The volume editor's introduction – a survey that puts the archaeologist's contributions in context, discusses its importance, and shows how it connects with work by other scholars.
- The main part of the book will consist of a series of chapters by different scholars that link their work to that of the archaeologist being celebrated. These should include relevant biographical and contextual material that demonstrates the way in which archaeological understandings have developed particular historical and social contexts, linked to the celebrated archaeologist. A feschrift-style volume with diverse chapters by those who know the archaeologist but which do not link the contributions to their biography will not be accepted.
- Response to the contributions by the archaeologist to whom the volume is devoted, or if they are deceased by the volume editor.
- List of publications by the archaeologist being celebrated.

For more information on submitting a proposal to the series, please contact the Series Editor, Harold Mytum (h.mytum@liverpool.ac.uk), or the Publishing Editor, Christi Lue (christi.lue@springernature.com).

Arkadiusz Marciniak • Michał Pawleta
Wlodzimierz Rączkowski
Editors

The Poznań School of Archaeology

The Origin, Growth and Significance

Springer

Editors
Arkadiusz Marciniak
Faculty of Archaeology
Adam Mickiewicz University
Poznań, Poland

Michał Pawleta
Faculty of Archaeology
Adam Mickiewicz University
Poznań, Poland

Wlodzimierz Rączkowski
Faculty of Archaeology
Adam Mickiewicz University
Poznan, Poland

ISSN 3059-3093 ISSN 3059-3107 (electronic)
Outstanding Contributions to Archaeology
ISBN 978-3-032-07847-6 ISBN 978-3-032-07848-3 (eBook)
https://doi.org/10.1007/978-3-032-07848-3

This Springer imprint is published by the registered company Springer Nature Switzerland AG
The registered company address is: Gewerbestrasse 11, 6330 Cham, Switzerland

Contents

List of Figures

List of Tables

The Poznań School of Archaeology: An Outline

Arkadiusz Marciniak, Michał Pawleta, and Włodzimierz Rączkowski

Abstract The chapter serves as an introduction to a volume dedicated to the Poznań School of Archaeology (PSA), which is the most distinctive, original, and rigorous school of archaeological thought in Poland, developed after the Second World War. Emerging in the 1970s from the Poznań School of Methodology, the PSA has been continuously practiced by three generations of archaeologists. The chapter details the core philosophical and methodological tenets defining the PSA such as a rigorous epistemology, a sistematic and explicitly designed methodology, rejection of simplistic functional explanations. It stresses the School emphasis on the leading role of theory in scientific research, and promotion of interdisciplinary collaboration. The chapter further presents several key concepts and influences from its intellectual roots in the Poznań School of Methodology, including idealizational theory of science, humanistic interpretation, socio-regulative theory of culture, functional-genetic explanation, and logical reconstruction. The chapter concludes by presenting major figures responsible for the emergence of Poznań School of Archaeology, in particular Jan Żak, Stanisław Tabaczyński, Danuta Minta-Tworzowska, and Anna Pałubicka.

Keywords Poznań School of Methodology · History of archaeological thought · Methodology · Epistemology · Socio-regulative theory of culture · Idealizational theory of science

A. Marciniak (✉) · M. Pawleta · W. Rączkowski
Faculty of Archaeology, Adam Mickiewicz University in Poznań, Poznań, Poland
e-mail: arkadiusz.marciniak@amu.edu.pl; michal.pawleta@amu.edu.pl; wlodzimierz.raczkowski@amu.edu.pl

A. Marciniak et al. (eds.), *The Poznań School of Archaeology*, Outstanding Contributions to Archaeology, https://doi.org/10.1007/978-3-032-07848-3_1

1 Introduction

Schools of research are defined by distinct and coherent ways of understanding the reality under study, an explicitly defined epistemology and corresponding methodologies. In Knud E. Jørgensen's (2015, 1) words, they also imply "broad trajectories of scholarship, distinct dialectics of continuity and change as well as the gradual establishment of a scholarly infrastructure". The milieu of their promoters and practitioners is characterised by their shared research questions, the means of their analysis, including the choice of methods, the character of the research process and the nature of the explanations sought. Various research traditions imply different assumptions and diverse ways of looking at measurement and social science research. Research traditions provide the general framework within which more specific analytical interventions take place, thereby constituting a field of study.

The Poznań School of Archaeology is the most distinctive, original and rigorous school of archaeological thought in Poland that developed after the Second World War. It developed in the milieu of academic archaeology in Poznań in the 1970s and has been continuously practised by three generations of archaeologists. It emerged in the tradition of the Poznań School of Methodology (hereinafter referred to as PSM), adopting its various elements as manifested in the works of its main founders, including Jerzy Kmita, Leszek Nowak and Jerzy Topolski. It is characterised by the rigorous epistemology developed within the framework of Leszek Nowak's idealised conception of science and Jerzy Kmita's humanistic interpretation and socio-regulatory theory of culture. It is characterised by a rigorous and explicitly designed methodology as an intrinsic element of any archaeological endeavour, and an explicit and multifaceted interest in the nature of the archaeological process. It proposed an insight rooted in Marxism but offered an expanded epistemology. It rejected the simplistic functional and rigid deductive-nomological explanations but remained structural. It focused on historical epistemology, non-Marxist historical materialism and the original concept of culture. In the following decades, it also incorporated various elements of major research schools, especially from the Anglo-Saxon area, and tried to integrate elements of these schools into the tradition that originally constituted the Poznań School of Archaeology.

Among the most important facets of the Poznań School of Archaeology are the intellectual roots and the origin of the discipline of archaeology, the character of archaeological evidence, the conceptualisation of spatial behaviour and the religious and symbolic character of human groups, the social context of practising archaeology, and the meta-reflection of archaeology's presence in the contemporary world and its future. It has also been institutionalised, in particular in the form of the Department of History and Methodology of Prahistory at the Institute of Prahistory, student conferences organised by members of the Poznań School of Archaeology, seminars, conferences and most recently the Poznań Academic Archaeology Seminars (PASA). These achievements translated themselves into a commonly held opinion about the methodological strength of academic training at the Adam Mickiewicz University in Poznań. Over the decades it developed into an original school of archaeological thought, embedded in the research traditions of cultural studies, philosophy, history and the natural sciences. It became the only distinct school of archaeological thought in Poland. It made Poznań the dominant research centre in archaeological theory and methodology, continuously setting the agenda in the country.

One of the foremost proponents of the Poznań School of Archaeology is Danuta Minta-Tworzowska. She has played a pivotal role in shaping and advancing the School's intellectual direction. By attending seminars conducted by Jerzy Kmita, she became well-versed in the foundational ideas of the PSM, which profoundly influenced her approach to archaeology. Minta-Tworzowska not only absorbed these ideas but also innovatively applied them to the context of Polish archaeology. Her extensive works on the theory and methodology of archaeology (Żak and Minta-Tworzowska 1991; Minta-Tworzowska 1996, 2000a, 2002, 2006, 2015, 2017, 2021a, 2022), classification and typology as reflections on research procedures (Minta-Tworzowska 1994), the conceptualization of archaeological sources (Minta-Tworzowska 1998a, 2000b), and the study of symbols and symbolism in archaeological research (Minta-Tworzowska 2000c, 2008, 2019) have led to the development of a unique approach to archaeological methodology, significantly strengthening the theoretical foundations of Polish archaeology. Her rigorous scholarship and dedication embody the values of the Poznań School of Archaeology, establishing her as a central figure in its history and ongoing success.

The Poznań School of Archaeology is one of the European traditions of archaeological thought, relatively unknown to the wider research community. Evžen Neustupný (1997/1998) almost 30 years ago would classify it as a minority archaeology, mainly because of its negligible impact on world archaeology. The potential impact has also been undermined by several factors: a relatively small community of its practitioners, the use of the Polish language in publications, a relatively low level of institutionalisation, and the dominance of the cultural-historical school imposed on the structure of Polish archaeology, including the educational system. Additionally, political impositions and the use of the German tradition as a means of developing and strengthening the role of some archaeologists have also played a role.

However, the division of research traditions into minority and majority, as proposed by Neustupný, is no longer viable in the period of the post-paradigm phase in the history of archaeological thought and an increasing democratisation of academic practices. Furthermore, the Poznań School of Archaeology contradicts Neustupný's thesis that archaeology in the countries of Central Europe historically developed within the milieu of the German tradition. The School itself, as well as other research traditions in Polish archaeology, such as bioarchaeology (Marciniak 2015, 2018) and ethnoarchaeology (Kobyliński 2012; Marciniak and Yalman 2013), developed independently of German intellectual currents. The Poznań School of Archaeology must therefore be seen as a distinct, original and powerful mode of archaeological practice, and as such an important element of European archaeological thought.

2 The Theoretical and Methodological Context of the Poznań School of Archaeology

The Poznań School of Archaeology developed within the framework of the Poznań School of Methodology. The latter is a philosophical and scientific movement known for its significant and original contributions to the philosophy and methodology of science. It emerged as a post-war trend in Polish philosophy and methodology that rejected the so-called "humanistic interpretation of Marxism" and aimed to incorporate elements of positivism and the achievements of the Lviv-Warsaw School into Marxist philosophy. The PSM focused on epistemological and methodological issues of science, with a particular interest in historical epistemology, the methodology of the humanities and their philosophical foundations, considered from a Marxist point of view using analytical methods.

The PSM has made significant contributions to the philosophy of science, particularly in the areas of theory confirmation, scientific explanation, and the structure of scientific theories. It has investigated how scientific knowledge is developed, justified and structured. Since its theoretical foundations have been discussed in detail elsewhere (e.g. Kotowa 2010; Zamiara 2010; Musiał 2016), and in the chapters of this volume (Brzechczyn, chapter "Poznań School of Methodology: Institutional History – Research Program – Main Achievements"; Grad, chapter "The Poznań School of Methodology and Its Continuations"), we will only briefly highlight the elements that are important from the perspective of the development of the Poznań School of Archaeology.

The PSM emerged at the turn of the 1960s and 1970s at the Adam Mickiewicz University in Poznań and was centred around a group of eminent researchers. Its first pillar was Marxist philosophy, but not in the form of official Marxism, as a reference to the writings of Marx himself. The second pillar was the philosophy of science, combining the influences of Karl Popper and Kazimierz Ajdukiewicz. The third was the German anti-naturalist methodology of the humanities, such as Wilhelm Dilthey and Max Weber. These different currents, previously seen as antagonistic, became a source of originality for the school, but eventually led to its division. In the 1970s, some researchers around Leszek Nowak focused on the idealizational theory of science (Nowak 1980). At the same time, other researchers associated with Jerzy Kmita concentrated on the methodological foundations of the humanities, developing the concept of humanistic interpretation and later the socio-regulatory concept of culture (e.g. Kmita 1971, 1973). These were further strengthened by Jan Such and his collaborators, who studied the Marxist-dialectical characteristics of the scientific worldview.

Although the various facets of the PSM developed differently, it maintained the original line of thought and methodological project, based on several theoretical and cognitive foundations. As Jan Grad (chapter "The Poznań School of Methodology and Its Continuations") notes, the most distinctive elements of this research tradition included the application of cognitive categories such as the premise of rationality, humanistic interpretation, idealization, logical reconstruction, and functional-genetic explanation.

Firstly, the PSM emphasises the importance of empirical and formal methods in scientific research and advocates a rigorous approach to the study of scientific methods and the logical structure of scientific theories. Secondly, it focuses on the logical analysis of scientific language and theories. This involves studying the logical syntax and semantics of scientific statements and theories to clarify their meaning and structure. Thirdly, it pursues 'logical reconstruction', which involves the identification of various assumed and often unrecognised premises and theoretical underpinnings of the research process, and recognising their impact on the results obtained. It has been recognised as a powerful methodological tool for understanding and evaluating theories across disciplines. By focusing on the rational core and ensuring systematic coherence, logical reconstruction enhances epistemological clarity and promotes interdisciplinary collaboration. For this reason, the PSM promotes an interdisciplinary approach to methodological problems, integrating various fields of science such as philosophy, logic, mathematics and natural sciences. Fourth, the Poznań School of Methodology advocates the concept of humanistic interpretation, which seeks to bridge the gap between the natural sciences and the humanities. This concept represents a sophisticated and nuanced approach to understanding human culture. Finally, the School introduced the socio-regulatory theory of culture, which provides a comprehensive understanding of how culture functions as a regulatory system within society. It emphasises the importance of norms and regulations in maintaining social order and the adaptive nature of culture in response to changing circumstances of its persistence. This concept has been successfully adopted by archaeologists.

The concept of humanistic interpretation is one of the main achievements and characteristic features of the PSM. It emphasises the importance of context, meaning and intentionality, and advocates a methodology that respects the unique qualities of human cultural phenomena. It takes the form of a subjective-rational explanation. It assumes the rationality of human action and relies on deduction, where the conclusion must follow logically from the premises. It challenges the dominance of positivist methodologies and offers a richer, more empathetic understanding of human action and creation. While it departs from heuristic relativism, it is similar to inference in the natural sciences. The mechanism of objectifying knowledge itself is also similar to that of the natural sciences.

The PSM has had a lasting impact on the philosophy of science in the Polish social sciences and humanities. It has influenced various fields, including the methodology of empirical sciences, the theory of language, and the philosophy of social sciences. Its emphasis on rigorous analysis and interdisciplinary research continues to inspire contemporary philosophical and scientific inquiry. It has also influenced archaeology and inspired archaeologists.

3 The Emergence of the Poznań School of Archaeology

The PSM had a profound influence on Polish archaeologists, mainly based in Poznań and working at the Adam Mickiewicz University and the Polish Academy of Sciences. Its influence on the discipline of archaeology was mainly inspired by the work of the philosophers Jerzy Kmita and Leszek Nowak and the historian Jerzy Topolski.

The foundations of the Poznań School of Archaeology were laid by the distinguished prehistorian and medievalist Jan Żak (1923–1990). He was strongly opposed to positivist archaeology in its evolutionist-diffusionist version, which was associated with the dominant cultural-historical approach. Consequently, he turned to the Poznań School of Methodology, which was growing dynamically at the time, as the only alternative to positivism in Poland at the time. He aimed to bring the insights and findings of the philosophers of this school into the field of Polish archaeology. This included the main tenets of the Poznań Methodological School, such as humanistic interpretation, functional-genetic explanation and the idealizational concept of science. He aimed to modernise archaeology and establish it as an academic discipline based on a rigorous methodological framework and an explicit and in-depth theoretical foundation (Minta-Tworzowska 2001, 2023).

Żak was particularly inspired by Jerzy Kmita and wanted to introduce a concept of humanistic interpretation into archaeology. He advocated its heuristic potential for explaining various phenomena of the past, both synchronically and diachronically. His main research interests were the Early Middle Ages in Scandinavia, the ethnogenesis of the Slavs and settlement studies, a field in which empirical research dominated the archaeology of his times. In particular, he attempted to apply this concept to the study of the continuity or discontinuity of the Slavs' presence concerning their economic position in the fifth and sixth centuries AD (Żak 1974, 1975b, 1985). Żak also stressed the importance of the functional genetic method in explaining the historical consciousness of prehistoric societies (Żak 1975a). He used this framework to explain this way of thinking among hunter-gatherer societies in the Mediterranean area. He saw the consciousness of these communities as both a reflection of their living conditions and a regulator of their lives, intertwined with clan and tribal institutions, and described it as naïve-naturalistic.

Żak was not only a vivid advocate of the ideas developed by the PSM, but also made an important contribution to the institutionalisation of these efforts. As a result, in 1982 the Department of History and Methodology of Prehistory was established at the Adam Mickiewicz University in Poznań—the only such department and university organisational and educational unit of its kind in Poland (Mamzer 2020, 110). Since then, the methodology of archaeology and/or prehistory has become an important aspect of the archaeology curriculum at the Poznań centre (Minta-Tworzowska and Rączkowski 1996, 2007; Minta-Tworzowska 2001, 2002, 2017, 2021b, 2023; Minta-Tworzowska and Pawleta 2013).

The ideas, scientific problems and questions addressed within the department were continued and developed by J. Żak's colleagues and successors, both during his lifetime and especially after his death. His students, and later generations of researchers, sought to answer the question explicitly posed by Henryk Mamzer (2020): why do we need the methodology and theory of archaeology? Specifically, what is 'theory' in archaeology, what is its relationship to the practice of the discipline, to the humanities and the intellectual world in general, and what role has it played in the development of Polish archaeology in recent decades?

Another important proponent of the PSM and influential archaeological theorist was Stanisław Tabaczyński (1930–2020) (Cyngot et al. 2020). He graduated from the university in Poznań, but worked throughout his academic career at the Institute of Archaeology and Ethnology of the Polish Academy of Sciences in Warsaw. He developed an original concept of the archaeological process, largely embedded in the Poznań School of Archaeology, and further blended with the Annales School and inspirations from the Italian academic archaeological milieu. He borrowed the concept of culture from Jerzy Kmita and debated with Jerzy Topolski on the concept of archaeological sources.

Tabaczyński's views were eloquently expressed in a co-authored work with the Poznań philosopher Anna Pałubicka, herself a member of the Poznań School of Archaeology (Pałubicka and Tabaczyński 1986). In this publication, they presented an original perspective on the subject of archaeological research and its methodological foundations. In particular, they advocated issues of historical explanation, humanistic interpretation and functional-genetic explanation as indispensable elements of archaeological practice. Their approach to the issues of social development and cultural change involved an analysis of two conceptual frameworks: the systemic approach and the Marxist approach, and illustrated the criteria for interpretive validity within these frameworks.

Tabaczyński also developed an original concept of culture (1987). Among its key elements were the process of creating material remnants of the past, long-term processes, ethnogenetic processes and the polysemantisation of culture. The possibility of achieving the proposed research goals was to be ensured by a coherent methodology of archaeological research. This methodology was characterised by the integration of two intertwined research perspectives: historical and anthropological. This led to the development of a particular way of understanding archaeology as the anthropology of the prehistoric past, stretching between anthropology and history.

The works of Jerzy Topolski, a historian and co-founder of the Poznań Methodology School, also significantly influenced the concept of archaeological research in Poznań (Topolski 1983, 1984, 1996). His main interests concerned the nature and practice of studies of the past (Drozdowski et al. 1988; Pomorski 1988; Wrzosek 2013; Domańska 2016; Domańska and Topolska 2022). Inspired by new interpretations of historical materialism, he advocated the primacy and primordiality of ontology, defining the object of research, over the research itself. It means that the research objectives define the selection of epistemological and methodological solutions. Hence, historical materialism provided a theoretical model of the historical process, both as a means of selecting and categorizing facts and as a research agenda. Its essential premise was an 'activist mode of the historical process', which broke with the emphasis on deterministic factors external to man. In this view, the historical process is made of humanistic components—intentional and conscious human actions. The dualistic nature of social reality, which consists of subjective human actions and their objective effects, prompted the adoption of 'humanistic interpretation' as a model to explain human actions. A further step aimed at strengthening our understanding of the research process involved the introduction of an 'integral explanatory model' that is a fusion of 'humanistic interpretation' and a deductive-nomological explanatory model. It made it possible not only to answer the question, 'Why did X act in this way and not in that way?', but also the question, 'How did X get to know certain knowledge and norms'. This approach made it possible to draw attention to two important aspects: (i) the explanation of the motivation of people's actions in the past (i.e., a definite departure from the concept of even history) and (ii) the modes of carrying out research by historians and archaeologists alike. As regards the latter, J. Topolski emphasised that the effect of research conducted by a historian depends not only on the knowledge of sources but also on 'extra-source knowledge', i.e. the non-verbalised, unconscious knowledge (e.g., the vision of the world and man, the understanding of science, etc.). To explain the phenomena of the past, it is required to have a coherent vision of the past (theoretical knowledge), and a comprehensive understanding of the complex nature of the research process. The so-called 'conscious control' requires the historian to be continuously aware on the becoming of the historical process and the role of the human being in this process. Hence, Topolski argued that one cannot talk about a single historical science (the same applies to archaeology) and that the criteria of their scientific character are not given but continuously changing.

The ideas pursued by J. Topolski have been well received by archaeologists attempting to move away from a positivist, cultural-historical practice of archaeology. Their impact can be identified in at least several research areas. From the beginning, critical reflection on the theoretical background of archaeological pursuits influenced the ways of researching the past (e.g., Żak 1974; Minta-Tworzowska 1994; Rączkowski 2002). These works showed how different modes of doing archaeology (posing questions, applying explanatory procedures, selecting modes of inference, etc.) were dependent on the theoretical assumptions advocated (consciously or not). The archaeologists were also inspired by Topolski's approach to historical sources. This led to the development of the new concept of archaeological source (e.g., Minta-Tworzowska 1998b; also Marciniak 1996) as interpreted in the context of contemporaneity, defined and specified by the archaeologist (see also Pawleta, chapter "Danuta Minta-Tworzowska and Her Role in Conceptualizing the Poznań School of Archaeology and Advancing the Methodology of Archaeology in Poland"). The collaboration of J. Topolski and J. Żak significantly influenced a new approach to settlement pattern studies in the form hitherto conceptualized in Polish archaeology (Żak 1977, 1985). This new proposal was in line with the 'activist conception of the historical process' and unequivocally dissociated itself from the environmental determinism inspired by the anthropogeography of Friedrich Ratzl that has dominated archaeological practice to date (Rączkowski 2001). No less important was J. Żak's and J. Topolski's collaboration, which aimed at the development of a new type of synthesis of the past. Consequently, the theoretical assumptions concerning the historical process shaped the character of the first, consciously theoretically designed, synthesis of the prehistory of the Polish lands (Żak 1975b).

Jerzy Topolski's original concepts, shaped at the time of the Poznań School of Methodology, were subject to change as a result of a deepening reflection on the research practice of historians, partly as a result of facing new views emerging in world history methodology (e.g., Georg Iggers, Hayden White). Without questioning the foundations of his thinking, J. Topolski introduced new themes. And this freshness of his thought constantly inspired archaeologists from the Poznań School of Archaeology to consciously take up new issues developed within the framework of new theoretical trends.

Another important contribution to the adoption of some elements of the Poznań School of Archaeology came from the philosophy of biology. In particular, it was inspired by the work of Krzysztof Łastowski, the Poznań biologist and philosopher, whose work focused on merging and comparing the theory of evolution with historical materialism. Łastowski is the author of a synthetic and naturalistic theory of evolution, and he was particularly concerned with its methodological aspects (Łastowski 1987). The work was inspired by the ideas of Leszek Nowak and his research environment. In particular, it was developed in the context of the idealizational theory of science.

4 The Research Traditions of the Poznań School of Archaeology

The archaeologists of the Poznań School of Archaeology follow two main methodological orientations: humanistic and naturalistic. Jan Żak's school greatly influenced his students, most notably Danuta Minta-Tworzowska (see Pawleta, chapter "Danuta Minta-Tworzowska and Her Role in Conceptualizing the Poznań School of Archaeology and Advancing the Methodology of Archaeology in Poland"), Andrzej Weber (1987, 1988), Włodzimierz Rączkowski (2002), Arkadiusz Marciniak (2005), as well as Anna Pałubicka (1990), and Andrzej P. Kowalski (1999). The most prominent representatives of the next generation are Adriana Ciesielska (2002), Rafal Zapłata (2005), Michał Pawleta (2016) and Lidia Żuk (2019). There are also several younger researchers. The intellectual paths of individual scholars are diverse, reflecting their search for paradigms and self-reflection on the role of theory in archaeology. The quest for interdisciplinarity has emerged as crucial, not only in theory but also in archaeological practice. This is manifested in an engagement with the methodology of the humanities, philosophy of science, methodology of history, cultural anthropology, inferential procedures and empirical foundations. The proponents of the Poznań School of Archaeology demonstrated in their work the indispensability of archaeological theory in every aspect of the research process. In the 1970s and 1980s, they attempted to apply theories directly derived from the PSM. However, from the 1990s onwards, after the fall of the Iron Curtain, they attempted to apply theories from different schools of research, mostly from the Anglo-American tradition, while maintaining other elements of the research agenda inherited from the Poznań School of Archaeology (e.g. Minta-Tworzowska 2015, 2017, 2022).

The philosophical status of archaeology, as understood by the Poznań School of Archaeology, has been extensively explored by Henryk Mamzer (2004, 2021). In his book *Archaeology and Discourse. Metaarchaeological Considerations* (2004), he explores philosophical reflections, termed metaarchaeological considerations, on the nature of the object of archaeological inquiry, the means of creating the past, and the value-oriented study of the past. He argues for an integrated humanistic, hermeneutic and constructivist perspective within the discipline. He sees the humanities as distinct from the natural sciences in terms of methodology, and engages in debate with proponents of empiricism and scientific methods typical of the natural sciences within archaeology. While Mamzer argues for archaeology as a humanistic discipline, he opposes its traditional practice stance dominated by realism, naturalism and universalism. At the same time, he promotes the idea of archaeology as a dialogue that operates at both the heuristic and the worldview-educational level, intending to promote understanding through agreements between scholars from different disciplines.

Irrespective of whether archaeologists in Poland have correctly grasped the principles of the PSM, its application to archaeology has encountered numerous difficulties (Minta-Tworzowska 2022, 19–20). One of these was the logical orientation of the school, which was in strong contrast to positivism. Another difficulty stemmed from the demands of humanistic interpretation, which, according to J. Kmita, required deductive reasoning, which was very different from the common types of reasoning used in archaeology (inductive, reductive, statistical and *per analogiam*). There was also an ongoing debate as to whether Marxism, as espoused by the PSM, had a chance of becoming a theory that would significantly change the character of Polish archaeology, particularly in terms of its methodology (Minta-Tworzowska 2000a; Tabaczyński 2000). Interestingly, from today's perspective, reinterpreted Marxism, unlike neo-Marxism in Western Europe, did not significantly influence the development of Polish archaeological theory. However, its development outside of philosophy was significantly hampered by external factors, such as the general societal aversion to Marxism following the collapse of the communist state in Poland in 1989. More generally, the opening up to the outside world led to the abandonment of the interpretive potential of non-orthodox Marxism in archaeology and other humanistic disciplines (Burda 2004, 343–344). Interestingly, other elements of the school remained very much in place.

Many ideas and proposals advocated and developed within the Poznań School of Archaeology have had a significant impact on archaeologists in Poland, leading to the creation of a unique and distinct form of theory and methodology of Polish archaeology. The scholars representing the school are mainly associated with the academic centre in Poznań and, to a lesser extent, Warsaw (mainly Stanisław Tabaczyński and his research milieu) and Gdańsk (Andrzej P. Kowalski and his collaborators). This school, based at the Adam Mickiewicz University, became the dominant research centre for methodology and theory in Polish archaeology. It has played a crucial role in the training of generations of archaeologists, emphasising methodological rigour and interdisciplinary collaboration. These aspects have been incorporated into academic programmes and archaeological fieldwork training initiatives throughout Poland, shaping the skills and perspectives of future archaeologists. Many scholars trained at the Poznań School of Archaeology have made significant contributions to archaeological research both in Poland and internationally.

The distinctiveness, uniqueness and specificity of the Poznań School of Archaeology can be seen in several factors. First, the School is known for its rigorous and conscious methodological approach to archaeological research. It has advanced methodological precision in archaeology by setting high standards for data collection and interpretation, which has helped to ensure the reliability and accuracy of archaeological research and interpretation. Secondly, in addition to its methodological impact, the Poznań School of Archaeology has contributed significantly to the development of archaeological theory in Poland. It has emphasised the leading role of theory in scientific research, which was not evident or even recognised within the historical-cultural paradigm of archaeology. Scholars associated with the School have engaged in theoretical debates within the discipline, challenging established paradigms and proposing new interpretive frameworks. This intellectual engagement has significantly enriched archaeological discourse and stimulated critical reflection on the nature of the archaeological process. Thirdly, the School promotes interdisciplinary collaboration by encouraging archaeologists to work with specialists from other scientific disciplines such as anthropology, history, physical anthropology, zoology, geology and environmental sciences. Interdisciplinarity enriches archaeological investigations by providing different perspectives and analytical techniques, allowing researchers to address complex questions about past human societies, their interactions with the environment, and their cultural practices.

5 The Structure of the Volume

The volume aims to provide a systematic overview of the Poznań School of Archaeology, in particular its foundations, current developments and impact on archaeology practised in Poland and beyond. It is the first book to examine the impact of this original school of thought on archaeology and the archaeological process in Poland in a broad intellectual context. It places the development of Polish archaeological thought and theory within the social and intellectual framework of Europe. The contributors to this volume are archaeologists, philosophers, historians and cultural anthropologists, who provide a multi-scalar analysis of its unique and distinctive character and together shed light on the trajectory of the development of theory and methodology of archaeology in the Poznań School of Archaeology. The volume has a syncretic character, bringing together different voices into one whole.

The volume is divided into three interrelated parts and consists of 12 chapters. The parts are: (1) THE EMERGENCE AND INTELLECTUAL CONTEXT OF THE POZNAŃ SCHOOL OF ARCHAEOLOGY, (2) The Poznań School of Archaeology—CONTENT, APPLICATIONS AND DEVELOPMENTS, and (3) The RECEPTION AND SIGNIFICANCE of the Poznań School of Archaeology.

The first part of the volume consists of three chapters, including the introduction written by the editors. Krzysztof Brzechczyn in his chapter "The Poznań School of Methodology: Institutional History—Research programme—Main achievements" provides an overview of the origins and subsequent development of the Poznań School of Methodology, its sources of inspiration (Popperism, Marxism, German anti-naturalism) and its research programme (anti-positivist naturalism, [Popperian] hypotheticalism, essentialism, idealisation and rationalism [of the human individual]). It outlines its three main components, including (i) idealization theory of science and non-Marxist historical materialism (Leszek Nowak), (ii) humanistic interpretation and social-regulatory theory of culture (Jerzy Kmita), and (iii) dynamic theory of historical source, concept of integral explanation and theory of historical narrative (Jerzy Topolski). In the next paper, "The Poznań School of Methodology and ITS CONTINUATIONS", Jan Grad further clarified the theoretical context in which the Poznań School of Archaeology developed and flourished. He outlined the character of the naturalistic and anti-positivist methodological model of the humanities proposed by Jerzy Kmita in the form of humanistic interpretation. Another important factor was a procedure of idealization and specification formulated by Leszek Nowak and embedded in his interpretation of Marxism. Grad has also highlighted the historical epistemology and socio-regulatory theory of culture proposed by Kmita as a third major contribution.

The second part of the volume "The Poznań School of Archaeology—CONTENT, APPLICATIONS AND DEVELOPMENTS" consists of six chapters. Henryk Mamzer in his introductory chapter "An *Apologia* of Methodology of Archaeology at Adam Mickiewicz University, Poznań" provides an overview of the main achievements of the representatives of the Poznań School of Archaeology at Adam Mickiewicz University in Poznań. He captured their academic standing as embedded in three different traditions within the Poznań School of Methodology, namely those of Jerzy Kmita, Leszek Nowak and Jerzy Topolski. He also outlines the characteristic features of the 'methodology of archaeology' developed in this research milieu. In the second chapter, "Danuta Minta-Tworzowska and Her Role in Conceptualizing the Poznań School of Archaeology and Advancing the Methodology of Archaeology in Poland", Michał Pawleta presents the academic achievements of Danuta Minta-Tworzowska—the current leading figure of the Poznań School of Archaeology. He outlined the main themes pursued by this scholar, such as the reflection on the theory and methodology of archaeology, the reception of the socio-regulatory concept of culture, the conceptualisation of archaeological sources, classification in archaeology, the social role of contemporary archaeology, and issues of heritage protection and management. He also discussed the impact of her work on the development of archaeological thought and practice in Poland and beyond. Anna Pałubicka's chapter "Did/Do Early Traditional Peoples Have a History?"

discusses the understanding of time among early traditional peoples and the emergence of historical thinking. She argues it appeared in a culture in which linear time and logic were constituted. It is illustrated by the emergence of philosophy in ancient Greece, which introduced a dualism of attitudes towards the world: the attitude of detachment (theoretical) and the attitude of involvement. In the next chapter, "'Archaeology Is Anthropology or It Is Nothing': But What Kind of Anthropology?", Andrzej P. Kowalski considers the impact of ethnology on archaeological interpretations of prehistoric culture. The main claim of the article is that archaeology has already become the anthropology of archaic societies. The study of the past of these cultures will never disappear, at least from the field of interest of archaeologists. They also face the challenge of developing their own theories of primitive societies. Jacek Woźny, in the following chapter "Philosophical Reflections on the Sources of Prehistoric Symbolism", examines the impact of Danuta Minta-Tworzowska's works on the interpretation of prehistoric symbolism. In this view, reductive and constitutive hermeneutics are applied to the conceptualisation of symbolic meaning. The former includes a semantic aspect of symbols and studies them synchronically and historically, while the latter refers to their non-semantic character and addresses their essence in a diachronic way. In the final chapter of the part of volume "Shaping Archaeological Awareness and the Idea of Socialising Archaeology", Violetta Julkowska presents the heuristic potential of the category of cultural landscape concerning the content of historical and archaeological education. The combination of this category with aerial photography aims to use the visualisation of archaeological artefacts present in space to promote their mindful protection. The essence of this approach is to show the way for adding new content to historical education to promote archaeological awareness.

The third part of the volume "The Reception and Significance of the Poznań School of Archaeology" consists of three chapters. Ludomir R. Lozny's chapter "Philosophy of Archaeology and the Poznań School of Methodology" discusses the contribution of the Poznań School of Methodology, which developed out of isolation from the West, to a theory of scientific cognition suitable for archaeological reasoning. It also discusses the process of acquiring archaeological knowledge and how this affected a larger question about the nature of archaeology, which falls within the realm of the philosophy of science. The main part of the chapter is a comparison of the methodological proposal of the Poznań School of Archaeology with Western thought at the time, placing it between those who see archaeology as a scientific discipline and those who see it as a humanistic field. He sees it as a Marxist-inspired epistemology, unique in Eastern Europe, which is fodder for philosophical analysis and understanding. Jacek Kowalewski in his next chapter "Ethnoarchaeology in Toruń and the Influence of the Poznań School of Methodology" discusses a distinct academic phenomenon of ethnoarchaeology developed at the Nicolaus Copernicus University in Toruń. The paper explores the relationship between Toruń archaeologists and representatives of the Poznań School of Methodology from various disciplines. The final chapter in the volume entitled "Late Mesolithic and Early Neolithic Hunter-Gatherer Cultural Processes in Cis-Baikal, Eastern Siberia: A Case Study in Darwinian Evolutionary Archaeology" was written by Andrzej W. Weber. It discusses substantial spatio-temporal variations in the adaptive strategies of Late Mesolithic and Early Neolithic hunter-gatherers in the Cis-Baikal region of Eastern Siberia in the context of different theoretical positions of the author. In particular, he examines the influence of the PSM, especially the intellectual links with Jerzy Kmita, Leszek Nowak, Anna Pałubicka, and Jan Żak, some of the more ardent proponents of this school of thought. He emphasised that the representatives of the Poznań School of Archaeology significantly contributed to the shift from the dominant positivist approaches in the empirical sciences (including archaeology) with the anti-positivist paradigm and, consequently, from inductive to deductive reasoning.

References

Burda, T. 2004. Reinterpretacja marksizmu przez metodologiczną szkołę poznańską i próba jej zastosowania w archeologii. In *Archeologia, kultura, ideologie*, ed. B. Gediga and W. Piotrowski, 331–346. Biskupin/Wrocław: Muzeum Archeologiczne w Biskupinie, Polska Akademia Nauk, Oddział we Wrocławiu.

Ciesielska, A. 2002. *Elementy teorii społecznej w archeologii: koncepcje grup, instytucji i struktur społecznych.* Poznań/Kalisz: Uniwersytet im. Adama Mickiewicza, Wydział Pedagogiczno-Artystyczny.

Cyngot, D., H. Kowalewska-Marszałek, A. I. Zalewska, and D. Minta-Tworzowska. 2020. Profesor Stanisław Tabaczyński (1930–2020) – wybitny uczony i intelektualista. *Folia Praehistorica Posnaniensia* 25:9–36.

Domańska, E. 2016. (Ed.) Jerzy Topolski. *Teoretyczne problemy wiedzy historycznej. Antologia tekstów.* Poznań: Wydawnictwo Nauka i Innowacje.

Domańska, E., and A. Topolska. 2022. (Eds.) *Theory and methodology of historical knowledge: An anthology.* Faculty of History Press, Adam Mickiewicz University in Poznań.

Drozdowski, M., A. Wyczański, J. Pomorski, and A. Zybertowicz. 1988. Jerzy Topolski – for a new shape of the historical science. In *Między historią a teorią. Refleksje nad problematyką dziejów i wiedzy historycznej. Studia ofiarowane profesorowi Jerzemu Topolskiemu w sześćdziesiątą rocznicę urodzin*, ed. M. Drozdowski, 50–64. Warszawa/Poznań: Państwowe Wydawnictwo Naukowe.

Jørgensen, K. E. 2015. Introduction: Research traditions. In *Sage handbook of European foreign policy*, ed. K. E. Jørgensen, Å. Kalland Aarstad, E. Drieskens, K. Laatikainen, and B. Tonra, 3–13. Los Angeles/London/New Delhi: Sage Publications.

Kmita, J. 1971. *Z metodologicznych problemów interpretacji humanistycznej.* Warszawa: Państwowe Wydawnictwo Naukowe.

Kmita, J. 1973. *Wykłady z logiki i metodologii nauk: dla studentów wydziałów humanistycznych.* Warszawa: Państwowe Wydawnictwo Naukowe.

Kobyliński, Z. 2012. Etnoarcheologia. In *Przeszłość społeczna. Próba konceptualizacji*, ed. S. Tabaczyński, A. Marciniak, D. Cyngot, and A. Zalewska, 721–731. Poznań: Wydawnictwo Poznańskie.

Kotowa. 2010. Od filozofii humanistyki do filozofii jako dziedziny kultury. Główne kierunki badawcze szkoły poznańskiej od połowy lat siedemdziesiątych. In *Filozofia na Uniwersytecie w Poznaniu*, ed. T. Buksiński, 309–334. Poznań: Wydawnictwo Naukowe IF UAM.

Kowalski, A. P. 1999. *Symbol w kulturze archaicznej.* Poznań: Wydawnictwo Naukowe Instytutu Filozofii UAM.

Łastowski, K. 1987. *Rozwój teorii ewolucji: studium metodologiczne.* Poznań: Wydawnictwo Naukowe UAM.

Mamzer, H. 2004. *Archeologia i dyskurs. Rozważania metaarcheologiczne*. Poznań: Instytut Archeologii i Etnologii Polskiej Akademii Nauk.

Mamzer, H. 2020. "Mowa obrończa" metodologii archeologów Uniwersytetu im. Adama Mickiewicza w Poznaniu. *Folia Praehistorica Posnaniensia* 25:109–148.

Mamzer, H. 2021. *Jak człowiek konstruował świat i siebie samego. Archeologia w procesie humanizacji świata*. Poznań: Instytut Archeologii i Etnologii Polskiej Akademii Nauk.

Marciniak, A. 1996. *Archeologia i jej źródła. Materiały faunistyczne w praktyce badawczej archeologii*. Warszawa/Poznań: Wydawnictwo Naukowe PWN.

Marciniak, A. 2005. *Placing animals in the Neolithic: Social zooarchaeology of prehistoric farming communities*. London: UCL Press.

Marciniak, A. 2015. Mainstream and minority archaeologies. The case of the beginnings of the Polish bioarchaeology. In *Paradigm found. Archaeological theory—present, past and future. Essays in honour of Evžen Neustupný*, ed. K. Kristiansen, L. Šmejda, and J. Turek, 137–146. Oxford: Oxbow.

Marciniak, A. 2018. Archeologia i antropologia w polskiej tradycji. Pomiędzy współpracą, konfrontacją a wzajemną fascynacją. In *Stary materiał – nowe spojrzenie. Funeralia Lednickie – spotkanie 20*, ed. W. Dzieduszycki and J. Wrzesiński, 33–44. Poznań: Stowarzyszenie Naukowe Archeologów Polskich.

Marciniak, A., and N. Yalman. 2013. Non-anglophone ethnoarchaeologies in the past and today. In *Contesting ethnoarchaeologies. Traditions, theories, prospects*, ed. A. Marciniak and N. Yalman, 1–13. New York: Springer.

Minta-Tworzowska, D. 1994. *Klasyfikacja w archeologii jako sposób wyrażania wyników badań, hipotez oraz teorii archeologicznych*. Poznań: Wydawnictwo Naukowe UAM.

Minta-Tworzowska, D. 1996. Culture and society in traditional and processual archaeology. *Archaeologia Polona* 34:59–81.

Minta-Tworzowska, D. 1998a. Jerzego Topolskiego koncepcja źródeł historycznych a ujęcia źródeł archeologicznych. In *Świat historii. Prace z metodologii historii i historii historiografii dedykowane Jerzemu Topolskiemu z okazji siedemdziesięciolecia urodzin*, ed. W. Wrzosek, 329–340. Poznań: Instytut Historii Uniwersytetu im. Adama Mickiewicza.

Minta-Tworzowska, D. 1998b. Continuity and change of classification in archaeology. In *Theory and practice of archaeological research. Dialogue with the data: The archaeology of complex societies and its context in the '90. V. III*, ed. S. Tabaczyński, 191–223. Warszawa: Institute of Archaeology and Ethnology, Committee of Pre- and Protohistoric Sciences, Polish Academy of Sciences.

Minta-Tworzowska, D. 2000a. Kwestia przełomu metodologicznego w archeologii i prahistorii polskiej. In *Archeologia i prahistoria Polska w ostatnim półwieczu*, ed. M. Kobusiewicz and S. Kurnatowski, 527–537. Poznań: Wydawnictwo Poznańskiego Towarzystwa Przyjaciół Nauk.

Minta-Tworzowska, D. 2000b. Świat archeologii w świetle źródeł archeologicznych. *Acta Historica et Museologica Universitatis Silesianae Opaviensis* 5:49–59.

Minta-Tworzowska, D. 2000c. Symbole i symbolika w perspektywie badań archeologicznych. In *Kultura symboliczna kręgu pól popielnicowych epoki brązu i wczesnej epoki żelaza w Europie Środkowej*, ed. B. Gediga and D. Piotrowska, 45–54. Warszawa/Wrocław/Biskupin: Polska Akademia Nauk, Oddział we Wrocławiu, Muzeum Archeologiczne.

Minta-Tworzowska, D. 2001. Profesora Jana Żaka poszukiwanie paradygmatu. In *Archeologia. Paradygmat. Pamięć*, ed. D. Minta-Tworzowska and W. Rączkowski, 109–122. Poznań: Wydawnictwo Poznańskie.

Minta-Tworzowska, D. 2002. Between a community of inspiration and the separateness of archaeological traditions. In *Archaeologies of Europe: History, methods and theories*, ed. A. Gramsch, P. Biehl, and A. Marciniak, 53–64. Münster: Waxman.

Minta-Tworzowska, D. 2006. Interdisciplinary research and archaeology. *Archaeologia Polona* 44:41–51.

Minta-Tworzowska, D. 2008. Teoretyczne problemy interpretacji "sztuki" pradziejowej. In *Sztuka pradziejowa i wczesnośredniowieczna jako źródło historyczne*, ed. B. Gediga and W. Piotrowski, 23–47. Wrocław/Biskupin: PAN, Oddział we Wrocławiu, Muzeum Archeologiczne.

Minta-Tworzowska, D. 2015. Czy współczesna i przyszła archeologia będzie "potrzebowała" teorii? *Archeologia Polski* 60:21–37.

Minta-Tworzowska, D. 2017. Tradycja i inspiracje teoretyczne we współczesnej archeologii. *Filo-Sofija* 17 (1):267–286.

Minta-Tworzowska, D. 2019. The program of "new" aesthetics regarding the expectations of archaeologists. *Folia Praehistorica Posnaniensia* 24:149–163.

Minta-Tworzowska, D. 2021a. Are we where we wanted to be? Modernist tendencies versus the postmodern reality of archaeology. Some remarks on the methodology of archaeologists at Adam Mickiewicz University in Poznań. In *Treasures of time. Research of the Faculty of Archaeology of Adam Mickiewicz University in Poznań*, ed. D. Żurkiewicz, 452–473. Poznań: Wydawnictwo Naukowe UAM.

Minta-Tworzowska, D. 2021b. Metodologia archeologii/prahistorii profesora dr. hab. Jana Żaka (1923–1990) po 30. latach. Między modernistycznymi dążeniami a postmodernistyczną rzeczywistością. *Folia Praehistorica Posnaniensia* 26:145–169.

Minta-Tworzowska, D. 2022. Teoretyczne i metodologiczne podstawy archeologicznego dyskursu w Polsce. Zarys problematyki. *Przegląd Archeologiczny* 70:5–40.

Minta-Tworzowska, D. 2023. Nowoczesna archeologia Profesora Jana Żaka (1923–1990). W 100. rocznicę urodzin. *Folia Praehistorica Posnaniensia* 28:231–253.

Minta-Tworzowska, D., and M. Pawleta. 2013. Teoria i metodologia archeologii na Uniwersytecie w Poznaniu – dlaczego właśnie tu? *Folia Praehistorica Posnaniensia* 18:217–233.

Minta-Tworzowska, D., and W. Rączkowski. 1996. Theoretical traditions in contemporary Polish archaeology. *World Archaeological Bulletin* 8:196–209.

Minta-Tworzowska, D., and W. Rączkowski. 2007. Standardy teoretyczne we współczesnej polskiej archeologii. In *Pół wieku z dziejów archeologii polskiej (1939–1989)*, ed. J. Lech, 219–247. Warszawa: PWN.

Musiał, G. 2016. Poznańska szkoła metodologiczna—jej rozwój i znaczenie. *Przegląd Filozoficzny – Nowa Seria* 25 (2):553–563.

Neustupný, E. 1997/1998. Mainstream and minorities in archaeology. *Archaeologia Polona* 35–36:13–23.

Nowak, L. 1980. *The structure of idealization. Towards a systematic interpretation of the Marxian idea of science*. Dordrecht/Boston/London: Reidel.

Pałubicka, A. 1990. *Kulturowy wymiar ludzkiego świata obiektywnego*. Poznań: Wydawnictwo Naukowe UAM.

Pałubicka, A., and S. Tabaczyński. 1986. Społeczeństwo i kultura jako przedmiot badań archeologicznych. In *Teoria i praktyka badań archeologicznych. Przesłanki metodologiczne, t. 1*, ed. W. Hensel, G. Donato, and S. Tabaczyński, 57–183. Wrocław/Warszawa/Kraków/Łódź: Zakład Narodowy im. Ossolińskich, Wydawnictwo Naukowe IAE PAN.

Pawleta, M. 2016. *Przeszłość we współczesności. Studium metodologiczne archeologicznie kreowanej przeszłości w przestrzeni społecznej*. Poznań: Wydawnictwo Naukowe UAM.

Pomorski, J. 1988. Metodologiczne boje o historię. In *Między historią a teorią. Refleksje nad problemami dziejów i wiedzy historycznej*, ed. M. Drozdowski, 28–38. Warszawa/Poznań: PWN.

Rączkowski, W. 2001. Między Jankuhnem a Łowmiańskim: w poszukiwaniu koncepcji studiów osadniczych. In *Archeologia. Paradygmat. Pamięć*, ed. D. Minta-Tworzowska and W. Rączkowski, 163–175. Poznań: Wydawnictwo Poznańskie.

Rączkowski, W. 2002. *Archeologia lotnicza: metoda wobec teorii*. Poznań: Wydawnictwo Naukowe UAM.

Tabaczyński, S. 1987. *Archeologia średniowieczna: problemy, źródła, metody, cele badawcze*. Wrocław: Zakład Narodowy im. Ossolińskich.

Tabaczyński, S. 2000. "Wymuszony przewrót metodologiczny" w archeologii polskiej i jego nieoczekiwane skutki. In *Archeologia i prahistoria Polska w ostatnim półwieczu*, ed. M. Kobusiewicz and S. Kurnatowski, 513–525. Poznań: Wydawnictwo Poznańskiego Towarzystwa Przyjaciół Nauk.

Topolski, J. 1983. *Teoria wiedzy historycznej*. Poznań: Wydawnictwo Poznańskie.

Topolski, J. 1984. *Metodologia historii*. Poznań: Państwowe Wydawnictwo Naukowe.

Topolski, J. 1996. *Jak się pisze i rozumie historię. Tajemnice narracji historycznej*. Warszawa: Oficyna Wydawnicza Rytm.

Weber, A. 1987. *Megalityczny obrządek pogrzebowy w kulturze pucharów lejkowatych. Próba ujęcia teoretycznego*. Unpublished PhD thesis, Institute of Prahistory of AMU.

Weber, A. 1988. Idealizacyjny model procesu neolityzacji. In *Szkice z antropologii ogólnej: propozycje teoretyczno-metodyczne badań społeczeństw pradziejowych*, ed. J. Piontek, 43–62. Poznań: Wydawnictwo Naukowe UAM.

Wrzosek, W. 2013. Jerzego Topolskiego projekt metodologii historii. *Studia Metodologiczne* 31:15–23.

Żak, J. 1974. Słowianie i Germanie w prahistorii polskiej i niemieckiej. In *Stosunki polsko-niemieckie w historiografii, cz.1*, ed. J. Krasucki, G. Labuda, and A. Walczak, 21–149. Poznań: Instytut Zachodni.

Żak, J. 1975a. Próba przedmiotowego modelu kultury społeczeństw młodszego i schyłkowego paleolitu (strefy śródziemnomorskiej). *Fontes Archaeologici Posnanienses* 25:28–59.

Żak, J. 1975b. Ziemie polskie w starożytności i we wczesnym średniowieczu. In *Dzieje Polski*, ed. J. Topolski, 14–76. Warszawa: PWN.

Żak, J. 1977. Uwagi o studiach osadniczych. *Archeologia Polski* 22 (2):421–424.

Żak, J. 1985. Uwagi metodologiczne nad problemem osadnictwa. *Studia Metodologiczne* 24:77–87.

Żak, J., and D. Minta-Tworzowska. 1991. Problematyka prahistorii pozytywistycznej. *Folia Praehistorica Posnaniensia* 4:9–25.

Zamiara, K. 2010. U początków poznańskiej szkoły metodologicznej. In *Filozofia na Uniwersytecie w Poznaniu*, ed. T. Buksiński, 283–308. Poznań: Wydawnictwo Naukowe IF UAM.

Zapłata, R. 2005. *Archeologiczne studia na przestrzenią: zastosowanie Systemów Informacji Geograficznej w badaniach nad wczesnośredniowiecznym osadnictwem grodowym na Pomorzu*. Unpublished PhD thesis, Institute of Prahistory of AMU.

Żuk, L. 2019. *Pamięć o krajobrazie jako kategoria analizy archeologicznej*. Unpublished PhD thesis, Faculty of Archaeology of AMU.

The Emergence of the Poznań School of Archaeology

Poznań School of Methodology: Institutional History—Research Program—Main Achievements

Krzysztof Brzechczyn

Abstract The chapter presents the main achievements of the Poznań School of Methodology against the background of the postwar history of Polish philosophy. In its first part, the history of Polish philosophy in the twentieth century is outlined. In the second part, the institutional history of the Poznań School is sketched, and its main sources of inspiration and determinants of the School's research program are identified. In the third part the main scientific achievements of Jerzy Kmita, Leszek Nowak and Jerzy Topolski—three major advocates of the School—are presented.

Keywords Idealization · Poznań School of Methodology · Jerzy Kmita · Leszek Nowak · Jerzy Topolski · Socio-regulative theory of culture

1 Poznań School of Methodology Against the Background of Postwar Polish Philosophy

The history of Polish analytic philosophy dates back to the end of the nineteenth century when Kazimierz Twardowski (1866–1938) became (in 1895) a professor at the University of Lviv. Twardowski began intense organizing, didactic, and popularizing work, the effects of which rippled through the other partitioned regions of Poland. In 1904, he established the Polish Philosophical Society in Lviv, and in 1911—"Ruch Filozoficzny" which he edited until 1936.

Before World War I, Twardowski mentored a number of students, who were later active in the fields of humanities and social science, from psychology and pedagogics to art history and literary studies. Among the philosophers, there were: Kazimierz Ajdukiewicz (1890–1963), Tadeusz Czeżowski (1889–1981), Tadeusz Kotarbiński

K. Brzechczyn (✉)
Faculty of Philosophy, Adam Mickiewicz University in Poznań, Poznań, Poland
e-mail: krzysztof.brzechczyn@amu.edu.pl

A. Marciniak et al. (eds.), *The Poznań School of Archaeology*, Outstanding Contributions to Archaeology, https://doi.org/10.1007/978-3-032-07848-3_2

(1886–1981), Jan Łukasiewicz (1878–1956), Władysław Witwicki (1878–1948), and Zygmunt Zawirski (1882–1948). In 1919, Twardowski's student, Jan Łukasiewicz, became the Minister of Religious Denominations and Public Enlightenment in Ignacy Paderewski's government. In the 1920s, Twardowski initiated the organization of Polish philosophical conferences, and he created—in cooperation with his students—curricula for teaching philosophy in high schools and at universities. The analytic style of philosophizing had an impact not only on philosophy but also on other fields of science and culture. That development of Polish philosophy was interrupted by the outbreak of World War II, during which about six million Polish citizens lost their lives, including 615 scientists, and among them—32% of doctors and 38% of professors of pre-war universities and institutions.

After the 1945 Yalta Conference, Poland became a part of the Eastern Bloc. Because of that, many philosophers and other scholars decided to remain in the West, and some left Poland just after the war. They were, among other scientists: Józef Maria Bocheński (1902–1995), Henryk Hiż (1917–2006), Zbigniew Jordan (1911–1977), Czesław Lejewski (1911–2000), Jan Łukasiewicz, Henryk Mehlberg (1904–1979), Bolesław Sobociński (1906–1980), Alfred Tarski (1901–1983), and Florian Znaniecki (1882–1958).

We can distinguish six periods of the post-war history of Polish philosophy (other periodizations see: Stępień 1990, pp. 22–23 and Skoczyński and Woleński 2010, 526):

1. 1945–1948—conditional tolerance,
2. 1949–1956—Stalinization,
3. 1956–1968—limited pluralism,
4. 1968–1980—normalization,
5. 1980–1989—gradual emancipation,
6. since 1989—restitution of freedom.

The Conditional Tolerance Period (1945–1948) After 1945, four pre-war universities resumed operations: Poznań University, Warsaw University, Jagiellonian University, and the Catholic University of Lublin. Four new universities were established: in Lublin (in 1944), Toruń, Łódź, and Wrocław. Polish Philosophical Society was reactivated, and several pre-war magazines were published again: "Przegląd Filozoficzny", "Kwartalnik Filozoficzny", "Ruch Filozoficzny", and "Studia Philosophica". Books written just before the war or during the war were published then as well, including Roman Ingarden's *Spór o istnienie świata* (*Controversy over the Existence of the World*; 1947–1948), Władysław Tatarkiewicz's *O szczęściu* (*On Happiness*; 1947), and Maria Ossowska's *Podstawy nauki o moralności* (*The Foundations of the Science of Morality*; 1947).

At the beginning, philosophy practiced within the framework of phenomenology, Thomism, or neo-positivism was tolerated, partly because of the dearth of academics knowledgeable about Marxism, and partly because there were more important—from the authorities' point of view—social spheres to subjugate. The full ideological offensive in science began after the falsified 1947 elections, the elimination of legal opposition, and the merger of the Polish Socialist Party with the Polish Workers' Party in December 1948.

The Stalinization Period (1949–1956) In April 1949 the Central Committee of the Polish United Workers' Party organized a plenary session during which an "ideological offensive on the scientific front" was discussed. In the consequence, the policy on the personnel of universities and research institutes changed because elections for university authorities was replaced with the principle of nomination. The Polish Academy of Learning (*Polska Akademia Umiejętności*) was substituted, in October 1951, by the Polish Academy of Sciences.

As noted by Józef Tischner (1981, 18):

> (...) dialectical materialism does not enter our culture by way of people's free choices; it is promulgated, and sometimes simply imposed with the use of political methods. Before the war, there was not one significant representative of Marxism in Poland. It is only after the war that Marxists are becoming abundant.

"Przegląd Filozoficzny" was closed in 1949, "Kwartalnik Filozoficzny"—in 1950, and "Ruch Filozoficzny"—in 1951. In 1951, "Myśl Filozoficzna" was established, as the only central philosophical journal, which only published Marxist texts. The aim of the magazine was:

> the fight to permeate the whole Polish science with the Marxist-Leninist world view, creative development of Marxist philosophy in strict connection with practical needs, popularization of Marxist scientific thinking and action among the broadest masses, and the fight against the remnants of idealism as they hinder social development, against bourgeois philosophy, and against the destructive philosophy of imperialism ("From the editors," *Myśl Filozoficzna*, 1951, no. 1–2, 11–12).

The journal contained articles which unmasked 'bourgeois' philosophic movements, with personal attacks against their representatives. Leszek Kołakowski (1927–2009) criticized neo-Thomism, while Roman Ingarden (1893–1970) and phenomenology were criticized by Tadeusz Kroński (1907–1958) who accused Ingarden of methodological idealism, lack of appreciation for Marxism as a breakthrough in philosophy, antihistoricism, of presenting philosophers' views as a search for truth rather than an expression of class interest, and of separating philosophy from science.

The Lvow–Warsaw School received plenty of attention. Adam Schaff (1913–2006) unmasked Kazimierz Ajdukiewicz's views, Bronisław Baczko (1924–2016)—Tadeusz Kotarbiński's, and Józef Chałasiński 1904–1979)—Twardowski's. Ajdukiewicz and Kotarbiński responded to that critique, and those texts were published in "Myśl Filozoficzna". The Catholic philosophers and Ingarden did not reply, and, it was not clear, as noted by Jan Woleński, "if they did not want to, or if it was clear from the start that their answers would not be published" (Skoczyński, Woleński 2010, 537).

The publication of the third volume of Tatarkiewicz's *Historia filozofii* (*History of Philosophy*—1958) and of Ajdukiewicz's *Zagadnienia i kierunki filozofii* (*Problems and Theories of Philosophy*—1949) was halted. Non-Marxist philosophers were denied administrative and didactic positions. That happened to Ingarden, Tatarkiewicz, Tadeusz Kotarbiński, Tadeusz Czeżowski, Izydora Dąmbska (1904–1983), Maria Ossowska (1896–1974), Stanisław Ossowski (1897–1963), and Henryk Elzenberg (1887–1967).

In 1978, Marxist philosopher Kazimierz Ochocki (1978, 198) evaluated that campaign as follows:

> First, it was very loud, and carried out with the use of scathing – sometimes outright aggressive – terms. Second, the main sources were Stalin's and Zhdanov's texts, with no understanding of many issues under discussion. The critique was baseless, it consisted in assigning pre-defined labels. Fifth, philosophical, epistemological, and methodological views were equated with political views.

The Limited Pluralism Period (1956–1968) After 1956, philosophical studies and classes were restored in all courses of study, and professors could teach philosophy again. In 1957–1962, Kotarbiński become the president of the Polish Academy of Sciences. "Ruch Filozoficzny" was reactivated, and "Myśl Filozoficzna" was replaced with "Studia Filozoficzne" (1957–1990). In 1956, the Institute of Philosophy and Sociology of the Polish Academy of Sciences was formed.

At that time, Lublin Philosophical School developed fully, and the Lvow–Warsaw School and phenomenology regained their influence. In Marxism itself, the dogmatic style of the Stalin era was condemned, and alongside the official version, two new orientations were formed: scientistic and humanistic. In Woleński's (1993, 65) words:

> Polish Marxist proposed a program of 'open Marxism,' by which they meant: open to problem and even to solution taken from 'bourgeois' philosophy. This movement split itself into two schools, namely 'humanistic' and 'naturalistic.' The former, influenced by existentialism in its various forms, concentrated its activities on philosophical anthropology; the latter, influenced by the Lvow-Warsaw school and by logical empiricism, was mainly involved in the philosophy of science.

Humanistic Marxism was best represented by the Warsaw School of Historians of Idea. It was founded by Kołakowski, the head of the Chair of the History of Modern Philosophy of the University of Warsaw, and by Bronisław Baczko, the head of the Chair of Modern Philosophy and Social Thought of the Polish Academy of Sciences. The seminars of the Polish Academy of Sciences, organized at regular intervals, were attended by, among other people, Andrzej Walicki (1930–2020), Jerzy Szacki (1929–2016), Krzysztof Pomian, Nina Assorodobaj-Kula (1908–1999), Henryk Hinz (1929–1992), Adam Sikora (1927–2011), Jan Garewicz (1921–2002), Ryszard Panasiuk, and Zbigniew Kuderowicz (1931–2015).

At that time, Marxist philosophy was under the influence of the analytical philosophical style. According to Nowak (1997, 29–30):

> (…) in the spiritual culture dominated by the Lwów–Warsaw School, one of the initiators of 20th-century analytic philosophy. The Polish style of even the most official Marxist philosophy, shaped mainly by Adam Schaff's writings, was decidedly different than that of the USSR and quite close to the clarity requirements imposed on philosophy by the Lwów-Warsaw School.

Skolimowski claimed that because of the influence of analytic philosophy, "after 1956, Poland witnessed incredibly interesting developments of Marxist philosophy" (Skolimowski 1969, 42).

The Normalization Period (1968–1980) In the 1960s, the representatives of the Warsaw School of Historians of Idea sympathized with revisionism which was critical of real socialism. As a result of the repressions after the pacification of the March 1968 protests of students and the youth, philosophers of Jewish descent, including Kołakowski and Zygmunt Baumann, had to emigrate.

In 1968, the University of Silesia was opened (in Katowice), and in 1970—the University of Gdańsk. Also, the administration of universities was reorganized. Independent chairs were closed, and institutes with departments were formed. Despite the efforts of the authorities to put universities in order, Marxist philosophy began to erode at that time. Some of its representatives entered a dialog with Catholic philosophy or incorporated theses of analytic philosophy. Importantly, there appeared research teams which studied Polish philosophical tradition, the popularization of which had earlier been banned for political reasons. In 1979, Karol Wojtyła (1920–2005) became Pope John Paul II. In his scientific career, Wojtyła combined Thomism and phenomenology. Philosophers started to engage in opposition activity. Stefan Amsterdamski (1929–2005), Bohdan Chwedeńczuk, and Krzysztof Pomian signed the Letter of 59 to protest against the introduction of friendship with the Soviet Union into the Constitution of the Polish People's Republic.

The Emancipation Period (1980–1989) In the emancipation period, the official Marxist philosophy started to fade away, while non-Marxist movements gained influence. More and more philosophers, including Mirosław Dzielski (1941–1989), Nowak, Tischner (1931–2000), Klemens Szaniawski (1925–1990), Amsterdamski, and Barbara Skarga (1919–2009), engaged in the actions of Solidarity and the opposition.

Restitution of Freedom (Since 1989) After downfall of communism in 1989 all political barriers were removed. The new phenomenon was development of the higher education. New universities were founded, in Opole (1994), Białystok (1997), Olsztyn (1999), Rzeszów (2001), and Zielona Góra (2001) where philosophical studies were initiated. In this period theological faculties were included in structure of some universities. There were founded new philosophical journals and publishing houses specializing in editions of philosophical books. In this period old philosophical orientations were restored and also new trends appeared: postmodernism, cognitive science, eco-philosophy, bioethics, and gender studies.

2 Poznań School of Methodology

2.1 Institutional History

The term "Poznań School of Methodology" was introduced by Janusz Kuczyński, the editor of "Studia Filozoficzne" in 1972. At that time that is in the fourth and fifth periods of history of Polish postwar philosophy Poznań School of Methodology was the most influential. It has never been a homogenous milieu because in the 1970s, as will be shown, there emerged three research teams gathered around Kmita, Nowak, and Topolski. However, the beginnings of this scientific milieu backed to the 1960s of the previous century that is to the third period in history of postwar Polish philosophy. It gradually declined in the sixth period, with the death of its founding fathers and of some of the first generation of their students.

In 1945, Ajdukiewicz, Twardowski's student and—privately—his son-in-law, was employed at Poznań University as the head of the Chair of the Theory and Methodology of Science (Gan-Krzywoszyńska and Leśniewski 2016). In 1948 Ajdukiewicz became the rector of the university, but in 1954, he moved to the University of Warsaw where he became the head of the Chair of Logic in the Institute of Philosophy. The scientific seminars at the university were attended by Kmita who traveled to Warsaw from Poznań to participate in them.

After his graduation from Polish language studies, Kmita was employed in the Chair of Logic of the Institute of Philosophy and History in 1957 (see: Grad 2011). In 1962, he defended his doctoral dissertation titled *Znaczenie jako wykładnik sposobu rozumienia wyrażeń (Meaning as an identifier of the manner of understanding of phrases*). The first supervisor of that work was Adam Wiegner (1889–1967), but as he retired in 1960, Jerzy Giedymin (1925–1993)—Ajdukiewicz's student and the head of the chair until 1967—became the new supervisor. In 1968, Kmita presented his habilitation thesis, *Problematyka terminów teoretycznych w odniesieniu do pojęć literaturoznawczych* (*Theoretical Terms and the Concepts of Literature Studies*), and he received the scientific degree of *doktor habilitowany*. Since January 1, 1969, Kmita became also the head of the Chair of Logic of the Institute of Philosophy and History, and then, after the Institute of Philosophy was formed in October 1969, the head of the Department of Logic and Methodology of Science. In 1977, he became the director of the newly founded Instutute of Cultural Studies, and he held that position until his retirement in 2002.

In 1946–1950, Jerzy Topolski (1928–1998) studied history in Poznań University, where he attended Jan Rutkowski's seminar (see: Domańska 2021, Domańska 2022). After Rutkowski's death in 1951, Topolski received a doctor's degree in historical sciences at Nicolaus Copernicus University in Toruń, on the basis of his work titled *Rozwój latyfundium arcybiskupstwa gnieźnieńskiego od XV do XVIII w.* (*The Development of the Latifundium of the Gniezno Archbishopry from the fifteenth to the eighteenth century*) and written under supervision of Stanisław Hoszowski. In 1951–1956, Topolski worked in the Institute for Scientific Staff Education. Since 1956, he was employed in the Institute of History of the Polish Academy of Sciences, and since 1959, he also worked in the Institute of History of Adam Mickiewicz University in Poznań. At the turn of the 1950s and 1960s, he cooperated with Andrzej Malewski. In 1962, they published *Studia z metodologii historii* (*Studies on the Methodology of History*). In 1965, Topolski founded and edited "Studia Metodologiczne. Zeszyty Poświęcone Integracji Nauki". In 1981–1987, Topolski was the head of the Institute of History of Adam Mickiewicz University in Poznań, and then, until his retirement, he was the head of the Department of Methodology of History.

Leszek Nowak (1943–2009) started in 1960 study of law in Poznań University, while also doing distance studies in philosophy at the University of Warsaw (see biographical details: Brzechczyn 2022). In 1966 he defended his master thesis titled *Model, prawda względna, postęp nauki* (*Model, Relative Truth, Progress of Science*) written under supervision of Janina Kotarbińska. In the fall of 1965, he was employed in the Department of Legal Applications of Logic of the Chair of the Theory of the State and Law. In 1967, he defended his doctoral dissertation titled *Problemy znaczenia i obowiązywania normy prawnej a funkcje semiotyczne języka* (The problems of the meaning and validity of a legal norm and the semiotic functions of language), under the supervision of professor Zygmunt Ziembiński. In October 1, 1970, Nowak moved to the new Institute of Philosophy, and he began working in the Department of Logic and Methodology of Science. In December 1970, Nowak presented his habilitation thesis *U podstaw metodologicznych "Kapitału" Karola Marksa* (*On the Methodological Foundations of Carl Marx's Capital*). In 1974, he became the head of the newly created Department of the Dialectics of Cognition, where his doctoral students could find employment.

In the second half of the 1970s, Nowak began working on his theory of socialism which was critical of the actualities of real socialism. The first part of the large typescript *U podstaw teorii procesu historycznego* (*The Foundations of the Historical Process*) was published in 1981 by the oppositional independent student association of the Agricultural University in Poznań. The whole was published in English as *Property and Power*, by the Reidel publishing house. On December 13, 1981 Nowak was interned because of his engagement in the Solidarity movement. He was released from the internment facility on December 9, 1982, and he resumed his work at the university as well as his underground publicist activity. In October 1984, the Minister of Science and Higher Education decided to suspend Nowak's academic teaching rights. In 1985, Nowak was expelled from the university. In 1988, he went abroad to lecture.

After his return to the Adam Mickiewicz University in Poznań, he was appointed a professor ordinary, and he did scientific, organizational, and editorial work. In 1991, his three-volume book titled *U podstaw teorii socjalizmu* (*The Foundations of the Theory of Socialism*, Nowak 1991b) was published, and it also contained his most accurate articles which had earlier been published in underground papers. Since the second half of the 1990s, as his health declined, he gradually stepped back from his duties at the university. In 1999, he resigned from his position of the head of the Institute of Epistemology, and he limited his didactic and scientific activity while focusing on the creation of a system of unitarian metaphysics.

2.2 *The Program Sources of Inspiration and Determinants of the Research of the Poznań School of Methodology*

The intellectual sources of inspiration for the Poznań School of Methodology include:

- **analytic philosophy of Lviv-Warsaw school** with its emphasis on methodological clarity, explication of notions, and posing of research questions;
- **Popperian hypothetism** with its emphasis on consciously formulating of scientific problems and looking for ways of its solutions;
- **Marxism**—Carl Marx's and Friedrich Engels's methodological and theoretical views were reconstructed and explicated in accordance with methods of analytic philosophy.[1]

It is also possible to distinguish following determinants of the research program of the Poznań School of Methodology that made it unique. There are:

- **naturalism**—it is assumed that the same methodological principles are adhered to in the humanities and social science as in natural science;
- **hypothetism**—it is assumed that in natural science and the humanities, it is crucial to formulate hypothesis which influence the course of empirical studies;
- **essentialism**—it is assumed that there are main and secondary factors of the studied reality;
- **idealization**—scientific theories are formed on the basis of modeling of the studied reality, which consists of the elimination of secondary factors and of focusing on the main factors; and
- **rationalism**—human actions and their products are explained and interpreted based on the assumption of rationality which has an idealizational status.

3 An Overview of the Main Achievements of the Founders of the Poznań School of Methodology

3.1 *Jerzy Kmita*

Point of departure of Kmita's research was conviction about methodological unity of humanities, social and natural sciences. However, this unity of science was perceived in different way from positivism philosophy of science. According to Kmita, these main domains of sciences shared some basic methodological principles. Those are:

[1] According to Zamiara (2010, 297), the basic sources of inspiration were: methodology of science (Ajdukiewicz), philosophy of science (Popper), and German anti-naturalist philosophy of humanities (Wilhelm Dilthey, Heinrich Rickert, Max Weber), see also Woleński (2011, 567).

1. intersubjective communicability of formulated statements,
2. intersubjective controllability of those statements,
3. empirical applicability of descriptive concepts (that is, non-logical and non-mathematical statements) (Kmita and Nowak 1968, 4–5).

These assumptions were present in Kmita's most important theories which were:

- concept of humanistic interpretation,
- historical epistemology,
- social-regulatory theory of culture.

Humanistic interpretation. The starting point for humanistic interpretation was an explication of the principle of rationality. According to Kmita (1971, 28):

> (Z_1) If X (in the time t) is to undertake one of actions $C_1, \ldots, C_n$, which, to his knowledge (in time t) exclude one another and (jointly) complement one another, and inevitably lead, respectively, to results $S_1, \ldots, S_m$, and results $S_1, \ldots, S_m$ are ordered via a relation of preference characteristic for X (in time t), then X (in time t) will undertake action C_i leading to preferred result S_j.

From the assumption of rationality (Z_1) and two premises:

Z_2: S_j is the result preferred the most by X_t; and
Z_3: it is a result of knowledge X_t that action C_i inevitably leads to result S_j,

there results the *eksplanandum*:

X will undertake, in time t, action $C_{i.}$.

The assumption of rationality was the basis for the concept of humanistic interpretation as Kmita distinguished a subclass of rational actions which was directed at interpretation and gave the examples of saying "Good morning" to someone or tipping one's hat. The aim of those actions will be achieved when someone interprets their meaning correctly. Actions like hammering a nail into a wall in order to hang a picture do not belong to that subclass. The aim of such an action will be *achieved* regardless of whether someone understands the reason for hanging the picture.

Actions directed at interpretation refer to the concepts of the rules of cultural interpretation and of cultural group. The rules of cultural interpretation are a set of hermeneutic principles—which are usually not fully codified and of which the individual is usually not fully conscious—which make it possible to capture the meaning of actions directed at interpretation. A collective of people who are mutually aware of their knowledge of cultural interpretation is called a cultural group.

Historical Epistemology Kmita (1976, 1980, 1988) formulated so-called theoretical history of science. According to him, the most important factors for development of science are external ones. In Kmita's reconstruction of Marxian historical materialism, development of different kinds of sciences resulted from the response for needs of different realms of social practice. The most fundamental social praxis is productive one that functionally determines the development of natural sciences. Humanities are directly depended on the legal-political practice (the superstructure in Marx's terminology) and indirectly on economical social practice (the base).

A new set of scientific views/theories become widespread because it responds better on the demand of a given realm of social practice in a given time. However, a new theory does not appear in an intellectual vacuum. This demand of a given realm of social practice operates on a set of hitherto existing scientific theories ("thought material") which may be selected, transformed and improved. In this way functional determination should be supplemented by genetic determination. According to Kmita development of science and "all forms of social consciousness is determined not in purely functional, but in functionally-genetic way" (Such 1996, 160). Scientific practice was seen by Kmita as a realm of social practice subjectively regulated by a specific set of beliefs—methodological consciousness of scientists. It is consisted of two kinds of beliefs: normative and instructive. The first one points out the kind of values (e gr. truth, utility) should be achieved, the second one tells how to achieve chosen values.

The Social-Regulatory Theory of Culture This theoretical approach to the history of science was developed into the social-regulatory theory of culture (Kmita 1982, 1985; Kmita and Kostyrko 1983; Banaszak and Kmita 1991). Culture was perceived a system of social consciousness consisted of normative and directive judgements. The first kind of judgements determines values that should be realized whereas second kind of judgments points out how these values should be realized. This system of beliefs (a) belongs to the given community *S* distinguished in a functional manner, (b) is generally respected in that community, and (c) intentionally regulates functional actions with respect to the global structural context of this these actions performed by this community (Kmita 1996, 582).

The starting point in Kmita's theory was the distinction between acceptance and respecting beliefs. He writes:

> The acceptance of belief *p* by a given individual consists in – we assume – that person consciously deeming it to be true, so the person is, in any way, aware of that belief. On the other hand, we speak of a person respecting belief *p* when the individual acts, undertakes the action as if that person accepted it, but we have not determined if that person indeed accepts the belief (Banaszak and Kmita 1991, 39).

Every type of social practices is subjectively regulated by an appropriate set of beliefs which determines what is worth doing and in what manner. Therefore, normative beliefs determine particular values which are the positive goals of an action, while directive beliefs point to actions which should (in particular circumstances) be undertaken in order to realize the assumed values (goals).

Social consciousness is not a sum of the consciousnesses of particular individuals but a set of beliefs consisting of directive and normative beliefs functional with respect to a type of social practice which is generally respected in a given collective. Social consciousness acts as a subjective regulator of social practice of a given type. The consciousness of particular individuals may consist of beliefs which are either functional or dysfunctional with respect to the social practice of a given type.

Kmita distinguished two types of normative beliefs: beliefs reachable in practice and beliefs unreachable in practice. For example, saving a certain amount of money or going on holidays are values reachable in practice. Values unreachable in practice cannot be realized because of the lack of sufficient means (at a given time and in a given place) or because they are unrealizable in the actual world. Values were also divided into final and instrumental. The latter are a means for realizing the former; the former are an end in themselves.

Those divisions formed the basis for distinguishing two spheres of culture: instrumental and symbolic. Kmita illustrates that with two examples:

1. a farmer plants seeds into the soil,
2. and a clergyman puts a metal ring onto the head of a standing person.

In order to achieve expected results determined by the sphere of instrumental culture, it is not necessary to observe normative and directive beliefs which would subjectively regulate them. In case of the coronation ceremony, the general adherence to the subjective beliefs which regulate the relevant set of actions is necessary.

The first kind of actions comprise a 'material' practice of the given society, which consists of the processes of production, exchange, and consumption. Kmita defines the production process as a transformation of a physical material (raw material) into objects which have usable features, that is, features which make those objects consumable without entailing the necessity of adhering to normative-directive beliefs regulating the production process.

As regards the condition of the effectiveness of the actions comprising a symbolic-cultural practice, normative-directive beliefs must be adhered to. Symbolic culture encompasses the sub-area of communication, that is: customs, language, and art, as well as the sub-area of non-communication, that is: science and political and legal consciousness, and the sub-area of worldviews.

The worldview sub-area determines a group of 'final' positive values (commonly called the meaning of life) and a group of 'final' negative values, and it characterizes the manner of the implementation of those final values with the help of practical (everyday life) values constitutes a holistic vision of the world (Kmita and Kostyrko 1983, 63–64).

Those sub-areas of symbolic culture have a socially integrating-distinguishing function, which means that they subjectively integrate particular individuals into certain collectives. Moreover, the communication sub-area maintains the information-communication contact inside the group through language. Social practice is mainly practical-cognitive, but it also has the additional didactic-cultural function which engages particular individuals in the cultural life.

Kmita also distinguished the world-view function of particular types of symbolic-cultural practice based on providing justification for beliefs belonging to other forms of social consciousness which regulate other social practices. There are three historical forms of a world views: magic, religion, and lay metaphysical philosophy.

3.2 *Leszek Nowak*

Leszek Nowak, the youngest father-founder of Poznań School authored six theories of the great scope (see: Brzechczyn 2017, 2022; Borbone 2021). However, in this chapter I presented only three connected with the topic of the volume. There are idealizational theory of science, the adaptive reconstruction of historical materialism and non-Marxian historical materialism.

The Idealizational Theory of Science The main idea of that concept is the assumption that a scientific theory is neither a generalization of facts nor a hypothetical-deductive system (Nowak 1980; Nowak and Nowakowa 2000). According to Nowak, a theory resembles a caricature more than a photograph because the construction of a scientific theory begins from a decisive deformation of reality, in the first, most idealized model. The idealizational law has the form of a conditional sentence. The antecedent contains counterfactual assumptions on the basis of which the influence of factors considered by the researcher to be secondary are omitted. The consequent describes how the studied phenomenon depends on its main factor. Thus, the obtained idealizational statement is only valid based on those idealizing assumptions—all of them.

Idealization is but one side of a scientific process. The other side is the concretization procedure, in the course of which the idealizing assumptions which have been made are canceled and the initial idealizational law is modified. It is shown how the studied phenomenon depends on secondary factors. Idealizational statements which are closer and closer to empirical data are obtained. Concretization takes place in a predefined order. First, those idealizing assumptions are canceled which pertain to the secondary factors with the greatest influence on the studied phenomenon. Then, secondary factors which exert a smaller influence are taken into account. Finally, all idealizing assumptions are canceled and a factual statement is obtained.

In research practice, though, final concretization is never carried out. Usually, after a series of concretizations has been made, the influence of the remaining, less important secondary factors are determined by way of an approximation, however, on condition that the level of the admissible divergences between the empirical data and the result obtained from the theoretical formula is not greater than the one generally accepted in the particular field of knowledge.

The functioning of a phenomenon is explained by showing the dependencies between that phenomenon and its main factor. Next, taking into account secondary factors, more and more realistic concretizations are derived from the idealizational law obtained in that way. The explained sentence is the results of the factual statement formulated on the basis of the procedure of concretization and of the initial conditions.

The Adaptive Interpretation of Historical Materialism One of the main applications of the idealizational theory of science was historical materialism. When Marxism is interpreted, it is commonly understood as saying that there are cause and effect relationships between global productive forces and relations of production, a social base and a legal and political superstructure, social and economic conditions and particular states of social consciousness. That has been giving rise to well-known interpretive difficulties as it has not been clear how productive forces are to cause the appearance of particular relations of production, a social base—of a legal and political superstructure, and social and economic conditions—of particular states of social consciousness.

One solution to that problem was the adaptive understanding of those dependencies proposed by Leszek Nowak (1982). According to him from set of states of affairs α that state of affairs A becomes widespread, which, in conditions B will lead to a result characterized by k to the greatest degree. This property k is called the criterion of adaptation.

Nowak illustrated his interpretive suggestion with the example given below. Let us assume that a person has invented a new device (conditions B) which improves work efficiency and, consequently, makes it possible to increase the surplus product (criterion of adaptation k). As they compare various systems of the organization of production (traditional, invented by specialists, etc.), owners select such a system (state of things A) as they believe to be capable of ensuring the greatest surplus product. However, if the expected increase does not take place, the owners—thinking that they have made the wrong choice—will continue looking for an optimal—for the given level of productive forces—system of the organization of production. If any of the owners delays the reorganization of production, that owner's profits from the surplus product will shrink and, in the end, he or she will go bankrupt. After a sufficiently long time, by trial and error, and the elimination of those who have not learned fast enough, an optimal system of the organization of production, with respect to the newly introduced tool, will become common in the observed economic sector.

There are similar adaptive dependencies between the level of productive forces and the system of the organization of production, between the superstructure and the economic base, and between social consciousness and social being. In this case, the interests of the owners' classes, guaranteed by the legal and political system, function as a selector of individual ideas. On a mass scale, such ideas become common as, in given social and economic conditions, ensure the durability of a political system.

During the adaptive-idealizational reconstruction of historical materialism, Nowak noticed the contradiction between the philosophical assumptions and the social theory of Marxism. That contradiction, called the paradox of historism. Namely, the Marxist dialectics presupposes the changeability of the main factors of studied phenomena (Nowak 1977, 89–90), while the Marxist historical materialism presupposes that productive forces and the relations of production are the main determining factors in social life, in all societies and in all historical periods. This presupposition gives rise to the question of whether the repertoire of main factors is or is not subject to change in historical process. If the answer is positive, then another question arises about the nature of those non-economic but still material factors.

Within the framework of Adaptive Interpretation of Historical Materialism researchers have been trying to deal with the aforementioned paradox of historicism by creating an idealizational-adaptive interpretation of Engels's historical materialism. For the proponents of that approach—which has been developed primarily by Jolanta Burbelka (1982)—the main factor in primitive societies is the reproduction of life and not economic production. They claim that the production of goods only becomes the main factor in the epoch of class societies.

Distinguishing between two types of materialism: Engels's and Marx's entailed the problem of generalizing them and of constructing such a third form of materialism as would be applicable to systems of real socialism. Piotr Buczkowski, Andrzej Klawiter, and Leszek Nowak (1982) tried to construct a third version of materialism. The authors assumed that the structures of the political and cultural spheres were analogous to that of the economic sphere. In the political sphere, we could distinguish the means of coercion and a system of the organization of power which constituted a political base; the political base and the system of political institutions combined to produce social and political conditions, and there was also political consciousness. The cultural momentum of a society was reconstructed in a similar way, by distinguishing the means of knowledge production and a system of the organization of knowledge production making up the cognitive base of a society. The cognitive base with the system of the organization of knowledge created cognitive and social conditions having an impact on meta-cognitive consciousness. The distinguished components of the political and spiritual momentums were considered to be linked by means of adaptive dependencies.

Non-Marxian Historical Materialism For Nowak, attempts at applying the generalized form of the adaptive interpretation of historical materialism to the construction of a theory of socialism appeared to be unconvincing, which inspired him to construct non-Marxian historical materialism (Nowak 1983, 1991a). The theory overtakes the AIHM view of the isomorphic structures of three realms of social life: politics, culture, and economy, and it strengthens the antagonistic nature of historical materialism. Consequently, the concept of a social class is generalized. Nowak assumes that social classes exist not only in economy but also in politics and culture. In political life the rulers' class, which has at its disposal the means of coercion, increases the global sphere of influence and restricts citizens' autonomy. In economic life the owner' class, which has at its disposal the means of production, maximizes their additional product at the cost of producers' direct income. In culture, the monopoly of the means of spiritual production allows the priests caste to augment their spiritual authority and restrict believers' autonomy. The social antagonisms based on unequal access to material social means (means of coercion, production, and indoctrination) in each of the three realms of social life are, then, autonomous.

Class divisions can also cumulate, and, for example, one social class, in order to boost its social power, can overtake the means of coercion and of production or the means of coercion and of indoctrination, etc. Hence, apart from class societies (with three separate classes), there also occur supraclass societies in which one social class controls culture and politics (fascism) or culture, politics, and economy (communism). Non-Marxist historical materialism contains an interpretation of 2500 years of the history of class societies of the European civilization, from the slave societies of ancient Greece and Rome to the contemporary Western European capitalist societies and East European socialist societies (see Brzechczyn 2007).

3.3 Jerzy Topolski

Jerzy Topolski, the eldest of the founding fathers of the Poznań School of Methodology, developed Nowak's and Kmita's concepts creatively within the framework of the methodology of historical research and theoretical history (collection of Topolsk's papers in English, see: Brzechczyn 2009; Domańska 2022). Some of Topolski's most important achievements are:

- the dynamic concept of the historical source,
- the theory of historical narration,
- the concept of integral explanation.

The Theory of a Historical Source In Topolski's (1968, 324) words:

> a historical source is any source of historical cognition (direct and indirect), that is, any information (within the meaning given to that term in the theory of information) about the social past, wherever they are, together with what transmits that information (the information channel).

Topolski distinguishes between a potential and effective source. A potential source is any source of a historian's knowledge about the past, while an effective source is the collection of information currently used by a historian.

In that concept, the credibility of a historical source is relative—dependent on the question. For one research question, the source can be quite credible, while for another question—not so credible. According to Topolski (1983, 257):

> Thus, the key which triggers the mechanism of searching for source information (that is, searching for sources) is the research question posed for the source. Without asking such a question, the source will not 'tell' the historian anything, it will remain silent even if the historian can read it. By asking those questions, based on – obviously – the historian's other knowledge (and methodological consciousness), the historian undertakes the initial modeling of the reality research.

When he characterized the information structures of the source, Topolski was inspired by Jerzy Kmita's concept of sign, in which a sign is defined as an activity or an outcome of a human activity which is a result of intentional communication of a particular state of things. Apart from signs, there are symptoms, which can be divided into humanistic and natural. Humanistic symptoms are human activities or their outcomes not intended by their creator to be used for communication. For example, smoke coming out of a chimney is a symptom of the human presence in that house and of the preparation of a meal, but it is not intended to communicate anything (Kmita 1973, 23–33, 210–211).

Topolski (1983, 263) distinguished three layers of the information structure of a source. The surface layer of the information structure of source is a set of

> (…) the information which can be extracted from the sources in the simplest way, that is, which is legible without the necessity of asking questions other than questions concerning directly the information in the information set.

The questions are of the simplest type: what, who, and when. The sign layer of the information structure of a historical source is revealed by asking what the information extractable with those questions signifies. We reach the symptom layer of the information structure when we ask what the information in the source is a symptom of.

Topolski also introduced an original division into addressed and unaddressed sources. He claims that addressed sources were created with the intent to communicate a state of things, while non-addressed sources were not. In this case, the criterion of division is the fact of the existence or non-existence of the information relationship of the type: information sender—information channel—recipient between the creator of the source and the historian (or another recipient). The division does not concern the material layer of sources (the carrier of information) but the information included in the sources (Topolski 1983, 260). The addressed sources can be divided into certain categories, depending on whether they are addressed to:

1. recipients of a source contemporary to the author—for example, letters, reports, announcements,
2. posterity in general—inscriptions on headstones, documents which confirm someone's rights,
3. a historian—diaries written with that intention (Topolski 1983, 260).

Because of the addressed source authors' intention to communicate something, the sources have an element of persuasiveness directed toward their recipient, and they already contain an interpretation of the information they impart. For that reason:

> the historian who wants to extract information from such sources must interpret the sources, that is, carry out the interpretation already included in the source. Examining the credibility of the informant – that is, considering whether the informant is telling the truth in the given case does not suffice here (Topolski 1996, 341)

The active role of a historian in the search for source information was called—controversially—the ontological and epistemological creation of the past, that is, including the knowledge about the past in the individual and social consciousness. In Topolski's words (1977, 45):

> Obviously, the past exists objectively as something that happened, but without its reconstruction it would not exist for people, and it would not be able to influence human actions through people's consciousness of their history. In such a case, there would only exist an influence of the past of which they would not be aware.

According to Topolski, searching for sources encompasses the discovery of new types of sources possibly containing the information looked for by the researcher, the constitution of sources by creating a set of question, and interpreting the source information extracted in that way.

The Theory of Historical Narration The results of historical research are presented in the form of narration which contains both idiographic and theoretical layers. Topolski formulated the paradox of historical narration. Seemingly, narration is treated as a conjunction of its constituent sentences. As we know, a conjunction is true when all its constituents are true. However, in the case of a historical narrative:

1. the truthfulness of all constituent sentences does not warrant the truthfulness of the whole narrative,
2. a historical narrative may remain true even if some of its constituent sentences are false, and
3. a greater proportion of true constituent sentences in a narration in comparison to another narration does not guarantee a greater degree of the truthfulness of the first narration.

According to him, the truthfulness of narration depends on its theoretical layer which includes the basic relations and dependencies. Topolski assumes that a historical narration consists of historical sentences, sequences of historical sentences, and historical images. Historiographical sentences and sequences of historiographical sentences comprise articulated parts of a historical narrative, while a historiographical image is a non-articulated part of such a narrative which contain theoretical assumptions (Topolski 1987a).

Topolski distinguished the following layers of narration: surface (factographic), persuasive-communicative and theoretical-ideological. The theoretical-ideological layer determines the two other layers of a narrative:

> The deep layer of a narrative is a dynamic mechanism which imparts that narrative its structure and determines the choice and the configuration of the item information which the historian has at his disposal (Topolski 1994, 37)

The surface layer of narration is maneuvered through the theoretical layer by means of three mechanisms:

- selection of source information and non-source information,
- hierarchization of source and non-source information, and
- comparison of the information selected and hierarchized in that way in narrative wholes (sequences and images).

The Concept of Integral Explanation According to Topolski, the historical process is a global result of conscious and intentional actions of people (individuals and groups). Those global results, however, only partially overlap with the people's intentions and goals. People set their goals of action based on particular motives and taking into account environmental and social conditions, which they can evaluate with varying degrees of accuracy. Thus, such subjective factors as knowledge, motivation, values, or preferences are linked to objective factors (environmental and social conditions of being) through people's knowledge and valuing system. The efficacy of people's actions depends on the adequateness of their knowledge about the conditions in which they operate, including the environmental and social laws as well as the popularized value systems which determine the goals. Therefore, integral explanation should encompass people's motivational structures as well as (by a reference the laws) the objective results of people's actions.

One example of such explanation was presented in book devoted the rise of capitalism in Europe (Topolski 1987b). According to him, in the fourteenth century, the income of the gentry in Western Europe decreased, as a result of various external factors (the plague, inflation, the increase in the numbers of the gentry, etc.). Consequently, there was growing disparity between the lifestyle dictated by the contemporary culture and the revenue which would allow for such a way of life. The gentry was forced to counteract its relative impoverishment. Topolski lists four basic methods of increasing the income of this social estate: appropriating more income at the cost of the peasants, overtaking a part of the income of the bourgeoisie, changing the division of income within the feudal class, and undertaking business activity (Topolski 1987b, 72). The attempts at a greater exploitation of the peasantry triggered a series of rebellions and uprisings in the whole Europe, from the second half of the fourteenth century to the beginning of the sixteenth century. It was not easy to limit the income of the bourgeoisie, either, as it was a strong class the finances of which could only be restricted through family relationships. Overtaking church assets, with the help of the state, was only a short-term solution. It turned out that the only valid long-term method of increasing the income was for the gentry to undertake business activity. The particular forms of this entrepreneurship depended on the local conditions. In England, it was raising sheep, in Bohemia—fish farming in ponds, and in Poland—grain production. In Western Europe, the power of bourgeoisie and the existing capitalist system directed the business activity of the gentry toward commodity money economy with the use of mercenary workforce, which was conducive to the appearance of capitalism. In Central Europe, the economic activity of the gentry contributed to functional serfdom.

4 Conclusions

Poznań School of Methodology was a scientific milieu that in the 1970s and 1980s of the twentieth century exerted a great impact on the development of various kinds of natural and social sciences and humanities, including archeology, in Poland. On the one hand, the word "Poznań" in the name of school suggests a limitation to one city. In fact, proponents of the school were employed in universities in Toruń, Lublin, Łódź, Szczecin or Gdańsk. On the other hand, the word "school" suggests a strict set of uniform views and an existence of a one Master. In fact, this milieu was created by three charismatic personalities of Kmita, Nowak and Topolski followed by teams of their pupils and collaborators. Very often, alternative explanations of a given phenomenon were offered in the school. It was the case of philosophy of science. E. gr. Nowak in his idealizational theory of science focuses on internal factors of development of scientific activity, whereas Kmita in his historical epistemology emphasized the influence of external factors. The common feature was a style of research based on methodological self-consciousness and logical clarity inspired by analytical tradition of doing philosophy.

In the time of its heyday, the school gathered about several hundred people. They edited several journals, published hundreds of books and collections of papers. The most important journals were *Studia Metodologiczne* and *Poznańskie Studia z Filozofii Nauki*. Their works have been published in English in the *Poznań Studies in the Philosophy of the Sciences and Humanities* publication series which is still being published today by the Brill.

One can risk putting a thesis that factor contributing to the rise of the school was relative isolation of Polish science from the Western ones. Relative because polish humanities and social sciences had some contacts with the most important achievements of their Western counterpart. However, these limitations imposed by the political system forced to critical and original reception of Western output. After 1989 the opening of Polish philosophy (and general Polish culture) on Western philosophy and humanities contributed to uncritical acceptance of concepts and theories coming from the better (scientifically) developed West. In order to get on in Polish science, it is enough to popularize what was invented in the first scientific world (Nowak 2012). This provincial attitude can hamper the further development of the school which was an important part of history of the humanities in Poland.

References

Ajdukiewicz, K. 1949. *Zagadnienia i kierunki filozofii: teoria poznania, metafizyka*. Warszawa: Czytelnik.

Banaszak, T., and J. Kmita. 1991. *Społeczno-regulacyjna koncepcja kultury*. Warszawa: Instytut Kultury.

Borbone, G. 2021. The relevance of models. Idealization and concretization in Leszek Nowak. München: Grin Verlag.

Brzechczyn, K. 2007. On the application of non-Marxian historical materialism to development of non-European societies. In *The courage of doing philosophy: Essays dedicated to Leszek Nowak*, ed. J. Brzeziński, A. Klawiter, T. A. F. Kuipers, K. Łastowski, K. Paprzycka, and P. Przybysz, 235–254. Amsterdam/New York: Rodopi.

Brzechczyn, K. (Ed.) 2009. *Idealization XIII: Modeling in history. Poznań Studies in the Philosophy of the Sciences and the Humanities*. Vol. 97. Amsterdam/New York: Rodopi.

Brzechczyn, K. 2017. From interpretation to refutation of Marxism. On Leszek Nowak's non-Marxian historical materialism. *Hybris. Internetowy Magazyn Filozoficzny* 37:141–178.

Brzechczyn, K. 2022. Preface. In *New developments in the theory of the historical process: Polish contributions to non-Marxian historical materialism. Poznań Studies in the Philosophy of the Sciences and the Humanities*, vol. 119, ed. K. Brzechczyn, vii–xxvii. Leiden/Boston: Brill.

Buczkowski, P., A. Klawiter, and L. Nowak. 1982. Historical materialism as a theory of social whole. In *Social classes action and historical materialism. Poznań Studies in the Philosophy of the Sciences and the Humanities*, vol. 6, ed. L. Nowak, 236–280. Amsterdam: Rodopi.

Burbelka, J. 1982. Historical materialism. General theory and forms. In *Social classes action and historical materialism. Poznań Studies in the Philosophy of the Sciences and the Humanities*, vol. 6, ed. L. Nowak, 211–235. Amsterdam: Rodopi.

Domańska, E. 2021. Jerzy Topolski's Marxist anthropocentrism. *Journal of the Philosophy of History* 15 (3):361–377.

Domańska, E. 2022. Jerzy Topolski's theory and methodology of historical knowledge. In *Theory and methodology of historical knowledge: An anthology*, ed. E. Domańska and Anna Topolska, 17–48. Poznań: Faculty of History Press.
Gan-Krzywoszyńska, K., and P. Leśniewski. 2016. *Kazimierz Ajdukiewicz (1890–1963)*. Poznań: Wydawnictwo Naukowe UAM.
Grad, J. 2011. Jerzy Kmita – uczony, nauczyciel akademicki, organizator życia naukowego, popularyzator nauki. *Filo-Sofija* 11 (12):19–68.
Ingarden, R. 1947–1948. *Spór o istnienie świata*, Vol. 1–2. Kraków: Polska Akademia Umiejętności.
Kmita, J. 1968. *Problematyka terminów teoretycznych w odniesieniu do pojęć literaturoznawczych*. Poznań: Poznańskie Towarzystwo Przyjaciół Nauk.
Kmita, J. 1971. *Z metodologicznych problemów interpretacji humanistycznej*. Warszawa: Państwowe Wydawnictwo Naukowe.
Kmita, J. 1973. *Wykłady z logiki i metodologii nauk*. Warszawa: Państwowe Wydawnictwo Naukowe.
Kmita, J. 1976. *Szkice z teorii poznania naukowego*. Warszawa: Wydawnictwo Naukowe PWN.
Kmita, J. 1980. *Z problemów epistemologii historycznej*. Warszawa: Państwowe Wydawnictwo Naukowe.
Kmita, J. 1982. *O kulturze symbolicznej*. Warszawa: COMUK.
Kmita, J. 1985. *Kultura i poznanie*. Warszawa: Państwowe Wydawnictwo Naukowe.
Kmita, J. 1988. *Problems in historical epistemology*. Warszawa/Dordrecht: Wydawnictwo Naukowe PWN, Reidel.
Kmita, J. 1996. Towards cultural relativism "with a small 'R'". In *Epistemology of history. Poznań Studies in the Philosophy of the Sciences and the Humanities*, vol. 47, ed. A. Zeidler-Janiszewska, 541–615. Amsterdam: Brill.
Kmita, J., and T. Kostyrko. 1983. *Elementy teorii kultury: wykład dla studentów kulturoznawstwa*. Poznań: UAM.
Kmita, J., and L. Nowak. 1968. *Studia nad teoretycznymi podstawami humanistyki*. Poznań: Wydawnictwo UAM.
Malewski, A., and J. Topolski. 1962. *Studia z metodologii historii*. Warszawa: Państwowe Wydawnictwo Naukowe.
Nowak, L. 1966. *Model, prawda względna, postęp nauki*. Unpublished MA thesis, UAM Poznań.
Nowak, L. 1977. U podstaw dialektyki Marksowskiej. Próba interpretacji kategorialnej. Warszawa: PWN.
Nowak, L. 1980. *The structure of idealization. Towards a systematic interpretation of the Marxian idea of science*. Dordrecht/Boston/London: Reidel.
Nowak, L. 1982. The theory of socio-economic formation as a theory of adaptation process. In *Social classes action and historical materialism. Poznań Studies in the Philosophy of the Sciences and the Humanities,* vol. 6, ed. L. Nowak, 110–121. Amsterdam: Rodopi.
Nowak, L. 1983. *Property and power. Towards a non-Marxian historical materialism*. Dordrecht/Boston/Lancaster: Reidel.
Nowak, L. 1991a. *Power and civil society. Toward a dynamic theory of real socialism*. New York: Greenwood Press.
Nowak, L. 1991b. *U podstaw teorii socjalizmu*. Vol. 1–3. Poznań: Nakom.
Nowak, L. 1998. The adaptive interpretation of historical materialism: A survey. On a contribution to polish analytical Marxism. In *Marx's theories today. Poznań Studies in the Philosophy of the Sciences and the Humanities*, vol. 60, ed. L. Nowak and R. Panasiuk, 201–236. Amsterdam/Atlanta: Rodopi.
Nowak, L. 2012. The structure of provincial thought. Half essay, half thesis. In *Thinking about provincialism in thinking. Poznań Studies in the Philosophy of the Sciences and the Humanities*, vol. 100, ed. K. Brzechczyn and K. Paprzycka, 51–66. Amsterdam/New York: Rodopi.
Nowak, L., and I. Nowakowa. 2000. Idealization X: The richness of idealization. In *Poznań Studies in the Philosophy of the Sciences and the Humanities*, vol. 69. Amsterdam/Atlanta: Rodopi.
Ochocki, K. 1978. *Wokół sporów o filozofię*. Warszawa: Wiedza Powszechna.

Ossowska, M. 1947. *Podstawy nauki o moralności*. Warszawa: Spółdzielnia Wydawnicza Czytelnik.

Skoczyński, J., and J. Woleński. 2010. *Historia filozofii polskiej*. Kraków: Wydawnictwo WAM.

Skolimowski, H. 1969. *Polski marksizm*. Londyn: Odnowa.

Stępień, A. 1990. Rys rozwoju filozofii marksistowskiej i problem dialogu filozoficznego z marksizmem od roku 1945. In *Wobec filozofii marksistowskiej. Polskie doświadczenia*, ed. A. Stępień, 21–59. Lublin: TN KUL.

Such, J. 1996. Types of determination vs. development of science in historical epistemology. In *Epistemology of history. Poznań Studies in the Philosophy of the Sciences and the Humanities*, vol. 47, ed. A. Zeidler-Janiszewska, 157–168. Amsterdam: Rodopi.

Tatarkiewicz, W. 1947. *O szczęściu*. Kraków: Wiedza-Zawód-Kultura, Tadeusz Zapiór i S-ka.

Tatarkiewicz, W. 1958. *Historia filozofii. T. III. Filozofia XIX wieku i współczesna*. Warszawa: Państwowe Wydawnictwo Naukowe.

Tischner, J. 1981. *Polski kształt dialogu*. Paryż: Editions.

Topolski, J. 1968. *Metodologia historii*. Warszawa: Państwowe Wydawnictwo Naukowe.

Topolski, J. 1977. *Marksizm i historia*. Warszawa: Państwowy Instytut Wydawniczy.

Topolski, J. 1983. *Teoria wiedzy historycznej*. Poznań: Wydawnictwo Poznańskie.

Topolski, J. 1987a. Historical narrative: Towards a coherent structure. *History and Theory* 26:75–86.

Topolski, J. 1987b. *Narodziny kapitalizmu w Europie XIV-XVII wieku*. Warszawa: Państwowe Wydawnictwo Naukowe.

Topolski, J. 1994. A non-postmodernist analysis of historical narratives. In *Historiography between modernism and postmodernism. Contributions to the methodology of the historical research. Poznań Studies in the Philosophy of Sciences and the Humanities*, vol. 41, ed. J. Topolski, 9–86. Amsterdam/Atlanta: Rodopi.

Topolski, J. 1996. *Jak się pisze i rozumie historię. Tajemnice narracji historycznej*. Warszawa: Rytm.

Woleński, J. 1993. Marxism and the professionalization of philosophy. In *Philosophy and political change in Eastern Europe*, ed. B. Smith. La Salle, Ill: Hegeler Institute.

Woleński, J. 2011. From controlled liberalism to real pluralism. The development of philosophy in Poland at the end of the communist era. In *Twenty years after the collapse of communism. Expectations, achievements and disillusions of 1989*, ed. N. Hayoz, L. Jesień, and D. Koleva, 559–572. Bern: Peter Lang.

Zamiara, K. 2010. U początków poznańskiej szkoły metodologicznej. In *Filozofia na Uniwersytecie w Poznaniu. Jubileusz 90-lecia*, ed. T. Buksiński, 283–308. Poznań: Wydawnictwo Naukowe Instytutu Filozofii UAM.

The Poznań School of Methodology and Its Continuations

Jan Grad

Abstract The Poznań School of Methodology is a research milieu, which was created at Adam Mickiewicz University in Poznań from the mid-1960s until 1973. Its founder was Jerzy Kmita—a philosopher and methodologist. It developed an original, naturalistic, and anti-positivist methodological model of the humanities, in which a scientific explanation procedure—referred to as the humanistic interpretation—defines the specific character of the humanities and social sciences.

Between 1970 and 1973, the group analyzed Marx's research method, determining that it consisted of a procedure of idealization and concretization. After this period, the Poznań School of Methodology disbanded as a research group sharing common theoretical assumptions. Three academic teams emerged, of which the one considered a continuation of the School was the group of researchers centered around Jerzy Kmita. They pursued the work in the realm of the theory of scientific knowledge within the framework of the historical epistemology proposed by Kmita himself and later from the perspective of his socio-regulative theory of culture. These approaches formed the basis of what is known as the "Poznań School of Cultural Studies."

Keywords Poznań School of Methodology · Marxism · Idealization · Concretization · Historical epistemology · Socio-regulative theory of culture · Jerzy Kmita

The following reflections are not intended to be a detailed account of the history of the Poznań School of Methodology and its activities. This has already been the subject of various works published on the occasion of the 90th and 100th anniversaries of the founding of Adam Mickiewicz University in Poznań (Mackiewicz 2001;

J. Grad (✉)
Institute of Cultural Studies, Faculty of Anthropology and Cultural Studies, Adam Mickiewicz University in Poznań, Poznań, Poland
e-mail: gradjan@amu.edu.pl

A. Marciniak et al. (eds.), *The Poznań School of Archaeology*, Outstanding Contributions to Archaeology, https://doi.org/10.1007/978-3-032-07848-3_3

Kotowa 2010; Zamiara 2010; Grad 2019). This paper recognizes the scholarly importance of the Poznan school of methodology for the Polish humanities, as well as its original and innovative theoretical and cognitive insights into the study of culture. The achievements of the Poznań school and its methodological model of the humanities have had a significant impact on research practice in the humanities and social sciences. Although the period from the mid-1960s to the early 1970s is considered the heyday of the Poznań School of Methodology, its continuations have adhered to the school's original model, incorporating the concept of humanistic interpretation, or the Marxist approach, and they also include studies in historical epistemology conducted from the perspective of socio-regulatory cultural theory ("Poznań School of Cultural Studies").

1 The Development of the Poznań School of Methodology (from the 1960s to 1969) as a Naturalistic and Anti-positivist Model of the Humanities

The establishment of the Poznań School of Methodology was connected with the research and teaching activities of members of the Department of Logic at the Faculty of Philosophy and History of Adam Mickiewicz University in Poznań, which was headed in 1961–1967 by Professor Jerzy Giedymin (1925–1993). After Giedymin left for the United Kingdom in 1966 to continue his academic career there (until his retirement in 1990), Dr. Jerzy Kmita (1931–2012) (associate professor after 1968) became the head of the Department.

The research activities of the Department of Logic from the mid-1950s to the end of the 1960s constitute the first period of the Poznań School of Methodology witnessing the emergence of its research program and paradigms. In this period, according to Kmita (1977a):

> (…) a) research was carried out in the field of general methodology of empirical sciences, but mainly focused on individual disciplines (…); b) from the very beginning our research was characterized by an increasingly intense anti-positivist orientation (…); c) conceptual methodology of sciences, or theory of scientific cognition, was systematically developed as an empirical and humanistic discipline at the same time.

Thanks to the publications of Giedymin and Kmita, the Department of Logic pursued methodological research in an innovative way, which was readily recognized by the academic community of the time. The Department's research program was based on the concept of "empirical logic", i.e. the pragmatic methodology of Kazimierz Ajdukiewicz (1890–1963), which postulated a logical analysis of specific ways of research in various disciplines leading to cognitively valuable results. This was in contrast to overly formalized logical research in science, i.e. "a kind of formal graphomania", as Kmita called it, and to speculative studies based on ontological and theoretical assumptions of certain philosophical orientations, without any practical-cognitive implications and relevance to the actual research practice in the empirical sciences. The basic research procedure for analysis recommended by Ajdukiewicz in his descriptive methodology was "logical reconstruction", which was later acknowledged in the humanities as a typical analysis of theorems used by the Poznań School of Methodology. It involved the identification of tacitly accepted and even unrecognized premises, but above all, theoretical assumptions in the research procedure leading to the achievement of specific results. The cognitive effectiveness of this procedure in theoretical and methodological studies was noted by the reviewers of Kmita's postdoctoral dissertation, who emphasized the originality and novelty of his research results. The most significant for the naturalistic and antipositivistic logical reconstruction in the methodological analyses of the Department of Logic was the finding that research in the humanities assumes various theories of rational behavior, which makes logical reconstruction more appropriate than anti-naturalistic approaches. The idea of rational action was developed by Ajdukiewicz, adopted by Giedymin, and found its "ultimate" manifestation in Kmita's concept of "subjective rational action".

In addition to Ajdukiewicz's logological approach, the hypothetistic epistemological ideas of Karl R. Popper (1902–1994) inspired the new methodological research. They were transplanted into Polish philosophy of science and creatively developed (since the late 1950s) by Jerzy Giedymin, who worked on the general methodology of sciences and on the methodology of social and historical sciences in particular (Giedymin 1958, 1964). Wilhelm Dilthey's anti-naturalistic philosophy of the humanities was also of interest to Giedymin. Jerzy Giedymin examined the fundamental "dispute between naturalism and anti-naturalism in the understanding of social sciences" and the "concept of rational action" (1958/1959, 1961, 1964), which remained in the scope of interest of the already formed Poznań School of Methodology. These concepts became the main "slogans" of the school.

Equally important for the development of research in logical theory of science at the Department of Logic were Kmita's analyses of the "language of empirical sciences" contained in his doctoral dissertation: *Znaczenie jako wykładnik sposobu rozumienia wyrażeń* (*Meaning as an Exponent of Understanding Expressions*) (1961), in which he positively resolved the problem of whether the humanities, as sciences based on understanding (in accordance with Ajdukiewicz's approach), are empirical sciences in the strict methodological sense.

Kmita related general methodological analyses to empirical sciences (humanities and social sciences) and illustrated them with findings in specific disciplines from the works of Giedymin (1964), Kmita (1964a, b), and Zamiara (1974). Kmita's post-doctoral dissertation *Problematyka terminów teoretycznych w odniesieniu do pojęć literaturoznawczych* (*Problems of Theoretical Terms in Relation to Literary Concepts*) (1967), as the title suggests, generally focused on the methodology of literary studies, a discipline with which the author, as a graduate of Polish Studies, was most familiar.

Other scholars who also addressed methodological issues in the humanities included Teresa Kostyrko (born 1932), who focused on evaluative statements and the language of theoretical and historical description of art; Włodzimierz Ławniczak in his studies of research practices in art history (Ławniczak 1971); and Tadeusz

Zgółka (1976), who examined structuralist procedures for explaining linguistic phenomena. The crowning achievements of the Poznań school were *Studia nad teoretycznymi podstawami humanistyki* (*Studies on the Theoretical Foundations of Humanities*) by Jerzy Kmita and Leszek Nowak (1968), and Kmita's monograph *Z metodologicznych problemów interpretacji humanistycznej* (*On Methodological Problems of Humanistic Interpretations*, 1971). At the same time, the philosopher Jan Such(1968, 1969) published his works on general methodology of sciences. In 1969, the last year of the Department of Logic, its members were Jerzy Kmita, Teresa Kostyrko, Włodzimierz Ławniczak (1927–2012), Krystyna Zamiara (1940–2010), and Tadeusz Zgółka (1945–2021). This group of researchers, representing disciplines of the humanities (art history, psychology, linguistics, jurisprudence), reconstructed the methodological premises of research on the basis of their original disciplines. After all, according to Ajdukiewicz's pragmatic methodology, the "logic of science" should cover specific modes of research in certain extra-philosophical disciplines, and this task can be best performed by logically and methodologically competent representatives of various disciplines, i.e. by those with knowledge of the logical apparatus. Jan Such (born 1931) conducted his research on general methodological issues in the Department of History of Philosophy at the Faculty of Philosophy and History (since 1961). According to Kmita and Nowak (1968, 4), *Studia nad teoretycznymi podstawami humanistyki* constituted:

> (…) an attempt to counter the anti-naturalistic concepts of the humanities with a naturalistic position that, while respecting the valid intuitions of the anti-naturalistic approaches, would nevertheless be free of some of their flaws.

Among the anti-naturalistic approaches to logical reconstruction, Kmita and Nowak examined the following orientations: intuitionist-irrationalist (Wilhelm Dilthey), intuitionist-intellectualist (Eduard Spranger), and instrumentalist-antipositivist (Max Weber); and among the naturalistic approaches, the positivist (John Stuart Mill) and antipositivist (Karl Popper) orientations. Their cognitive confrontations led Kmita and Nowak to construct an antipositivist naturalistic model of the humanities using Popper's hypothetism. They replaced its inherent methodological and objective individualism with methodological structuralism, supplemented by an anti-psychological directive that humanistic research presupposes rational theories of behavior.

The finding that the research practice of the humanities tacitly applies theories of rational behavior is an important methodological discovery. It is reflected in the methodological model of the humanities of the Poznań school in its premise of rationality, which plays the role of the law of science in humanistic interpretation, understood as a kind of deductive explanation. Humanistic interpretation is precisely the methodological specificity of the humanities in relation to the natural sciences.

The outlined methodological model of the humanities was intended to provide a theoretical context for research analyses in the humanities and social sciences. It did not aim at solving various specific problems, let alone at demonstrating concrete research applications, which was eagerly awaited by representatives of these scientific fields. Jerzy Kmita, to a certain extent, took this into account in his work *Z metodologicznych problemów interpretacji humanistycznej* (1971), in which he exemplified theoretical and methodological considerations of explanatory procedures with reference to the conduct of research in the humanities. This problem has also been recognized by representatives of the humanities and social sciences, who have affirmed the usefulness of relevant methodological analyses in their research practice.

The second period of activity of the Poznań School of Methodology (1970–1973) is connected with the establishment of the Institute of Philosophy in 1969 after the reorganization of Adam Mickiewicz University and its institutional units: the Department of History of Philosophy, the Department of Marxist Philosophy, and the Department of Logic and Methodology of Sciences. The faculty members of these new departments came from the former Department of Logic. In 1970, the Department of Logic and Methodology of Sciences hired Dr. Leszek Nowak (1943–2009), who had previously attended Jerzy Giedymin's seminars as a graduate student. Nowak's intellectual and substantive contribution to the theoretical and methodological concepts formed in his logic seminars made him a co-founder of the research paradigm of the Poznan school of methodology.

2 The "Marxist Turn" in the Poznań School of Methodology (1970–1973)

The new period of activity of the Poznań School of Methodology is marked by a transition from "purely methodological" or "logistic" research, or "empirical logic" (in Ajdukiewicz's words), to Marxist social research, while maintaining its analytical premises (Kotowa 2010, 309). A number of references to the Marxist concepts of action and the determination of human activity by socio-economic relations in *Studia nad teoretycznymi podstawami humanistyki* (Kmita 1968) can be seen as harbingers of the "Marxist turn" in the research of Poznań methodologists. Undoubtedly, the historian and historical methodologist Jerzy Topolski (1928–1998) was very influential in bringing the Poznań methodological community to focus on Marxist methodology of the humanities, as he inspired a group of researchers to take an interest in Karl Marx's research on the socio-economic system of capitalism. Their early findings were published in a collection edited by Topolski (1970) entitled *Założenia metodologiczne 'Kapitału' Marksa* (*Methodological Foundations of Marx's Das Kapital*). By the time Leszek Nowak joined the Poznań School, he had already made some in-depth analyses in his postdoctoral dissertation, *U podstaw Marksowskiej metodologii* (*The Foundations of Marxist Methodology*, 1971). In the Marxist methodology practiced "in the Poznań way", "... the question of the role in the sciences... of laws and theories reconstructed as so-called idealization laws" and the question of the "explanation inherent in historical materialism" were particularly emphasized.

The Marxist turn in the Poznań School of Methodology began with the publication of the collective study *Założenia metodologiczne 'Kapitału' Marksa* (*Methodological Premises of Marx's Das Kapital*) (Janiec 1970) and ended with the publication of *Elementy marksistowskiej metodologii humanistyki* (*Elements of Marxist Methodology of the Humanities*) (Kmita 1973b). According to a representative of the Polish philosophical community, the research program of the Poznań methodologists was "(…) the first major attempt in Poland to develop a truly adequate Marxist methodology of the humanities" (Augustynek 1977). Studies reconstructing Marxist methodological directives for explanation in the humanities, including idealization and concretization, were confronted with opposing philosophical and methodological orientations: positivism, anti-naturalism, Popperism, and structuralism. The Marxist turn analyses demonstrated that,

> Marxist methodology, i.e., a corresponding system of (…) directives, dominates the various types of methodological systems opposed to it. It surpasses them primarily in terms of content, i.e., research procedures, especially in the humanities, according to its authentic recommendations, fulfill the necessary condition for achieving research results of cognitive value, which are otherwise unattainable when adopting a different methodological perspective (Kmita 1973a, 6).

The analyses and findings of the Marxist methodology of the Poznań School provoked wide-ranging discussions and polemics among Marxist philosophers and Marxist-oriented researchers in the humanities and social sciences for the next several years. The unorthodox "reading of Marx" by the Poznań philosophers of science became, in the eyes of their most ardent Marxist critics, a "simplification of Marxism", while representatives of non-Marxist philosophical orientations appreciated the originality of their research. Tadeusz Kotarbiński, referring to "the way Marxism is practiced in philosophy", stated that the studies of the Poznań methodologists "(…) come to the fore of genuine primary research in the methodology of the humanities in our country" (Kotarbiński 1974, 11). A similar view was expressed by Edward M. Swiderski (1984, 97), who argued that,

> The uniqueness of the Poznań school lies in the fact that, although influenced by Marxism, which it seeks to express and explain, it is grounded in the Polish tradition of logical and methodological inquiry, with particular attention to the method of the social sciences and humanities. The resulting "analytical" Marxism is superior to much theoretical Marxist analysis, which is too often conducted in the style of 19th-century philosophy.

During this period, the Poznań School of Methodology reached its greatest popularity, and the number of participants from Poznań and other academic centers in Poland who enrolled in seminars offered by the Department of Logic and Methodology of Science of the AMU Institute of Philosophy reached 40 people. This period of methodological research development of the Poznań school is often considered to be the years of the Poznań School of Methodology in general, if identified with the practice of Marxist methodological research (Kot 1984; Musiał 2016). In fact, the name "Poznań School of Methodology" was applied to the Poznań model of Marxist methodological practice by philosopher Janusz Kuczyński (1930–2017), at a philosophical and methodological conference in Zegrzynek (May 4–5, 1972). During the conference, the novelty of the Marxist methodological research carried out by the Poznań philosophical community was "ultimately" realized. However, the years 1970–1973 are usually considered the "heyday" of the Poznań School of Methodology, after which it disintegrated, although its rapid demise was not predicted by its founder at the time. In his "Foreword" to the volume of the conference proceedings, Jerzy Kmita (1974, 8) wrote:

> Within the framework of the views on the nature of the Marxist methodology of sciences, which could be perceived from the works of the Poznań school, the contributions of the representatives of the school in the present volume are a third link. The first one is the essays in *Metodologiczne założenia "Kapitału" Marksa* (Warsaw 1970), the second - *Elementy marksistowskiej metodologii humanistyki* (Poznań 1973). I would by no means claim that the third link is the last one.

In fact, however, the third link became the "last link", for by this time differences in the understanding of "Marxist method" (Nowak 1974) were becoming increasingly apparent among the Poznań methodologists, and the conference "Marxist Guidelines for the Study of Historical Development" revealed significant discrepancies in the theoretical analysis of historical materialism and the materialist views of social development derived from it. The post-conference publication *Założenia teoretyczne badań nad rozwojem historycznym* (*Theoretical Assumptions of the Study of Historical Development*—Kmita 1977b) was the last to include contributions from researchers who soon formed separate strands of Marxist philosophy of science. The subject of the dispute was "the Marxist way of analyzing science and its methods" (Kotowa 2010, 310–311), which led to the adoption of different concepts of the Marxist theory of scientific knowledge. It resulted in a theoretical split among scholars who had previously shared assumptions in the field of scientific cognition theory, and in the creation of separate research factions. Thus, the Poznań School of Methodology and its scientific paradigm disintegrated.

Contrary to the opinions that limit the functioning of the Poznań School of Methodology to the years 1970–1973, when Marxist methodological research was conducted by its members, it is reasonable to apply the name of the Poznań School of Methodology to the studies in the Department of Logic, which were conducted from the point of view of "empirical logic" and developed its basic methodological paradigm, without which the original and innovative analyses of Marxist research on capitalist social and economic aspects would not have been possible. Even Jerzy Kmita, the founder and organizer of the Poznań school, who dissociated himself from the name "Poznań School of Methodology", saw the two periods of research activity of the Poznań methodologists as a continuum.

3 Research Continuations of the Poznań School of Methodology

Since the mid-1970s, after the breakup of the relatively unified Poznań School of Methodology, or as Jerzy Kmita (1977a, 5) put it:

> (...) as a result of the further increase in the number of researchers in the theory of scientific knowledge at the Institute of Philosophy and the natural diversification of their interests, three methodological teams have been established to deal with different research problems

> (…). The team led by Professor Jan Such deals with the methodological problems of the natural sciences (…). Professor Leszek Nowak's team conducts research in the theory of scientific cognition, taking into account a wide range of disciplines: from the mathematical sciences to the humanities and social disciplines such as sociology or jurisprudence. This research is based on ontological premises that form, in this view, a Marxist categorical dialectic (…). Finally, the third team working with me aims to construct a theoretical system called historical epistemology. Science is treated here as a particular, historically variable type of social praxis, whose development is subject to, among other things, the patterns of historical materialism.

These three methodological teams were organized into three separate departments: the Department of Marxist Philosophy (later the Department of Philosophy of Science), headed by Jan Such; the Department of Dialectics of Cognition (later the Department of Epistemology), headed by Leszek Nowak; and the Department of Logic and Methodology of Science, headed until 1977 by Jerzy Kmita. Accordingly, the faculty members of the Institute of Philosophy, then headed by Kmita, who in their studies of the logic of science relied on the common general assumptions that defined the Poznań School of Methodology, formed three research groups that practiced—albeit from Marxist positions—the theory of scientific knowledge in different ways.

Of the above-mentioned research trends that originated in the Poznań School of Methodology, the theoretical and methodological problems of the humanities were most extensively developed by Jerzy Kmita's team. This research on the philosophy of the humanities and the research practices of its disciplines is a creative continuation of the ideas and problems of the Poznań methodological school. Today, researchers refer to the historical epistemology (theoretical history of science) developed by Jerzy Kmita since the mid-1970s, or to his social-regulatory theory of culture, as a characteristic of the "Poznań School of Cultural Studies". The latter was mainly associated with thc Institute of Cultural Studies at Adam Mickiewicz University, to which Kmita moved in early 1977, and was committed to establishing the theoretical and methodological foundations of cultural studies as an independent discipline of the humanities in accordance with the new directions in university education.

Krystyna Zamiara sees the development of Jerzy Kmita's cognitive interests as a "sequence of problem shifts" and distinguishes two major transitions in his field of research. The first is the transition "from the logic of science to the non-logicist theory and philosophy of science", while the second is the transition "from the non-logicist philosophy of science to specific science" (Zamiara 2011, 111).

4 Historical Epistemology: Theoretical History of Science

The group of researchers around Jerzy Kmita, who developed their theoretical-cognitive framework in the second half of the 1970s and the early 1980s, consisted of the following members of the Poznań School of Methodology: Teresa Kostyrko, Włodzimierz Ławniczak, and Tadeusz Zgółka, who continued their theoretical and methodological research in their original disciplines; Krystyna Zamiara, who contributed to the so-called historical epistemology; as well as those considered the "second generation" of the Poznań School of Methodology: Anna Grzegorczyk, Barbara Kotowa, Anna Pałubicka and Anna Zeidler (Zeidler-Janiszewska) (1951–2017). In this period Jerzy Kmita published his monographs: *Szkice z teorii poznania naukowego* (*Sketches from the Theory of Scientific Knowledge*) (1976) and *Z problemów epistemologii historycznej* (*On the Problems of Historical Epistemology*) (1980), as well as the collective work *Zagadnienie przełomu antypozytywistycznego w humanistyce* (*The Question of the Antipositivist Breakthrough in the Humanities*) (1978).

The new studies inspired by Kmita were to some extent creative developments of the earlier findings of the Poznań School, especially those in *Z metodologicznych problemów interpretacji humanistycznej* (Kmita 1971, 5). They led Kmita to the observation that epistemological cognition is historical cognition and, therefore, to the development of a certain version of epistemology that presupposes historical materialism, which he called *historical epistemology*.

Historical epistemology is a theory of the development of science as a field of social practice with a methodological consciousness that regulates it, producing knowledge systems that change with revealing technological demands and expand due to their increasing technological efficiency. In his historical epistemology, Kmita uses two basic concepts: social practice and social consciousness. He sees science as a kind of social practice (research practice) regulated by a supra-individual set of methodological norms and directives that constitute social methodological consciousness, the "verbalization of which … is addressed by the theory of scientific cognition, which always represents a certain philosophical orientation" (Kmita 1980, 31). The product of research practice is scientific knowledge (findings and concepts, scientific theories, etc.). Research practice is functionally linked to other fields of social practice, to which it provides certain premises that make it possible to predict the effects of actions taken in its field. It is historically variable, responding to the developmental demands of the entire social practice emerging at a given moment, reflected in the social methodological consciousness. This means that the development of science is determined by objective factors outside the sphere of science, especially in material" practice. Thus, it is determined, on the one hand, functionally and, on the other, genetically, by a given system of scientific knowledge that is undergoing a more or less radical transformation. Successive systems of knowledge can exist within the same theoretical-methodological framework, competing, so to speak, on the same plane. A later system may represent only a clarification, extension, or generalization of earlier research results, may have "greater exploratory power", or the systems may be incommensurable and logically incomparable. Any system of knowledge (scientific theory) has two types of object reference: literal and practical-objective. The latter can be identified only by the subsequent theory, which demonstrates its technological effectiveness. An exploratory (generalizing) correspondence or, in the latter case, a significantly corrective correspondence, which determines cognitive progress, can then take place between the later and earlier systems of knowledge. Furthermore, historical epistemology has been transformed into a theory of science outside of philosophy and has been defined as "theoretical history of science".

Jerzy Kmita's monograph *Kultura i poznanie* (*Culture and Cognition*) (1985) is still of great importance with regard to his epistemology in *Szkice z teorii poznania naukowego and Z problemów epistemologii historycznej*, as it introduces the concept of cultural epistemology, which considers the cultural conditions of scientific cognition in the context of the socially regulating concept of culture. Taking into account this primary circumstance that science is a cultural field (more precisely, a set of norms and methodological directives governing the practice of social research) and that research activities are cultural in nature, then—as a consequence—the philosophical, epistemological category of legitimate cognition also becomes a cultural category. This raises the problem of cultural relativism in scientific cognition and of historical-cultural limitations in declaring the cognitive legitimacy of the research results obtained. Its framing and resolution depend on the adopted theoretical-cultural perspective. It is therefore important to recognize that:

> (…) not only is the theory of cognition impossible without a theory of culture, but - moreover - at least some of the central problems of the theory of (scientific) cognition are determined by the relevant premises of the theory (philosophy) of culture, by means of which these problems can be solved (Kmita 1985, 7).

Within the framework of historical epistemology, these researchers have addressed the question of the role of philosophical (epistemological) orientations in the development of science (A. Pałubicka, K. Zamiara), art (A. Zeidler), and especially literature (A. Grzegorczyk, B. Kotowa).

5 From Methodological Research to Cultural Studies

The theoretical approach to science became the basis for Jerzy Kmita's concept of culture, which is a cultural continuation of certain ideas of the Poznań School of Methodology from its initial research activities in the Department of Logic. Cultural issues, more precisely artistic culture, in particular literature, were the original research areas of the school's founder. Kmita's doctoral dissertation was devoted to Jan Kochanowski's poetic art, and after graduating from the Polish Studies Department, he was temporarily employed at the Main Library of Adam Mickiewicz University and received a full-time position at the Faculty of Philology, where he cooperated with his supervisor the then associate professor Jerzy Ziomek (1924–1990). At the time, Kmita was actively involved in literary criticism, essay writing, and journalism, contributing to *Wyboje*, "a magazine of Greater Poland students and young intellectuals", as well as literary magazines and poetry almanacs. Kmita's interest in poetry stemmed in part from his own attempts at writing poetry and translating the works of prominent French poets during his middle and high school education and later during his Polish studies at AMU. Even after his employment at the Department of Logic at the Faculty of Philosophy and History and his fundamental shift of research focus to logic and methodology of empirical sciences, Kmita continued to pursue issues of philosophy, theory, and semiotics of art, mainly through essays in the Poznań-based social and cultural monthly *Nurt* and the quarterly *Sztuka*, as well as in his theoretical and methodological studies of literature, which was primarily the subject of his post-doctoral dissertation. As a result of his reflections on art since the end of the 1960s and his explication of various concepts of art theory, his regular research was expanded to include analyses of artistic culture. Jerzy Kmita's crowning achievement was the development of the "semantic concept of the artwork". It enabled researchers to further study art as an object of explanation in the course of humanistic interpretation.

For Kmita and Nowak, however, the most important factor was the need to consider "the world of culture" (the cultural character and context of cognitive activities) in a more general methodology or epistemology (since "all humanistic disciplines study cultural activities"). According to Kmita and Nowak (1986, 236), the methodological model of the humanities "(…) (1) presupposes certain theories of rational behavior, and (2) follows the directive of methodological structuralism". It therefore required a cultural conceptual framework. In their co-authored monograph *Studia nad teoretycznymi podstawami humanistyki* (*Studies on the Theoretical Foundations of the Humanities*—Kmita 1968)—the flagship work of the Poznań School of Methodology—the fifth chapter, "Cultural System, Cultural Activity, Performatives" included definitions of the titular concepts as well as the cultural product, cultural fact, cultural object, and the rules of cultural interpretation. It presupposed a certain understanding of culture from among several concepts of culture known in the humanities and social sciences. Kmita briefly pointed out the different semantic intuitions associated with the word "culture":

> The various definitions of culture emphasize that it is that part of the human environment that is produced by humans themselves, or that it is everything that humans learn, or that can be internalized by the individual because of its supra-individual origin. It is also emphasized that the elements of culture (cultural objects) are symbolic and that they are signs for something other than themselves, or that culture is everything that is capable of satisfying (or is produced to satisfy) all human needs (or only some, for example "spiritual" ones), etc. (…) It is difficult to provide a definition of culture that would satisfy any of these diverse intuitions. It remains, therefore, to choose an intuition that seems interesting for the reconstruction of the objects of interest in the humanities (Kmita and Nowak 1968, 254–255).

In view of the above, Kmita did not attempt to offer his own definition of culture at this point, but was content to explain the concepts of cultural activity and cultural product necessary for the methodological analysis of humanistic research. Starting from the basic idea of object-oriented anti-naturalism, which contrasts culture with nature, Kmita offers a broad (global or anthropological) understanding of culture, i.e. "a cultural fact is everything in which man 'has his fingers'" (Kmita and Nowak 1968, 256). In this view, "cultural activities are rational activities that are co-determined by the knowledge disseminated in a given community" (Kmita and Nowak 1968, 256). These are cultural activities in a broad sense, and so are their products, which range from the "material" to the "spiritual". Among the cultural activities understood in this way, one can distinguish those "oriented towards cultural understanding", for which "(...) it is characteristic that they are relative to the rules of cultural interpretation assumed in a given culture" (Kmita and Nowak 1968, 257). These are cultural activities in the narrow sense, which in the then concepts of culture were classified as "spiritual culture" or "social culture", and in later concepts were placed in the sphere of symbolic culture. These findings were specified in the monograph *Z metodologicznych problemów interpretacji humanistycznej* and allowed the introduction of new concepts. They were later used in the methodological description of humanistic interpretation "as a special case of deductively conceived explanation" (Kmita 1971, 7). In Kmita's framework, the following concepts were integrated into a coherent theoretical system: cultural activity, cultural object, cultural group, rules of cultural interpretation (qualifying and indirect), composition of a given cultural system, cultural competence, and sign—all used in the presentation of the varieties of interpretation in humanistic research. These findings and observations inspired Jerzy Kmita's later systematic theoretical studies of culture, which resulted in the socio-pragmatic (later renamed socio-regulatory) concept of culture.

Although in his further analyses Jerzy Kmita referred to findings from his monograph *Z metodologicznych problemów interpretacji humanistycznej*, he did not use any related theoretical concepts there. Some of them appeared only in passing in *Szkice z teorii poznania naukowego*, where Kmita (1976, 7–8) noted that:

> Concepts that are used for a closer analysis of the components of the explanans of humanistic interpretation - or those that can be defined by them - allow for the description of certain further concepts. Some of them, such as the concepts of sign, cultural system, or culture, serve to explicate concepts used in all the humanities, while others, such as representational structure, represented structure, "content" communicated by a work of art, reductive (realistic) type of communication, deductive (symbolic-metaphorical) type of communication, etc., serve to explicate concepts used in selected areas of the humanities, especially in art studies.

Jerzy Kmita's transfer to the Institute of Cultural Studies resulted in the concentration of his scholarly activities on the construction of the theoretical foundations of cultural studies as a scientific discipline in line with the direction of university education, since, according to academic tradition, "(...) without the existence of an appropriate, independent research discipline, there can be no separate teaching at the university level (...)" (Kmita 1985, 5). In undertaking this task, Kmita (1985, 6) was also mindful of the cognitive needs of the Polish humanities:

> There is (...) an undeniable research gap in the Polish humanities: a space for a general, but also empirically controllable science of culture, and for theoretically oriented research on the totality of culture, which constitutes a natural, so to speak, area of contact of individual humanities disciplines.

Cultural studies, understood as a general science of culture, must, in order to achieve its main goal, be equipped with an appropriate theory of culture and a research methodology that allows for the synthesis and integration of cognitive results from other disciplines of the humanities and social sciences. Cultural studies thus conceived become "integrated humanities".

Jerzy Kmita's innovative achievement, the socio-regulatory concept of culture, initially called socio-pragmatic, was to become the theoretical integrative platform of cultural studies. The socio-regulatory theory of culture was an intellectual achievement of the Polish humanities of the 1970s and 1980s. It began to be formulated at a time when many researchers realized the cognitive necessity of changing the theoretical and methodological model of cultural studies. It can be said that there was a demand for a new radically anti-positivist theory of culture in humanistic research, using, as it were, routinely popularized positivist research schemes. The epistemological reasons for the conviction that the theory in question brought cognitive progress were materialized in the research of those who made use of this theory.

The socio-regulatory theory of culture perceives culture as a "reality of thought", remaining in the mainstream of thinking and defining culture on the basis of modern cultural anthropology as signified by Ward Goodenough. Its dissemination in the academic community establishes culture as a set of forms of social consciousness constituted by systems of normative beliefs (determining the goals-values to be realized) and directive beliefs (determining the means of realizing these goals) that regulate in a subjective-rational way the corresponding types of social practices. Kmita's concept of culture has been modified over the years, as can be seen from its first published presentation (Kmita 1978), subsequent revisions (Kmita 1982a, b; Kmita and Kostyrko 1983; Banaszak and Kmita 1991), and the most recent interpretation, which represents "(…) a certain summary of the concept of culture that I have been developing in fragments since the mid-1970s, and mainly of that area of culture that I describe as symbolic culture" (Kmita 2007, 11).

This view of culture provides a more adequate perspective of the cultural sphere and incomparably greater opportunities to explain cultural phenomena and changes in cultural contexts with reference to the different varieties of determination of cultural phenomena discovered by Kmita, in particular: functional, functional-genetic, and subjective-rational. The socio-regulatory theory of culture is an effective methodological tool for the study of socio-cultural reality and its theoretically or philosophically diverse descriptions. From its perspective, the study of cultural phenomena and the ways of their scientific recognition in various disciplines of the humanities and social sciences has been carried out by the group of researchers representing it. In addition to the above-mentioned first and second generation of researchers of the Poznań School of Methodology, the third generation consists mainly of the research and teaching staff of the Institute of Cultural Studies: Grzegorz Banaszak, Zbigniew Gierszewski (1950–2003), Jan Grad, Ewa Kobylińska, Krzysztof Moraczewski, Paweł Ozdowski, Ewa Rewers, and Jacek Sójka; from the Institute of Ethnology and Cultural Anthropology: Michał Buchowski, Wojciech Burszta (1957–2021), and from other universities and research centers: Danuta Minta-Tworzowska, Dorota Angutek, Piotr Bauć, Artur Dobosz, Andrzej P. Kowalski, Henryk Mamzer, Andrzej Pluta, Barbara Pogonowska, Andrzej Szahaj and Wojciech Wrzosek. Their results were published in monographs and collective works. Today's representatives of the younger generation of researchers, who are interested in the achievements of the Poznań School of Methodology and use its theoretical and methodological premises, Andrzej Bełkot, Mateusz Bonecki and Maciej Musiał can be considered the fourth generation of researchers of the Poznań school.

This collectivity of scholars justifies the continued existence of the Poznań School of Methodology, best expressed by Barbara Kotowa (2010, 334), one of its most active participants:

> The Poznań School of Methodology, like any historical-social formation, has repeatedly changed its membership, but it is still a school in the usual sense of the word, because it fulfills the necessary conditions of a school: it is headed by an undisputed scholarly authority, a leader who inspires new generations of students of the humanities through his texts; and the members of the school are bound by a common theoretical and methodological framework of research. Within the school, new research teams are formed (…) the range of its influence is also expanded, both institutionally and territorially.

After Jerzy Kmita's death, Anna Pałubicka appears to have taken over the leadership of the school, bringing together a group of young students of the humanities and social sciences.

Among academic humanists, the criterion for attributing texts to the Poznań School of Methodology is the application of such cognitive categories as the premise of rationality, humanistic interpretation, idealization, logical reconstruction, functional-genetic explanation. It is also connected with the concepts and ideas of the later—after the dissolution of the Poznań School—research of its founders and participants, such as historical epistemology, theoretical history of science, pre-theoretical and theoretical stage of development of science, and socio-regulatory theory of culture. Sometimes, authors who are not sufficiently familiar with the relevant concepts quote or use only related keywords literally, without being aware of the thematic references and relevant findings. However, the theoretical way of thinking formed under the influence of the Poznań School of Methodology and socio-regulatory theory of culture is important and sometimes manifests itself in the analysis of cultural phenomena carried out in the field of cultural sciences.

Jerzy Kmita, during the ceremony of awarding him the Palmae Universitatis Studiorum Posnaniensis medal (at the inauguration of the 2011/2012 academic year), in his unpublished address, noted that his research was deeply rooted in the tradition of seeking the specificity of scientific knowledge in the humanities—a tradition initiated and developed by Florian Znaniecki (the idea of the "humanistic coefficient") and Kazimierz Ajdukiewicz (the concept of "understanding statements"). It is precisely in Ajdukiewicz's view that such knowledge is "capable of being a science", and this approach was further continued by Jerzy Giedymin and Jerzy Topolski (Grad 2012).

These explorations led to changes in the research problems and the respective different periods of activity of the Poznań School. Nevertheless, the basic theoretical and methodological premises and tools of analysis developed in the first period of its activity focused on the "logic of science" were preserved. The teams of researchers have changed, but the basic research paradigm has ensured the continuity of the Poznań School of Methodology as a significant scholarly formation.

References

Augustynek, Z. 1977. *Opinia w sprawie nadania prof. dr hab. Jerzemu Kmicie tytułu naukowego profesora zwyczajnego* (file reference 825/394, Archive of UAM).

Banaszak, G., and J. Kmita. 1991. *Społeczno-regulacyjna koncepcja kultury*. Warszawa: Instytut Kultury.

Giedymin, J. 1958. Model historycyzmu K. R. Poppera. *Studia Filozoficzne* 3:205–214.

Giedymin, J. 1958/1959. Koncepcja racjonalnego działania i charakterystyka metodologiczna opartych na niej teorii. *Rocznik Polskiego Towarzystwa Ekonomicznego* 6:61–81.

Giedymin, J. 1961. Spór między naturalizmem i antynaturalizmem w pojmowaniu nauk społecznych. *Rocznik Ekonomiczny Polskiego Towarzystwa Ekonomicznego* 13:173–191.

Giedymin, J. 1964. *Problemy – założenia rozstrzygnięcia. Studia nad logicznymi podstawami nauk społecznych*. Poznań: Państwowe Wydawnictwo Naukowe.

Grad, J. 2012. Jerzy Kmita (1931–2012) – wspomnienie. *Nauka* 3:177–188.

Grad, J. 2019. "Poznańska szkoła metodologiczna" – osiągnięcia, oceny i kontrowersje. *Nauka Polska. Jej Potrzeby, Organizacja i Rozwój* XXVIII (LIII):99–118.

Janiec, B. (Ed.) 1970. *Założenia metodologiczne 'Kapitału' Marksa*. Warszawa: Książka i Wiedza.

Kmita, J. 1961. *Znaczenie jako wykładnik sposobu rozumienia wyrażeń*. Unpublished Ph.D. thesis, UAM Poznań.

Kmita, J. 1964a. Przegląd zasadniczych koncepcji znaczenia wyrażeń. *Zeszyty Naukowe UAM. Filozofia – Psychologia – Pedagogika* 8:326–380.

Kmita, J. 1964b. Twierdzenia bezpośrednio oparte na rozumieniu wyrażeń. *Zeszyty Naukowe UAM. Filozofia – Psychologia – Pedagogika* 8:382–451.

Kmita, J. 1967. *Problematyka terminów teoretycznych w odniesieniu do pojęć literaturoznawczych*. Poznań: Poznańskie Towarzystwo Przyjaciół Nauk.

Kmita, J., Nowak, L. 1968. *Studia nad teoretycznymi podstawami humanistyki*. Poznań: Uniwersytet im. Adama Mickiewicza.

Kmita, J. 1971. *Z metodologicznych problemów interpretacji humanistycznej*. Warszawa: Państwowe Wydawnictwo Naukowe.

Kmita, J. 1973a. Wstęp. In *Elementy marksistowskiej metodologii humanistyki*, ed. J. Kmita, 5–20. Poznań: Wydawnictwo Poznańskie.

Kmita, J. (Ed.) 1973b. *Elementy marksistowskiej metodologii humanistyki*. Poznań: Wydawnictwo Poznańskie.

Kmita, J. 1974. Słowo wstępne. In *Metodologiczne implikacje epistemologii marksistowskiej*, ed. J. Kmita, 5–20. Warszawa: Państwowe Wydawnictwo Naukowe.

Kmita, J. 1976. *Szkice z teorii poznania naukowego*. Warszawa: Państwowe Wydawnictwo Naukowe.

Kmita, J. 1977a. O badaniach metodologicznych prowadzonych w ośrodku poznańskim. *Zagadnienia Naukoznawstwa* 1:4–5.

Kmita, J. (Ed.) 1977b. *Założenia teoretyczne badań nad rozwojem historycznym*. Warszawa: Państwowe Wydawnictwo Naukowe.

Kmita, J. 1978. O naukowo-teoretycznym ujęciu kultury symboliczne. In *Zagadnienie przełomu antypozytywistycznego w humanistyce*, ed. J. Kmita, 69–84. Warszawa: Państwowe Wydawnictwo Naukowe.

Kmita, J. 1980. *Z problemów epistemologii historycznej*. Warszawa: Państwowe Wydawnictwo Naukowe.

Kmita, J. 1982a. Dziedziny kultury. In *Przedmiot i funkcje teorii kultury*, ed. S. Pietraszko, 71–87. Wrocław: Wydawnictwo Uniwersytetu Wrocławskiego.

Kmita, J. 1982b. *O kulturze symbolicznej*. Warszawa: Centralny Ośrodek Metodyki Upowszechniania Kultury.

Kmita, J. 1985. *Kultura i poznanie*. Warszawa: Państwowe Wydawnictwo Naukowe.

Kmita, J. 2007. *Późny wnuk filozofii: wprowadzenie do kulturoznawstwa*. Poznań: Bogucki Wydawnictwo Naukowe.

Kmita, J., and T. Kostyrko. 1983. *Elementy teorii kultury: wykłady dla studentów kulturoznawstwa*. Poznań: UAM.

Kot, W. 1984. *Jak to jest z Poznańską Szkołą Metodologii Nauk?* "Gazeta Poznańska. Magazyn", 12 października 1984 r., 4.

Kotarbiński, T. 1974. Wystąpienie otwierające konferencję metodologiczną w Zegrzynku 4–6 maja 1972 roku. In *Metodologiczne implikacje epistemologii marksistowskiej*, ed. J. Kmita, 1974. Warszawa: Państwowe Wydawnictwo Naukowe, 9-11.

Kotowa, B. 2010. Od filozofii humanistyki do filozofii jako dziedziny kultury. Główne kierunki badawcze szkoły poznańskiej od połowy lat siedemdziesiątych. In *Filozofia na Uniwersytecie w Poznaniu. Jubileusz 90-lecia*, ed. T. Buksiński, 309–334. Wydawnictwo Naukowe Instytutu Filozofii UAM.

Ławniczak, W. 1971. *O uzasadniającej roli analogii na przykładzie wnioskowań z zakresu historii sztuki*. Poznań: UAM.

Mackiewicz, W. 2001. Szkoły i orientacje filozoficzne, 6. Szkoła poznańska. In *Polska filozofia powojenna*, ed. W. Mackiewicz, vol. I, 66–70. Warszawa: Witmark.

Musiał, G. 2016. Poznańska Szkoła Metodologiczna – jej rozwój i znaczenie. *Przegląd Filozoficzny. Nowa Seria* 2:553–563.

Nowak, L. 1971. *U podstaw Marksowskiej metodologii nauk*. Warszawa: Państwowe Wydawnictwo Naukowe.

Nowak, L. 1974. *Zasady marksistowskiej filozofii nauki: próba systematycznej rekonstrukcji*. Warszawa: Państwowe Wydawnictwo Naukowe.

Such, J. 1968. *O uniwersalności praw nauki. Studium metodologiczne*. Poznań: Wydawnictwo Naukowe UAM.

Such, J. 1969. *Wstęp do metodologii ogólnej nauk*. Poznań: Wydawnictwo Naukowe UAM.

Swiderski, M. 1984. Humanistic interpretation and historical materialism: The methodology of the Poznan School. In *Contemporary Marxism. Essays in honour of J.M. Bocheński*, ed. J. M. O'Rourke, T. J. Blakeley, and F. J. Rapp, 97–108. Dordrecht/Boston/Lancaster: Reidel Publishing Company.

Topolski, J., ed. 1970. *Założenia metodologiczne 'Kapitału' Marksa*. Warszawa: Książka i Wiedza.

Zamiara, K. 1974. *Metodologiczne znaczenie sporu o status poznawczy teorii. Z problematyki związków między metodologią nauk i teorią poznania*. Warszawa: Państwowe Wydawnictwo Naukowe.

Zamiara, K. 2010. U początków poznańskiej szkoły metodologicznej. In *Filozofia na Uniwersytecie w Poznaniu. Jubileusz 90-lecia*, ed. T. Buksiński, 283–308. Poznań: Wydawnictwo Naukowe Instytutu Filozofii UAM.

Zamiara, K. 2011. Droga twórcza Jerzego Kmity jako sekwencja przesunięć problemowych. *Filo-Sofija* 11 (12):109–127.

Zgółka, T. 1976. *O strukturalnym wyjaśnianiu faktów językowych*. Warszawa/Poznań: Państwowe Wydawnictwo Naukowe.

The Character and Different Facets of the Poznań School of Archaeology

An Apologia of Methodology of Archaeology at Adam Mickiewicz University, Poznań

Henryk Mamzer

Abstract This chapter provides an overview and interpretation of the intellectual activity of archaeology methodologists from Adam Mickiewicz University in Poznań, mainly Arkadiusz Marciniak, Danuta Minta-Tworzowska, Michał Pawleta, and Włodzimierz Rączkowski. The idea they pursue is, in my opinion, a development of thought: from the object to thinking about the object, addressed both to archaeologists who are not convinced by the methodology and to representatives of archaeology practiced in the field of cultural studies, in order to ultimately define their contribution to integrated humanities. The presented overview has the form of a Bakhtinian polyphony of Poznań methodologists from the Adam Mickiewicz University, expressed in one voice. It is a voice merged from two different research approaches, both shaped by the Poznań School of Methodology of Jerzy Kmita, Leszek Nowak, and Jerzy Topolski. We are referring to natural science on the one hand, and the humanities on the other. It was on this basis that Professor Jan Żak created the methodological foundations of Poznań archaeology. After all, the research of the aforementioned methodologists is not so much about research methods as about the subject of archaeological research. The chosen research methods are a consequence of the specificity of the subject of research. If the focus of archaeological research is materiality, then it necessarily draws attention to the methods of natural science. However, if we ask about the subject of research, i.e., "what should

The term *apologia* as used in this article has two meanings in literature. The first is a defense of an opinion, as in *The Apology of Socrates* by Plato or a speech in defense by Lucius Apuleius Madaurensis (Apulei Platonici Madaurensis pro se de Magia liber) delivered in 160 A.D. at the Roman basilica in Sabratha near Oea in present-day Libya before Claudius Maximus the proconsular governor of Africa. The other meaning is contained in the very title of Marc Bloch's work *Apologie pour l'histoire au Métier d'Historien* (1949). The two meanings are not mutually exclusive but rather they are mutually enhancing. The apology (apologia) of the ethical position of Socrates or Lucius Apuleius Madaurensis is included in the "speech in defense" delivered by each of them.

H. Mamzer (✉)
Institute of Archaeology and Ethnology, Polish Academy of Sciences, Poznań, Poland

A. Marciniak et al. (eds.), *The Poznań School of Archaeology*, Outstanding Contributions to Archaeology, https://doi.org/10.1007/978-3-032-07848-3_4

we study?" and not just "how?", it means that it is formulated on the basis of the archaeologist's interpretation of the reality in which he himself functions "here and now"; it is cultural rather than natural science in nature, especially since the second part of the question, "how?", is addressed to the community of archaeologists themselves functioning in the present. The present provides archaeology with conceptual frameworks that form the basis for interpreting the past. These determine the results of the research. Archaeology, in turn, communicates the results of its research to the contemporary cultural environment of which it is an emanation, confronting them with social expectations. Archaeology, therefore, not only co-creates the subject of its research, but also the reality surrounding it, in which it functions, and itself. In this way, the results of archaeological research provide the material for shaping contemporary culture, of which archaeology is also a part.

Keywords Archaeological theory · Methodology · Natural science · Humanities · Arkadiusz Marciniak · Danuta Minty Tworzowska · Michał Pawleta · Włodzimierz Rączkowski

1 Introduction

The following paper does not constitute a review of research activities of each of methodologists of archaeology from Adam Mickiewicz University in Poznań (AMU) separately; nor is it an analysis of the singularity of their individual views. It is a syncretic attempt to integrate the voices, or rather thoughts, of each of them into a coherent whole. One might say that this paper involves a Bakhtinian polyphony of AMU methodologists' voices combined into a single voice of two different research attitudes. The "sound" of that voice was produced in the Poznań methodological school of Jerzy Kmita, Leszek Nowak, and Jerzy Topolski. On the one hand, there is Kmita's social-regulatory concept of culture and Topolski's implementation of this concept on the ground of history (especially as far as humanistic interpretation is concerned). On the other hand, there is Nowak's idealizational theory of science and unitarian metaphysics. According to Jan Such (2017, 39–40), these two threads of the Poznań school of methodology are characterized by a rational compromise combining opposing methodological stances into, for example, a coherent notion of moderate methodological anti-naturalism, which can also be referred to as moderate methodological naturalism. A reflection of this "dichotomy" can be found among the present-day representatives of AMU methodologists of archaeology: Danuta Minta-Tworzowska, Włodzimierz Rączkowski, and Michał Pawleta; and on the other hand, Leszek Nowak's student Arkadiusz Marciniak, who draws on Janusz Piontek's and Krzysztof Łastowski's natural science orientation. Notwithstanding, each of them individually follows their own way of thinking, determined by the specificity of their research.

The above-mentioned Poznań academics working on methodological reflection in archaeology are students of Professor Jan Żak, one could say, the offspring of his interests and methodologically oriented research activity. It was the works of representatives of the Poznań school of methodology of the humanities, leaning towards Kmita's and Topolski' views, that fascinated Jan Żak, who laid the methodological foundations of Poznań archaeology. He was not only an advocate of the Poznań school of methodology in general, but also its propagator, or indeed its continuator in the field of archaeology. On Prof. Żak's initiative the first Polish Department of History and Methodology of Prehistory was established at the Adam Mickiewicz University in Poznań.

In an effort to outline the current research interests of Poznań methodologists, we should begin with the very question of the "methodology of archaeology" without delving into the specific history of methodological orientation of AMU archaeologists. This question has been the subject of numerous studies by the aforementioned researchers (Minta-Tworzowska and Rączkowski 1996, 2007; Minta-Tworzowska 2002, 2017, 2022).

2 What Do We Need Methodology For?

The actual meaning of the above title is not so much an articulated question. It is rather a form of palpable negative reaction, an ironic "shrug of the shoulders" by a significant number of Polish archaeologists to this type of research problem. It is—as one might think—a somewhat justifiable reaction, not so much to the question addressed to archaeologists-practitioners: Is methodology needed in archaeology? but addressing the interest in methodology of some (Poznań) archaeologists, for whom it is the main subject of their research. The very fact of posing this question is an expression of doubts expressed in opinions or texts written by archaeologists who are negatively inclined in this regard. This reaction signifies a negative answer: We do not need methodology for anything. Below is my understanding of the approach of archaeologists of the Poznań methodological community, of their reaction, not quite fully disclosed, to not always enthusiastic expectations in the texts they write. It is not uncommon to hear that methodology is not quite archaeology, as it once took place at a scholarly conference in Gdańsk devoted to the problems of the Pomeranian culture.

Methodology of archaeology, meta-archaeology, philosophy of archaeology[1] or theory of archaeology are seemingly synonymous terms in spite of some significant differences between. They testify to the different understandings of the concept of methodology. However, due to the fact that they do not enjoy overly high popularity in Polish archaeology, I will not go into the underlying subtleties between them for the time being. I will do so, if necessary, in the course of my argument. For it is not the definitions of individual terms that are at issue here, but the almost universal, yet entirely unreasonable, aversion of archaeologists to methodology itself that provides the direction for the argument in this article. My task is to demonstrate the

[1] Reflections on the methodological problems of sciences and the humanities can be found, for example, in Kmita (1977), Ajdukiewicz (1985a), or Woleński (1996).

rationale of AMU methodologists by exposing the intellectual poverty of works devoid of such considerations. I am referring here to continental archaeology, including, above all, Polish archaeology. After all, bearing in mind this aversion to methodology and doubts not infrequently appearing in Polish archaeological literature about the sense of theoretical reflection in archaeology, the wording of the title, which exposes the methodological aspect of archaeological research, disguises the answer to the question: What do we need methodology **FOR**? Certainly, a rhetorical question, as it conceals the answer: **FOR NOTHING**. This is one of the basic problems facing academic methodology of archaeology in Poznań. This problem is not limited to a "select" group of field archaeologists-practitioners constituting "the salt of the earth" of archaeology in recent times, as W. Rączkowski (2009, 26) observes mockingly, for whom theoretical considerations are completely dispensable. It also applies to researchers for whom synthesizing approaches are important.[2]

To avoid paying lip service, let me quote, by way of example, a remark by Jacek Poleski delivered at a conference in Puszczykowo in 1997. Poleski, in a polemical fervour against Jan Michał Burdukiewicz's comments, said: "I am a traditionalist, and I would be inclined to favour the eloquence of what follows from the analysis of an archaeological source rather than from conceptualization" (Poleski 2000, 257). This is by no means an isolated statement. We should note here the recently published polemical *Letter* by a group of archaeologists to the Editor of *Przegląd Archeologiczny* (vol. 64, 2016, 195) revealing the authors' satisfaction therein with the claim that, supposedly, Ian Hodder "abandoned the development of new theoretical concepts [which is not true - HM], ... taking [instead - HM] a highly knowledgeable approach to recorded sources".[3] The implication of the Letter's authors, harbouring approval of Hodder's changing attitude from the methodological to the source analysis aspect of archaeology (i.e. that Hodder has, supposedly, finally ceased to be interested in theory thus raising hopes that archaeology will cease to bother itself with methodological problems) fills one with concern. The signatories of the Letter are not only commercial researchers. The Kraków archaeological academic community seems to be significantly represented among them. This is also how M. Parczewski (1999, 2) expresses his skepticism regarding

> attempts to entangle archaeological research norms in some theoretical structures which are not part of this science (being sometimes odd ideological-theoretical configurations, once imposed by force, and now by global trends).

---

[2] Similar concerns are expressed by A. P. Kowalski's (2015, 147) platoregarding the conceptualization of anthropology. "This situation will not change if, at a minimum, the professionalization of the anthropologist remains a preference for field research in place of developing original interpretive concepts".

[3] It is a letter a group of archaeologists sent to the Editors of *Przegląd Archeologiczny* (Archeological Review) vol. 64, 2016, in protest to A. Ciesielska's response (*Przegląd Archeologiczny* vol. 63, 2015) to a critical review of her book *Przemiany kulturowe na obszarze między Odrą a Wisłą na przełomie starożytności i średniowiecza. Studium metodologiczne* (Cultural transformations in the area between the Oder and Vistula Rivers in the late antiquity and the early Middle Ages. A methodological study), Poznań 2012, written by M. Parczewski and published in *Slaviae Antiqua* 55, 2014. Cf. *From the Editors* in the same volume (pp. 196–198) containing an accurate and extremely balanced response by Editor-in-Chief Prof. Bogusław Gediga.

Parczewski appears to suggest as if Polish (continental) archaeology has had its own "theoretical structures", which is not at all obvious, and I will discuss it in a moment. At the same time, Parczewski does not mention any concept used in archaeology that has not been borrowed by it. The allegation about using foreign rather than one's own theoretical findings (regardless of whether this foreignness applies to non-Polish archaeology or non-archaeological disciplines) makes one wonder about its motivation. Regardless of whether the research premises guiding Polish archaeology are adopted by Polish archaeologists from outside or developed by themselves, the quoted author, assuming their existence, is correct in this regard. For there is no presuppositionless science even when Parczewski's discipline uses "theoretical structures" other than its own.

All activities of archaeologists involve theory (Rączkowski 2002, 13). Matthew Johnson (1999, 6) even claims that: "we are all theorists". So there is no point in talking about the need for theory, we use it whether we like it or not.[4] The problem of absence of theoretical involvement of Polish archaeology is aptly captured by Justyna Gabriel, who examines it using the example of the difference between Polish (or more broadly, continental) archaeology and Anglo-American archaeology, which in her view, are situated at the opposite ends of the methodological spectrum. In her work on the theoretical aspects of the study of origins of social differences in archaeology (2018) Gabriel relies primarily on the findings of British archaeology. She does so because, in her opinion, Polish archaeology has not developed sufficient conceptual foundations for such considerations. The British archaeological milleau, on the other hand, is one of the leading ones when it comes to the conceptualization of archaeological data and by far dominates theoretical research. Gabriel argues so, despite the fact that British archaeology has not developed any social research methodology on its own either, and has relied since its inception on methods and concepts from other sciences. However, the ties between British archaeology and anthropology are far greater than those of historically oriented Polish archaeology, and they have fostered the borrowing of research practice models and conceptualizations of archaeological data. The situation is no different when it comes to the recently published work by Arkadiusz Marciniak and Stanisław Tabaczyński (2017) *Geneza i rozwój nierówności społecznych w paradygmatach badawczych archeologii. Zarys problematyki* (*The origins and development of social inequalities in research paradigms of archaeology. An outline*), or Adriana Ciesielska's *Elementy teorii społecznej w archeologii. Koncepcje grup, instytucji i struktur społecznych* (*Social Theory Elements in Archaeology. The Concepts of*

[4] While discussing the question of presence of theoretical reflection among the AMU archeologists, it is worth noting that Johnson's book (2013) was translated into Polish on their initiative by Agnieszka Tokarczuk-Różanska, Dariusz Blaszczyk and Michał Szymkowiak. The same applies to the translations of I. Hodder's *Czytanie przeszłości* [*Reading the past*], Obserwator, Poznań 1995; and C. Renfrew's *Archeologia i język. Łamigłówka pochodzenia Indoeuropejczyków* [*Archaeology and Language. The Puzzle of Indo-European origins*], Warszawa, Poznań: Wydawnictwo Naukowe PWN, 2001.

Groups, Institutions and Social Structures), Poznań 2002. Ciesielska also underlines the significant role of British and American archaeology in this regard. This is equally true for the volume *Przeszłość społeczna* (*The Social Past*) edited by Stanisław Tabaczyński, Arkadiusz Marciniak, Dorota Cyngot, and Anna Zalewska, Poznań 2012. Marciniak in his chapter *Teoria archeologii* (*The Theory of Archaeology*) (2012) points out that the research assumptions of archaeology are adopted by archaeologists from outside and applied in the form of rules of conduct, conceptual schemes, methods and ways of interpretation derived from social, humanistic or biological theories. In the same article Marciniak mentions mainly British and American archaeologists such as L. Binford, D.L. Clarke, K. Flannery, M. Johnson, L. Meskell, C. Renfrew, Ch. Tilley, and J. Thomas who have drawn their inspiration from phenomenology, Hans-Georg Gadamer's philosophical hermeneutics, Martin Heidegger, Paul Ricœur, Ferdinand de Saussure's semantics and semiotics, Immanuel Wallerstein's world-system, Robin G. Collingwood's social action theory, Alfred Tarski's theory of truth, Anthony Giddens's theory of structuration, Leslie White's systems theory, or Hayden White's metahistory.

Bearing in mind the opinions of the Kraków archaeologists, we can ask the following question: What sort of intellectual poverty would archaeology entail, if it were deprived of the inspiration provided by the aforementioned theories? This question is obviously rhetorical. I will recall at this point W.V.O. Quine (1976, 76), for whom:

> Boundaries between disciplines are useful for deans and librarians, but let us not overestimate them – the boundaries. When we abstract from them, we see all of science – physics, biology, economics, mathematics, logic and the rest – as a single sprawling system, loosely connected in some portions, but disconnected nowhere.

After all we should point out that Justyna Gabriel's work quoted above as well as Andrzej P. Kowalski's publications contain opinions that archaeology does not utilize its own theoretical concepts, developed on the archaeological grounds, because of their absence. Instead it draws mainly from anthropology, history, or linguistics. Accordingly, A.P. Kowalski (2015, 143) advocates anthropological thinking and application of anthropological interpretation to historical and archaeological data. In his view, "... 'anthropology of prehistory' is first and foremost 'anthropology' and only by extension 'prehistory' understood as an image provided so far mainly by archaeology". On the other hand,

> (...) traditional archaeology remains effective for uncovering, documenting, classifying, preserving and protecting cultural property [...] it can successfully constitute a specialty within museum studies and heritage conservation. Its main domain remains the provision of expert opinions identifying any given collection of antiquities. Meanwhile, the study of cultures of prehistoric societies could fall within the framework of cultural archaeology, a discipline practiced as a specialty of philosophy of culture, cultural anthropology, sociology, art history, classical philology and linguistics" (Kowalski 2014, 40).

It would follow that since archaeology has no theoretical background of its own, borrowing exclusively from the theoretical realm of other disciplines, it thus does not meet the requirements of an independent academic discipline; it serves, at best, as an auxiliary science of history or anthropology.[5] The fact that archaeology provides sources not only for the eras preceding the appearance of writing, but also for the study of eras that are the domain of history *sensu stricto*, has become one of the important reasons for treating archaeology as a source science discipline of history, including prehistory. Particularly meaningful in this context is Witold Hensel's division of archaeology into archaeology as a method of source acquisition, and prehistory as the interpretation of sources. According to Hensel (1971, 449): archaeology "(…) is a typically analytical science, dealing with detection, selection, qualification and description of sources by means of archaeological methods. Archaeology in this sense is a typically descriptive, auxiliary science", i.e. an auxiliary science of prehistory, ancient history, medieval history, modern history, etc. Like sigillography, numismatics, or epigraphy, it falls into the domain of studies in historical sources. Thus, there is a need for the methodology of archaeology, if it is to be considered an independent discipline rather than an auxiliary branch of history or anthropology.

This raises another question: Where does this reluctance to theoretical thought in archaeology come from? This is a very interesting question, given that theoretical thought usually constitutes the "scientificity" of particular research disciplines. This applies not only to the mathematized natural sciences (molecular biology, theoretical physics, bio-organic chemistry, human genetics, etc.), but also to the theoretically advanced humanities (anthropology, history, economics, psychology, sociology or linguistics). Archaeology, on the other hand, and especially European continental archaeology, is one of exceptions in this regard. This does not apply, of course, to the AMU archeological community, which, following the withdrawal of the Warsaw archaeological community from this kind of research[6] is becoming a lonely island. Well, it is precisely the "loneliness" of the Poznań methodology of

[5]The Poznań university tradition involves J. Ostoja-Zagórski's opinion (2015, 7), following Jerzy Topolski, that "an archaeologist is also a historian, and archaeological sources are historical sources", thus denying archaeology its status as an independent academic discipline. This tradition was established by Jan Żak in his methodological works, by referring to archaeology as *prehistory*, or even *primeval history* (Żak 1966, 67–89) analogous to the terms *ancient history, medieval history, early modern history*, etc. This was also reflected in the institutionalized nomenclature of archaeological research centers, e.g., AMU Institute of Prehistory in Poznań, or the Committee for Prehistoric and Protohistoric Sciences [instead of Archeological Sciences—HM] of the Polish Academy of Sciences. This viewpoint is also represented by the organizers of the conference "Dorobek polskiej archeologii i prahistorii ostatniego półwiecza" ("The achievements of Polish archaeology and prehistory of the last half century") held in Puszczykowo near Poznań in 1997—see *Archeologia i prahistoria polska w ostatnim półwieczu* (*Polish Archaeology and Prehistory in the Last Half Century*), M. Kobusiewicz and S. Kurnatowski, eds., Wydawnictwo PTPN Poznań, 2000. The term *préhistoire* is commonly used in France, especially in reference to the oldest segment of prehistory.

[6]The Department of Archaeological Research Methodology first headed by Stanisław Tabaczyński was then dissolved by him. Tabaczyński was convinced that methodological reflection should be present and developed at all levels of research (Tabaczynski 2000, 259). However, his research attitude—undoubtedly valid—was not synonymous with the increased interest of archaeologists in how and why they are directed by their represented research activities.

archaeology, in other words, the absence of theoretical reflection in Polish archaeology that provokes yet another question: What is it about the methodology of archaeology, in general, that leads to the abandonment or even disregard of this type of research activity?

Let us return at this point to Justyna Gabriel's and Arkadiusz Marciniak's assertion that the level of advancement of Anglo-American archaeology differs significantly from that of Polish (continental) archaeology.[7] While continental, cultural-historical, archaeology mainly derived its ideas from the German-speaking cultural area (mainly from ethnology), the origins of processual archaeology lie in the Anglo-American cultural background, where, according to neo-evolutionists, archaeology and ethnology as identical disciplines were supposed to serve the reconstruction of cultural evolution processes. It should be remembered that Britain was the largest colonial power for many decades. Consequently, the knowledge of societies subordinated to the British Empire became an indispensable requirement of rational British policy. The problems of cultural diversity confronted by British anthropology required an ordering of phenomena often completely incompatible with those of European culture. This had a significant impact on the theoretical expansion of British anthropology, with which archaeology had an important relationship.

The same was true in the United States, the home of New Archaeology, where the problem of dominance of a technologically advanced civilization over indigenous societies is overtly obvious. Contact with societies at a different level of civilizational advancement than that of European culture became one of the reasons for reflection on the colonizing aspect of the emergence of anthropology, and thus of related archaeology. Thus, it was not only the resignation from or even opposition to the cultural-historical orientation, which contained expansionist tendencies saturated with "extermination", that was a highly desirable phenomenon. It was, first and foremost, about the axiological reorientation of culturally different societies,

[7]Although we fully agree with Justyna Gabriel and Arkadiusz Marciniak, in order to counterbalance the expressive force of the above-quoted argumentation with a slightly different point of view, let us pose the following question: To what extent does the unbelievable, even humiliating, pursuit of acceptance of our views in international circles today constitute a creative blockade of our domestic thought? Let us mention here Jerzy Kmita's (2015, 101) critical opinion towards the particularly devout attitude of Polish science towards anything Western, especially Anglo-American, not only when it comes to archaeology. According to Jerzy Kmita, "blocking thinking (clearly visible in our Polish archaeology as mentioned above—HM) is conditioned by a permanent association of the modern with the foreign. This kind of attitude, leading to intellectual impoverishment, is part "sorcery of the humanities". as Kmita once called it, and forces one to invoke "...ideas legitimized by their foreign origin, or by the fact that at least their carriers have been sanctified abroad". The currently practiced requirement to publish works in Western languages, and in Western journals is begging for recognition of our work in the eyes of the West. It shows not only our complexes, but also how far we have regressed to the 1970s, when critical works were published in the "humanist sorcery" or "terrestrial power of science" series. This requirement reveals an almost caricatured dimension of our evaluative attitude today, as evidenced by the notorious "running for points" (publish or perish) in foreign (Western) journals, which, in turn, defend themselves against the deluge of research papers.

and especially about changing their own attitudes towards themselves as researchers representing a "colonial" point of view, and thus situated at the opposite pole from the object of study itself, i.e. societies being the "subordinated" object. At this point, we should recall the views of Franz Boas, for whom "cultures" were of equal value and significant influence, and thus his and his generation's impact on American anthropology. Today, ethnoscience, also known as New Ethnography, whose representative was, among others, Ward Goodenough, popularized in Poland by W. J. Burszta and J. Kmita, is still referred to as "the last breath of cultural relativism". The awareness of expansionism and technological domination, on the one hand, and contact with cultures different from European civilization, with completely different preferences, on the other hand, had a significant impact on the conceptualization of anthropology and related Anglo-American archaeology involving an ongoing search for new theoretical developments. The theoretical aspects of Anglo-American archaeology therefore prove to be more prolific than continental archaeology, characterized by the constant, unchanging, cultural-historical premises from the period of its formation. This approach, which remained linked to the constantly emerging doubts and nagging questions arising from contact with other cultures, led to the emergence of anthropology as well as processual archaeology and then post-processual, interpretative, archaeology.

The implication is that the cultural factor is no less decisive than the epistemological perspective when it comes to the state of theoretical advancement of archeology. Secondly, the arguments above appear to indicate that in cultural-historical archaeology the object of its study was not only been formulated much earlier, but also standardized and conventionalized by recurrent cognitive games—as Zybertowicz (1995, 134) puts it, following A. Gurevich—to such an extent that the assumptions of archeological exposition become completely "invisible" (unconscious). Finally, methodology is rendered superfluous by the belief that "facts that speak for themselves" are entirely sufficient, and therefore completely without our participation, as is the case with the use of research-driven assumptions. I will return to this issue later.

In an attempt to at least partially come closer to AMU methodologists' answer to the question of the meaning of practicing methodology, I will quote here the opinion of Jerzy Kmita, one of the founders of the Poznań School of Methodology. In his view the methodology of science is a discipline that seeks an answer to the question: "how the various research activities should be performed (…) so that the most cognitively valuable results possible can be achieved" (Kmita 1977, 112). A similar observation is made by Jan Woleński (1996). However, he refers, like Kmita, to general methodology, and thus involving slightly different tasks than the methodology of a specific discipline such as archaeology. Methodology then provides rules of conduct to ensure research success. One might say, it formulates recommendations (e.g., in theoretical physics) for researchers-practitioners (experimental physicists) on how they should act to achieve it. As a result, methodology is sometimes treated by researchers (archaeologists-practitioners) as a discipline claiming to formulate **prescriptive guidelines** (not to say to harangue) on what and how we should do in archaeological practice. Such self-centered, in the eyes of the archaeologist-practitioner, placement of the methodologist in a privileged position vis-à-vis the researcher of the past appears to be (in addition to a different approach to the object of research) one of the reasons for the reluctance of field archeologists to methodology as well as to methodologists themselves. All the more so because the degree of generality of the aforementioned guidelines sometimes takes away the possibility of explaining or understanding, as well as resolving disputed problems, and does not encourage theorizing, especially since all authentic knowledge, according to the generally widespread belief, is based on facts. Hence is the question by representatives of "field archaeology": Why do we need methodology, if the facts discovered by the archaeologist speak for themselves? How was it really with this archaeologically studied past? Is this actually so?

The question "What do we need methodology for?" is intriguing simply because it is associated with an attitude of reluctance towards theoretical thinking that is limited to the "ascetic study of facts". Leszek Kołakowski (1984, 68) places archaeologists (next to paleographers, librarians, numismatists, sigillographers, or genealogists) among the meager craftsmen of the historical sciences. And he does so, among other things, because the attitude of the archaeologist, guided by the "fetish of facts", makes it impossible to "unveil" the meaning of past events, as it is the meaning given to the past by ourselves. Here I would like to bring up a remarkably straightforward definition of methodology by Ewa Domańska, a historian-methodologist with declared sympathies for archaeology and well acquainted with its research problems. According to her,

> methodology is a set of research practices, i.e. procedures and strategies that indicate how to build theories. Theory is then understood as a set of certain concepts and explanatory constructs that allow conceptualization and interpretation of the phenomena under study (Domańska 2010, 50).

It is theory that causes empirical research to acquire its meaning following the theory which designs and interprets it. We are dealing here with the theoretical interpretation of experience. Empirical knowledge (experience) would not articulate, as Jerzy Kmita puts it, any general knowledge about the world, about the past, if it did not presuppose a prior theory. Thus, we encounter a theoretical interpretation of experience when we expose the cognitive primitiveness of the theory, which is the opposite of the empirical interpretation of the theory placing experience (facts) in the foreground.

Therefore, if we accept the assertions of Minta-Tworzowska (1994), Marciniak (2012) or Rączkowski (2002, 14) that theory is the result of cognitive processes, while method is a way leading to research results, then we expose the idea inherent in those assertions that methodology is the science of ways of theorizing archaeology, and it is concerned not so much with the archaeologist's object of study itself (the past) but with ways of thinking about it. Thus, methodology of archeology is about the ways of learning about (pre)history, and not about (pre)history itself. In other words, according to theoretically oriented archaeology "… the object to be studied is not the event itself, but the thought expressed about it". This "understanding" attitude of theoretical archaeology most generally takes aim not so much at the past itself, but at ideas about the past guided by the ways of perceiving the world; the recognized, value-laden ideas, and the theoretical assumptions made in connection with them. At the same time, the ways of research are, as it turns out, imposed by the ideals of scientific cognition currently operating in the research community. According to K. Ajdukiewicz (1985b), methodology is a discipline of the humanities. Jan Woleński (1996, 242) believes similarly. Accordingly, he uses the notion of understanding, which is characteristic of the humanities.

It is, therefore, not entirely the case that methodology is established to make prescriptive statements. Its task is to 'reconstruct' the rules of the archaeologist's rational research conduct and way of thinking to try to understand and then explain the emerging variety of images of science formed following the application of these rules, which cannot be explained in the language of the discipline. Just as the methodology of science is a science of science, a science called upon to 'reconstruct' its rules, so too the methodology of archaeology is a science of the rules of archaeology investigating the past, not a science of the past itself. And since methodology of archaeology reveals the possibility of multiple and diverse viewpoints of the past, thus meta-archaeology that distances itself from them comes to the fore. For while archaeologists' opinions about the objects they study, states of affairs, or even about the past are statements in an object-oriented language, statements about ways of reflecting on the past require the use of a meta-language, i.e. a language that distances itself from the subject-oriented language, as Minta-Tworzowska (1991, 234) also reiterates.

Conclusion As methodology is the philosophy of science, so is the methodology of archaeology the philosophy of archaeology. Thus understood philosophy of archaeology is, according to Deleuze and Guattari (1991, 11–12) "not contemplation, nor reflection, nor communication. It is not contemplation, for contemplations are the things themselves as seen in the creation of their own concepts". This is an extremely important matter, not only in terms of archaeological study of the past, but also in terms of the co-construction of the present, through the "here and now" conceptual schemes it creates. Now, I will look closer at the problem of the reality of archaeological sources, which are the material remains of past societies.

3 *Archaeology i jej źródła* (*Archaeology and Its Sources*) and the Reality of the Past

Archaeologia i jej źródła (*Archaeology and Its Sources*) is the title of Arkadiusz Marciniak's book (1996), which, like Danuta Minta-Tworzowska's (1998, 2011) articles, discusses archaeological sources. Admittedly, Marciniak's work is about faunistic sources, but its very title directs the reader's attention to the tangibility of the remains of past societies. Following the generally common belief, archaeological sources, according to Jerzy Topolski, are direct sources and sources of true knowledge (as opposed to the always already interpreted historical written texts). They are simply facts. Fact is treated as a synonym for veracity. So when we talk about facts, there is no need to question their veracity. When D. Minta-Tworzowska and M. Pawleta (2013, 228) write about the strengths and assets of archaeology, they point, among other things, to the materiality of the past due to the materiality of its sources. They are guided by a scientific belief that archaeology is closer to the past than sciences with sources devoid of such materiality. According to this conviction materiality means reality. Reality, on the other hand, is necessarily, objectivity, almost at the border between culture and nature, devoid of any human interference. Naturality, on the other hand, is even universality. This is the process of mythicization of facts/sources suggesting their axiological neutrality, i.e. of sources from which all authentic knowledge of the past, devoid of bias, is derived.

Meanwhile, the very concept of "truth" is already a value. Consequently, the fact/material source synonymous with it also has a valuative character. Valuing invariably is related to the cultural context of whoever speaks about the veracity of facts. I will mention here a related archaeologists' belief that research is about making our statements about the past consistent with the past reality we are studying, which is the classical definition of truth. Such conformity is attained by way of corroboration by facts (sources), in other words, by way of experience. This is a thesis so obvious in the popular opinion of archaeologists-practitioners that one can hardly be surprised at the negative answer to the question about methodology, namely, that we do not need it for anything, if it is obvious that a significant part of archaeologists are convinced that facts are the source of all knowledge. For it is facts, according to this belief, that tell us themselves "how it really was" in the past we study.

The second reason for reluctance towards methodology of archeologists preferring that which derives much more from sources/facts, from experience, than from conceptualization is the almost universal belief that only "hard" facts matter. There is nothing in the mind that is not previously in the senses. "One must see in order to know". This is the philosophy of St. Thomas Aquinas that briefly characterizes epistemological empiricism, further exemplified by M. Rydlewski (2016) with the inverse of Ludwig Fleck's view: "to see one must know". Indeed, the point is that archaeologists believe that the essence of research remains the senses, experience, in other words, empiricism, especially when it comes to the context of justification (difficult, by the way, to separate from the context of discovery). At issue, then, is, bolstered by John Locke's metaphor, the conviction that the human mind is like a *tabula rasa*—a blank slate, on which only experience (facts or sources) can inscribe signs. The belief of a significant number of archaeologists becomes tantamount to this view that we, archaeologists, do not need methodology for anything.

The links between theory and experience (facts) determine the empirical foundation of archaeology. What usually happens is that the formulation of general claims (theory) "is based on the inductive generalization of collected archaeological data" (Rączkowski 2002, 13), i.e., as archaeologists quite often emphasize, on the enunciation of what directly follows from archaeological sources. According to this belief, it is the senses that are the source of experiential, empirical knowledge. Thus, a fundamental point of support in archaeology is the claim that "there are only facts". "Archaeology is about facts: if you want the truth, go next door to the philosophy department" wrote John Bintliff (1993, 100). J.C. Gardin (1998, 108) followed up with "facts established by science". Paraphrasing Ludwig Wittgenstein's lines in his 1922 *Tractatus logico-philosophicus* with quite a different meaning than the *Philosophical investigations* written more than 20 years later and published in

1953 (2 years after the author's death) we can say that the past (in Wittgenstein's case—the world) is a collection of facts. *Tractatus* … begins with a succinct passage: "The world is everything that is the case" ("Die Welt ist alles, was der Fall ist"). In other words, there is nothing in the world except facts. When reading this statement, an archaeologist undoubtedly comes to the conclusion that opinions about the past can only be true by virtue of such (empirical) facts. This, however, is already an interpretation from the perspective of epistemological empiricism. On the other hand, Wittgenstein himself does not claim that what we call facts, objects, or simple entities can be effectively isolated or known empirically. They appear as a logical necessity postulated by theory. It was the linguistic turn, linked to Wittgenstein's idea that the world (of facts) is mediated by language, that became one of the reasons for overcoming the dominance of empiricism. The linguistic turn was thus hinted at already in the *Tractatus* by the relation between language and the extralinguistic realm. The replacement of this relation in the *Investigations* is rendered as "(…) any view about the relation between language and non-language is bound to be nonsense" (Rorty 2007, 162).

Following the line of reasoning making methodology redundant for the archaeologist-practitioner, it is obvious that the obtained results should be consistent with reality as it really was. In other words, they should be empirically affirmed. The source of this conviction lies in the empiricism of the late 19th and early 20th centuries, which called for basing all knowledge on "pure experience" and eliminating anything beyond it. This is because experience is considered the source of all relevant knowledge about reality. Knowledge thus acquired is treated as cognitively reliable, as it arises from direct contact with objective reality. Here, we are dealing with the postulate of epistemological empiricism "according to which, all authentic knowledge is based - directly or indirectly - on experience, and only by experience can it be justified" (Stanosz 1991, 5). The distinguished role of experience (facts) means that only what is experienced, observable, is real. A theory, on the other hand, can adjudicate about the world, only if it is properly linked to experience.

Therefore, Friedrich Nietzsche's (1967, 267) statements from around the same time are perplexing:

> Against positivism, which halts at phenomena – 'There are only facts' - I would say: No, facts are precisely what there is not, only interpretations. We cannot establish any fact 'in itself': perhaps it is folly to want to do such a thing.

Exceptionally ironic in view of J. Bintliff's or J.C. Gardin's comments is an introductory quote from Wilkie Collins's *The Moonstone* used by Paul Feyerabend in his article "How to be a good empiricist: a plea for tolerance in matters epistemological": "'Facts?' he repeated. "Take a drop more grog, Mr Franklin (Gardin – HM), and you'll get over the weakness of believing in facts! Foul play, Sir!" (Feyerabend 1999, 78).[8]

---

[8]Today, however, as A.P. Kowalski (2018, 50) notes, there has been a return to the apologia of empirical experience, even though its limitations have already been the subject of methodological reflection on humanistic research. Philosophical positions appeared, such as that of Ian Hacking (1983), which not only questioned the primacy of theory over experience, but subordinated theoretical knowledge to practical action, or ignored it altogether (Pałubicka 2018b, 74–75).

The persuasion that facts are not enough is somewhat ironic. A more strongly directed and convincing suggestion in this regard is the title of one of the chapters of *Archaeological Theory: An Introduction, 2nd Edition* by Matthew Johnson (2010, 1), namely "common sense is not enough". It encapsulates an attempt to argue that the fact-bound "common sense" we use in resolving archaeological disputes requires moving beyond this fundamental tenet in archaeology that "there are only facts". For archaeology is the interpretation of facts from the perspective of the assumptions (theories) used in the various evolving orientations of archaeology. On the other hand, the common-sense ordering of individual facts, their systematization, and clarification of colloquial social experience, is a feature of pre-theoretical "crude and unreflective" science (archaeology) (Johnson 2010, 219; see also Kołakowski 1984, 68). Such archaeology turns out to be useless. It is suitable only for exploring, organizing, and recording data. In view of this, let us begin with the claim that archaeology is an empirical science.

Archaeological sources, the material remains of past societies, constitute the "core of archaeological research" also when we refer (like Rączkowski) to the significant role of the researcher/archaeologist in creating the source. This contradicts the claim that "the facts speak for themselves" since they happen thanks to the archaeologist's interference. The problem of the archaeologist's interference into the meaning of the source is also addressed by D. Minta-Tworzowska in her article "*Jerzego Topolskiego koncepcji źródeł historycznych a ujęcia źródeł archeologicznych*" ("Jerzy Topolski's notion of a historical source vs. approaches to archaeological sources") (1998) and, especially, in "*Człowiek i rzecz w perspektywie archeologicznej (czyli rzecz w perspektywie antropocentrycznej*") ("*Humans and objects in archaeological perspective (i.e., objects in anthropocentric perspective)*") (2011).

The problem of source material is approached somewhat differently by Arkadiusz Marciniak in his book *Archeologia i jej źródła. Materiały faunistyczne w praktyce badawczej archeologii* (*Archaeology and its Sources. Faunistic Materials in Archaeological Research Practice*) identifying the so-called objectifying planes, defining the limits of interpretation, inherent in the humanities. In any case, on their basis—according to D. Minta Tworzowska and M. Pawleta—conclusions are made (derived inductively from these sources) as well as interpretations about prehistory, oriented "here and now", and therefore they have little in common with induction. Thus, the basic activity of an archaeologist is the ordering and classification of material sources, to which D. Minta-Tworzowska devoted some of her significant studies, including *O metodzie systematyki wytworów kulturowych (na podstawie ceramiki naczyniowej z Kruszy Zamkowej, stan. 3* (*On the method of systematization of cultural artefacts (based on pottery from Krusza Zamkowa*) Minta-Tworzowska 1980 and *Klasyfikacja w archeologii jako sposób wyrażania wyników badań, hipotez oraz teorii archeologicznych* (*Classification in Archaeology as a Way of Formulating Research Findings, Hypotheses, and Archaeological Theories*) (1994). Obviously, classification using the criterion of similarity and difference is also underpinned by a value-oriented modern concept—the idea of progress/regression—the concept of change, which, together with the passage of time is the fundamental core of the notion of ordering material sources.

In this way, the above-mentioned authors, in particular Minta-Tworzowska, unveil the mechanism of constructing the "reality" of the past, driven by the nineteenth-century idea of nationalism, a "reality" that in research practice does not come off as constructed. It is then difficult to say that we are dealing here with something resulting from the sources, and with a preference of this outcome over conceptualization.

Conversely, the research subject as pursued by Minta-Tworzowska is a subject matter seemingly diverse to such an extent that it would be difficult to determine a common denominator (classification in archaeology, Teilhard de Chardin's philosophy of history, archaeological research on art/symbolism and magical interpretation). After all, it is dominated by the idea of order, which brings the thought out of chaos, even when it is devoted to criticism of Pierre Teilhard de Chardin's philosophy of history with its sympathies for French archaeology (Gardin, Gallay). Thus, methodology aims at ordering to bring out—as Wolenski puts it—the very essence of science from this chaos of multiplicity and diversity. It means creating concepts, beliefs, points of view, meanings. In short, it entails the creation of interpretive schemes that have varied throughout history. Let us recall at this point Theodor Lessing's work written during World War I *Geschichte als Sinngebung des Sinnlossen* (*History as giving meaning to the meaningless*) (2013). In it, Lessing argues in favour of the proposition that all representations of history as rational progress are essentially the forming of a formless state, stories of beginnings and endings, ups and downs, flowering and disintegration, which are generally subject to the rules of narrative. The problem is aptly summed up by Joseph Brodsky, who says, "As failures go, attempting to recall the past is like trying to grasp the meaning of existence" (1986, 3). It is worth mentioning at this point Ewa Bugaj's works devoted to the practice of the study of prehistoric and ancient art as a modern discipline of science. Especially when she cites Hans-Georg Gadamer's opinions on the concept of style in the study of Greek art (Bugaj 2018, 137):

> However let us consider whether we can use this concept of style as a historical category. Transferring the concept of style from history of art to history in general involves viewing historical events not in their own significance but in relation to a totality of forms of expression characteristic of their time. But the historical significance of an event does not have to be identical with its cognitive value as an expression of its time, and it is misleading to imagine that we have understood it, if we have understood it solely in this way, as an expressive phenomenon. If, in fact, we extend the concept of style to history in general [...] and expect this to yield historical knowledge, then we are compelled to assume that history itself obeys an inner *logos*. This may be true for particular lines of development that we pursue, but this kind of history is not really history. It is the construction of ideal types which [...] is legitimate only as a description. Viewing events in terms of history of style, like viewing art only in terms of history of style, fails to take account of the essential fact that something is taking place in it and we are not just being presented with an intelligible series of events (Gadamer 2004, 497).

On the other hand, remarks on the points of convergence between the French archaeological and historiosophical thoughts (Teihard de Chardin and the aforementioned archaeologists) can be found in the already cited works of Minta-Tworzowska.

Methodology, after all, is not just ordering as bringing out of chaos. It is an orderly uniformization of something that was never uniform. Thus, chaos is a dispersed multiplicity and diversity preceding order, subject to totalizing unification. Chaos always precedes order in a logical sense. Consequently, orderly reality is the corollary of the reality devoid of order, scattered, chaotic, it is the "reality" created by us from chaos that constitutes the preceding, authentic "reality", but to which we do not have access. The term "chaos" here has a metaphorical meaning. It is used to describe the past "reality" preceding the ordering, which is only a logical consequence of the preceding ordering situation. Thus, any interference by an archaeologist to interpret the past means giving up the belief that we are accessing the past reality as it actually was. As a result of our interference, it is a past "reality" already deformed by us. I would recall at this point Wittgenstein's opinion of extralinguistic "reality", which, according to him, exists, but it cannot be accessed without the mediation of language. It is language that organizes experience. It is, first of all, classification and ordering of the stream of sensory experience, yielding a certain order of the world. Let me quote here an extremely valuable remark by Heidegger (1962, 77):

> If an ordering principle is genuine, it has its own content as a thing, which is never to be found by means of such ordering, but is already presupposed in it. So if one is to put various pictures of the world in order, one must have an explicit idea of the world as such.

It therefore follows that at the root of the past "reality" we study, we are always ourselves. In any case, I can say that the same physical, material remnants of past societies will not necessarily lead all observers to the same image of the world. According to Kmita (2000, 59), empirical content is already organized by means of a conceptual scheme, applied to this content by our cultural teachers, starting with language. Concepts always precede rather than follow cognition. They guide human actions and impose a restrictive framework on them (Gellner 1988, 56). Their ultimate outcome—the result of the placement of these material remnants in the generated evaluative ordering in time and space—are archaeological cultures, identified with the significant aid of the modern criterion of similarity. It leads to the conclusion that the "same" products were made by the "same" producers. The similarity of material products is transferred to their makers. We can see the consequences of this way of thinking in the dispute between autochthonists and allochthonists, which is perfectly compatible with the answer to the question: What do we need methodology for? This dispute completely obscures the underlying substance. In the eyes of methodologists, the essence of this dispute is not about which side is right, but about revealing what kind of value-laden worldviews lie at its core, the consequences of which, in turn, go beyond the specifics of archaeology itself. The point is to expose the invisible but underlying presumptions, the consequences of which are—for example—tribalist, racist, xenophobic concepts in opposition to any otherness. The results of such convictional attitudes are to be found in the form of cultural, economic or, finally, military expansionism and, often in the "extermination" (to use Wawrzyniec Surowiecki's term) of cultures in the areas targeted for expansion. To the same extent this applies to cultural imperialism and the related idea of human subjugation and domination. More than aptly, this problem is illustrated by Benedict Anderson's examples of uses of archaeology to justify views that depend on the political attitudes of those in power (1997). Similar considerations can be found in Chapter II of Marian Kwapiński's recently published book (2020, 102ff.). It is one of the attempts of the Poznań methodologists (e.g. Rączkowski 1996) to answer the question: "What do we need methodology for?"—an answer that exposes the not-so-glorious value-oriented beliefs contained in the conceptual assumptions of the proponents of the ethnic qualification of archaeological sources: they reveal what the directly observable facts cannot reveal to us.

Conclusion The AMU methodologists of archeology are concerned with bringing to light the research premises, theories, that govern the interpretation of the facts uncovered by the archaeologist, and thus unveiling the value-oriented ideas that give direction to our archaeological interpretations, exposing illusions that often turn into a certainty that our study of the past is guided exclusively by "pure" facts. However, if we accept their view that they are brought to life by the archaeologist (Rączkowski 2009, p. 30), who "creates the source" then their objective character is no longer so obvious. This is because the above formulation contradicts the assertion that "the facts speak for themselves", since they do so thanks to the archaeologist's inteference.

Distinction should be made between at least two perspectives. The first is the epistemological perspective of scientistically oriented archaeology with claims to universality. It abstracts from the other perspective, i.e. all cultural and historical, contextual conditions, and it emphasizes its cognitive character. Cognition by this perspective should be identified exclusively with science. Its main measure is the rationality of cognition (Karl Popper's variant). It remains in opposition to the OTHER PERSPECTIVE, i.e. the view of the revolutionary nature of changes in archaeology as a science, i.e. Thomas Kuhn's view of its cultural and historical conditions (Amsterdamski 1983, 26–41). Before discussing the latter, I will turn to the materiality of the past and the related natural-scientific orientation of archaeology.

4 Materiality of the Past *Versus* the Natural-Scientific Nature of Archaeology

When discussing differences in the state of theoretical advancement of continental and Anglo-Saxon archaeology due to cultural rather than epistemological factors, I cannot ignore the dispute about the extent to what archaeology is a natural science or part of the humanities. According to Arkadiusz Marciniak, processual archaeology, closely related to New Archaeology, is characterized by a natural science orientation to a much greater extent than post-processual archaeology, which is oriented towards humanistic interpretation. Although both parts of the title: "Materiality of the past *versus* the natural-scientific nature of archaeology" appear to be significantly related to each other, I have contrasted them using the preposition *versus* rather than the conjunction *and*. This means that the materiality of the facts is not equivalent to archaeology being an *archaeological science*, despite adherence to such a view by a significant number of archaeologists. It can be said, therefore, that the first of the above-mentioned approaches remains—contrary to what is generally believed—in opposition to the other, that is, in opposition to epistemological empiricism, metaphysical realism, and methodological naturalism.

The desire for a natural scientific orientation of archaeology plays an important role in the views of many modern archaeologists, including Colin Renfrew and Paul Bahn (2000, 12), who claim that "archaeology, in short, is a science as well as a humanity". They express a conviction shared also by Stanisław Tabaczyński. They do so primarily because archaeology draws on the aid of natural sciences: biology, physics, chemistry. Using the results of scientific research does not yet make archaeology a natural science. For comparison, according to Aleksander Posern-Zieliński (2018, 324), criminology "whose techniques rely heavily on 'hard' data from physical, chemical, and biological analyses, does not thereby become an exact science". Such expectations are based on a view that natural sciences provide us with benchmark examples of authentic knowledge, and that sciences have a more direct access to reality, including past reality, than the humanities. On the grounds of archaeology represented by AMU methodologists this belief is shared, among others, by Arkadiusz Marciniak. The issue at stake here is the methodology used in archaeology. To what extent should these methods be drawn from natural sciences, and to what extent from the humanities? When, at one of the meetings of the Commission for Anthropology of Prehistory and the Middle Ages of the Polish Academy of Sciences, Marciniak asked "about the nature of differences between the results of isotope or neutron activity, on the one hand, and stylistic analysis, on the other, as a method of studying the provenance of objects and the raw materials from which they were made", he emphasized the significant differences between the humanities and natural sciences. In doing so, he demonstrates, as Jean-Claude Gardin and his colleagues do, a concern that archaeological methods should be closer to natural sciences, in other words, closer to science than to literature. This is because science-oriented disciplines provide representatives of empiricism with model examples of authentic knowledge. The method of natural sciences is the model scientific method (Woleński 1996, 47). The classic form of scientism emerges here, i.e. the view that science based on the scientific method is the best or the only way to render truth about the world. The scientific focus of archeological research not only suggests its association with positivism and scientism, but especially with empiricism.[9] Let us the consider closer the relationship between the humanities and natural sciences.

The claim to recognize archaeology as a natural science is also reflected in the positions of representatives of cognitive archaeology (e.g. Steven Mithen, Ezra Zubrow) or behavioural archaeology (Vincent M. LaMotta, Michael B. Schiffer).

[9] Opinions on this subject can be found in *Empiryzm współczesny* (*Contemporary Empiricism*), including essays selected, compiled, and preceded by an introduction by Barbara Stanosz, Wydawnictwa Uniwersytetu Warszawskiego, Warszawa 1991.

The desire to bridge the humanities and natural sciences can also be found in the research approach of historian Ewa Domańska, fascinated by archaeology, and Norwegian archaeologist Bjørnar Olsen, inspired, among others, by the views of Bruno Latour,[10] according to whom socio-nature is related with the failure to promote the idea of differences between natural science and the humanities. A view similar to Bruno Latour's is also expressed by Tim Ingold, particularly in his works on the cultural landscape. In Ingold's opinion the relationship between humans and nature does not form a dichotomous pair of opposites. In fact, it is characterized by dynamic ties, constant exchange of contents, and mutual transformations. The opposition of nature and culture turns out to be artificial, since both condition each other in many spheres of life, and humans' relations with nature, in his opinion, do not belong to culture (Ingold 2003, 74–84). This is a claim completely opposite to the one expressed by cultural studies scholars from Poznań, most eloquently articulated by Michał Rydlewski (2016, 31): "*Naturwissenschaften* are always *Kulturwissenschaften*" (Natural sciences are always cultural sciences) (see also Mamzer 2021, 65). According to Barbara Kotowa (2001, 35) the nature-culture opposition formulated from the position of cultural studies,

> (…) should by no means lead to the conviction that it considers man as an object of research in some absolute 'order of nature,' for both man as an object of research and the 'nature' that constitutes his biological context are considered within the framework of scientific practice already from the point of view of the 'order of culture' in which every researcher is entangled.

According to Leszek Kołakowski, man as part of the environment, when observing nature becomes an observer of himself as part of it. In Kołakowski's opinion, human contact with nature is not intelligible by itself, because it cannot identify its own observer's consciousness with its own consciousness as an object observed by itself. For man relates intelligibly not only to the natural world as the object of his investigation, and thus not only knows that he is a part of it, but also knows that he is conscious of it. Awareness of one's consciousness then requires going beyond the nature-culture order as an extension of nature requires going into the mythical (worldview) order, equipped with the memory of the past, as well as anticipatory foresight into the future (Kołakowski 1989, 134). Natural sciences will not provide full answers to the questions posed, especially questions about culture. "Thus, Husserl's reduction plan, which would suspend existence in indecision, is false, as is the scientists' hidden belief that they can, free from human partiality, interpret that human partiality" (Kołakowski 1989, 13). The objectivity of natural science, according to Kwapiński (2020, 124) citing Halina Walentowicz's opinion, is in this light a valuative belief, since it is co-created by the subject, which itself transforms

[10] Bruno Latour's ideas are also supported by Quentin Meillassoux (*After Finitude: An Essay on the Necessity of Contingency*, Ray Brassier, Continuum, 2008); and Graham Harman (*The Quadruple Object*, Zero Books, 2011; and *Prince of Networks: Bruno Latour and Metaphysics*, re.press, 2009). The cited works are significantly related to the idea of "democracy of objects" promoted by American philosopher Levi R. Bryant, which he presented at one of his meetings during his stay in Poznań in autumn 2019. See also Bryant (2011).

through feedback. This is also a cultural orientation, regardless of the verbal declaration and of the research attitude of Poznań methodologists of archaeology, including Arkadiusz Marciniak.

As indicated above the differentiation of the subject matter pursued by the AMU methodologists of archeology is manifested by a peculiar split between a humanistic orientation of archaeology (Minta-Tworzowska, Rączkowski), and a natural-scientific leaning (Marciniak). I am using the word "leaning" here, because it is by no means identical with the natural science orientation of Marciniak's research. Marciniak's perceptible nod towards natural science is most evident in his two major works on faunistic materials in archaeology (1996) and social relations with animals in the Neolithic (2006). It is not, then, precisely the scientific orientation of Marciniak's research that is at issue here, as is the case with processual archaeology. This is indicated not only by Marciniak's fascination with Paul Ricoeur's hermeneutics or René Girard's works on cultural studies (1996, 29, 145), but also by his articles in the volume *Przeszłość społeczna* (*Social Past*), especially those devoted to representation and narrativism in archaeology. The implications of Marciniak's fascination can be found, for example, in his interpretations of the outcome of his research at the well-known Čatalhöyük Neolithic site in Central Anatolia (Marciniak 2013).

It would seem that this kind of expectation lies solely in the search for research methods, in this case in the natural sciences. Meanwhile, first and foremost, it is about the object of study of archaeology, that is, in Marciniak's case, about developed prehistoric societies, especially the relationship between humans and animals, "social identity built through contacts with animals and also the symbolic and religious sphere associated with them" as in D. Minta-Tworzowska and M. Pawleta (2013, 226). Only in connection with the object of study of archaeology, do they relate to the study of applied research methods. It is also relevant for Marciniak, who seeks to link processual archaeology with contextual archaeology so as to combine the explanation inherent in natural sciences with the understanding characteristic of the humanities.[11]

Processual archaeology, according to Marciniak, is oriented toward a status similar to that of natural sciences. Post-processual archaeology, on the other hand, has strong ties with the humanities. Thus, it seeks to

[11] The notion of "understanding" inherent in the humanities and its opposite "explaining" characteristic of natural sciences are discussed by Marciniak (1996, 25ff) and Pałubicka (2018a), who, like Kmita (1971), present humanistic interpretation as a variant of causal explanation. According to Jerzy Kmita, humanistic interpretation has a cognitive role, corresponding to the term "explanation" characteristic of natural sciences. Anna Pałubicka explores this problem in detail in her article "*Hermeneutyczne i epistemologiczne koncepcje interpretacji*" ("*Hermeneutic and epistemological concepts of interpretation*") (2018). It is from Kmita's findings that Marciniak (1996, 24) derives information on this subject through Topolski's *Rozumienie historii* (*Understanding History*) (1978), when he writes about humanistic interpretation fulfilling the role of explanation in the humanities. See also Topolski (1978, 30ff), in whose opinion understanding is not some special scientific procedure different from explaining.

> (…) overcome the opposition that exists between archaeology as a science of human culture and the archaeological sciences as a set of disciplines growing out of exact and natural sciences. This integration, for the most part, must take place on the theoretical and methodological level. And this is the only way to achieve effective mergers of these two strands of research on the past. The proposed model is also an attempt to depart from the strong formalism, dogmatism and 'objectivity' of processual archaeology, and from the illusions and excessive expectations of sources of post-processual archaeology. It questions and attempts to overcome the relativism of the latter (Marciniak 1996, 33).

After all, overcoming relativism can only lead in the direction of realism, clearly evident in Marciniak's paper *Istota różnic pomiędzy ewidencją archeologiczną a przyrodniczą i ich artykulacja w praktyce badawczej (The significance of differences between archaeological and natural records and their articulation in research practice)*. According to Marciniak

> For I do not believe that knowledge of the past is a mere projection of the present, and I argue that the process of production of this knowledge is much more complicated than it is commonly believed. I therefore argue against indiscriminate constructivism,

thus leaning, one would like to say, toward Hilary Putnam's internal realism, more conciliatory than metaphysical realism.[12]

I will invoke here Richard Rorty citing Arthur Fine's acclaimed paper "The natural ontological attitude", which begins with the line: "Realism is dead". Fine, quoted by Rorty, who demonstrates significant parallels between realism and theism reveals that

> (…) realism is dying out among the philosophers, this is because of the attractions of a culture that is more deeply and unreservedly humanist than that offered by the arrogant scientism that was the least fortunate legacy of the Enlightenment (Rorty 2007, 136).

This kind of expectation, or even desire, by Olsen and Domańska to bring the humanities and natural sciences closer together can already be noticed in the views of Enlightenment thinkers, in particular, of Nicolas de Condorcet (1743–1794) who presented them more than 250 years ago in his inaugural lecture at the Académie

[12] The implication of the assertion that "our image of the past is a projection into the past of a perspective-oriented present" is worth noting. It is an expression, in the shortest way, of the idea of *Przeszłość w teraźniejszosci... (The Past in the Present...)* contained in the title of Robert Traba's book (2009), or *Przeszłość we współczesności...* (*The Past in the Present...*), which is the title of Michał Pawleta's book (2016). It is an idea in opposition to any juxtaposition of the present with what the past is. It is a warning against urging total oblivion of what once was, because looking backward is—as it is assumed—tantamount to the impossibility of looking ahead. Consequently, all the evils we faced, even in the recent twentieth century, should be eclipsed, precisely so that they do not obscure the view of the future. Future history, therefore, should not focus on negative history in order to preserve positive values for the future to enable the integration of the community in the present. Meanwhile, the quoted proposals to preserve the past in the present become a warning against repeating the mistakes once made. This is the meaning of the phrase "the past in the present", counteracting forgetting tragic past events, in any case proposing to, or at least consider, the problem of whether the awareness of past evils counteracts the shared values that bind the community together in the present? By limiting ourselves to focusing on this WARNING, we are overlooking the positive values that the past has passed on to the present.

Française. "The sciences"—as B. Suchodolski (1957, xvi) in the Introduction to the Polish translation of Condorcet's *Sketch for a Historical Picture of the Progress of the Human Mind* (*Esquisse d'un tableau historique des progrès de l'esprit humain*)—"whose object is man and whose task is the happiness of man" will develop as well in the future as natural science has become. However, they must, like the natural sciences, "be based on facts, use similar methods, acquire a language that is just as strict and accurate, and achieve a similar degree of certainty" (Suchodolski 1957, xvi). These expectations are almost entirely reflected in Condorcet's (1795) work cited above. I mention them for two reasons. First, Condorcet's philosophical writings, like Marciniak's, reveal a special concern to bring the level of the humanities close to that of natural science. Secondly, Condorcet's views anticipate the historiosophical idea of unification pursued in the twentieth century by Pierre Teilhard de Chardin, an author of particular interest to Danuta Minta-Tworzowska. Her examination of de Chardin's historiosophical concepts is the result of her practicing the archaeology of ideas. This is worth noting, if only because it illustrates the way of thinking from the objects dominant in archaeology to thinking about objects. After all, revisiting Teilhard de Chardin's thought I must point out that it involves concern for human development. It is a concept of evolution towards self-perfection with a tendency towards the unattainable perfection of the divine. As with the idea of progress, it has its origins in natural history, i.e. not only history, and not only history of man.

This universal naturalistic point of view on the origins of humans, their history, and evolution, which aims at unity, finds its distant in time but significant echo in Marciniak's views seeking to combine the natural and cultural standpoints. Marciniak's scientific leaning towards a processual archaeology of universal dimension can be noted in Minta-Tworzowska's (1986) critical assessment of Teilhard de Chardin's historiosophical concept, especially when she writes that "Teilhard de Chardin's phenomenology embraces the achievements of the empirical sciences, mainly geology, paleontology, biology, zoology, biogeography, and physics" (Minta-Tworzowska, 1991, 223) to lay the groundwork for the second stage of evolution associated with human history. Such an approach to the whole of history as a universal history can be seen as an attempt to unite the various natural sciences and humanities. Admittedly, Minta-Tworzowska herself does not share such Teilhard de Chardin's processual viewpoint. Her assertion, therefore, gives the impression of a polemic against A. Marciniak's idea, aiming to overcome the opposition between archaeology as a science of human culture, and archaeological sciences as a set of disciplines derived from exact and natural sciences. To some extent, Marciniak's idea resembles one of the theses of the Vienna Circle philosophers, namely that of unity of science, understood as the reduction of the conceptual apparatus of all science to the language of physics, i.e. physicalism. In any case, it is Teilhard de Chardin's concept of man and the world, which strives for unity, and contains, as in the case of Latour's and Ingold's views on nature and culture, the proposal to abandon the idea of fundamental difference between materialism and idealism.

Overcoming this difference, according to Minta-Tworzowska, is not entirely possible. It resembles Minta-Tworzowska's criticism of Teilhard's dream of eliminating the contradiction between materialists and idealists and merging the two points of view. However, "there is [...] no way to 'translate' the concepts of one discipline into those of another, and all of them into a versatile conceptual apparatus describing the universe" (Minta-Tworzowska 1991, 233). She recalls the interesting problem of confusion between the subject language and metalanguage at the time, leading to paradoxes such as, for example, the classical liar paradox, with which many thinkers struggled for at least more than 2500 years, until it was resolved by the Polish logician Alfred Tarski in his *Pojęcie prawdy w językach nauk dedukcyjnych* (*The concept of truth in formalized languages*).[13] The conclusion that emerges from Minta-Tworzowska's analysis of Teihard de Chardin's work is unequivocal. *Man* (vol. 1) and his place in Teilhard's model of evolution, is a work written from the deep humanistic perspective, even though it uses the achievements of the natural sciences. It shows a direction that merges a multitude of disparate elements into a single whole and heads toward the Omega Point, where all the developmental paths of the universe converge. This is a perspective full of contradictions, deterministic, and totalizing, as Danuta Minta-Tworzowska demonstrates in her analysis.

Conclusion Let us return to the humanities-sciences relationship and Arkadiusz Marciniak's research leaning particularly evident in the second publication mentioned above (Marciniak 2005). It seems to stem from the fact that his reserach activity is oriented towards the fusion of both currents of thought viewed from a humanistic perspective. Marciniak's natural-scientific attitude found its particular expression in the archaeozoology that he practices, and above all, in the human-animal relations, construction of social identity through contacts with animals, and the symbolic-religious sphere associated with them. There, his applied dialogic stance between natural science and the humanities in archaeology definitely begins to emerge. Thus, if there is a value-oriented research attitude of the natural scientist-archaeologist, as in the ordering of sources, and thus a classification devoid of such impartiality, then it is not only the natural sciences that are the consequence of a view from the perspective of cultural sciences. Also, Marciniak's attitude is that of a humanist, and it is from the humanist perspective that he views the above distinction. "Nature in this sense turns out to be an object that is just as historical as the subject that objectifies it" (Kwapiński 2020, 124).

5 "The Concept of the Source as a Cultural Construct", or the Linguistic Nature of the Past

The arguments presented so far constitute a way of thinking leading from the fact to Włodzimierz Rączkowski's "concept of the source as a cultural construct" (2009, 16), a phrase used in the title of this chapter, corresponding with Minta-Tworzowska's (1998) opinion on the cultural character of the fact/source as well as

[13] Ajdukiewicz's article in the volume *Język i poznanie* (*Language and Cognition*) (1985b) makes no reference to Tarski's work, as it was originally published in 1933. Ajdukiewicz's article is from 1931.

Kmita's and Pałubicka's (1992) exposition of "the cultural character of experience." The different images of the past that emerge from the application of models of its study reveal their value-oriented character leading to postulative judgments, to use Wolenski's term (1996, 241). They are parallel with Willard Van Orman Quine's naturalized epistemology, according to which all objects we talk or think about are postulated entities, not real objects, awaiting to be recognized. This means that our diverse perceptions of the past are not fixed once and for all by past events. A researcher's changing point of view over time means that a changing past, just as a diverse past, is in the eyes of concurrent, contemporary, representatives of different research orientations. The answer lies not only in the differences between them but also in the differences in the degree of theoretical advancement of continental (including Polish) and Anglo-American archaeologies. Hence, this objectivity of facts is, as it turns out, not quite what it seems. A thorough methodology of archaeology offers us the opportunity to become aware of what the facts do not tell us. Thus the negative answer implicit in the question: What do we need methodology for? that we do not need methodology for anything because the facts speak for themselves is not satisfactory to us, unlike to field archeologists who "prioritize whatever arises from the sources over conceptualization". What is called a fact, an event, is a most arbitrary construct, and each construct determines facts, not the other way around (Kołakowski 1984, 68).

Accordingly, the message contained in the quoted opinions remains in opposition to the conviction that "the facts speak by themselves" about "the way it happened" to use Leopold von Ranke's words. In Ranke's view history should reproduce what actually happened and thus avoid speculation that subordinates historical events to an arbitrary notion of truth. Let us recall at this point the common belief among archaeologists that scientific research is all about ensuring that our assertions about the past are consistent with the past reality we are studying. After all, statements doubting or even questioning the materiality of sources as reality, i.e. as remnants of the real past, mean that the classical definition of truth does not meet our expectations. For truth appears to be

> (…) a mobile army of metaphors, metonyms, and anthropomorphisms — in short, a sum of human relations which have been enhanced, transposed, and embellished poetically and rhetorically, and which after long use seem firm, canonical, and obligatory to a people. Truths are illusions about which one has forgotten that this is what they are; metaphors which are worn out and without sensuous power; coins which have lost their pictures and now matter only as metal, no longer as coins (Nietzsche 2019, 6-7).

This kind of stance, representing the cultural rather than the objective character of the source (facts or experience) is clearly manifested by Włodzimierz Rączkowski (2009, 16), who says that

> (…) archaeological source ceased to be a point of entry for the research proces and started to serve the verification/testing of hypotheses. (…) is the result of the archaeologist giving the object meaning. It is the archaeologist who brings the source to life.

Similarly, according to Minta-Tworzowska (2011, 46), "it is the archaeologist who determines what an artifact, an ecofact, a source is, and what it is not. It is the archeologist who creates the source". In other words, it is the present that gives meaning to the past, whereas what is called a fact, an event, is a most arbitrary construction.

Thus, the common-sense and, at the same time, positivist notion of fact-based experience constituting the hitherto unquestionable foundation of research results was challenged in the aftermath of the "linguistic turn" (Kmita and Pałubicka 1992; Rączkowski 2009; Minta-Tworzowska 2011). The AMU methodologists of archaeology emphasize this in almost all their works. It becomes evident in Rączkowski's *Archeologia lotnicza...* (*Aerial Archaeology*...), especially in Chapter 5: "Postrocesualna zmiana—nowe perspektywy" ("Post-processual change–new perspectives"); in Minta-Tworzowska's works, e.g. in *"Kwestia przełomu metodologicznego w prahistorii i archeologii polskiej" ("The question of methodological turn in Polish prehistory and archaeology")* (2000); or Marciniak's articles in a co-edited volume *Przeszłość społeczna* (*The Social Past*, 2012).

The linguistic turn, whose elements became noticeable at the end of the nineteenth century (Peirce, Frege), is most evident in the variability of Wittgenstein's views, who in *Tractatus*... declares that there is nothing in the world except facts, while in *Investigations*... demonstrates that language is the most essential component of human life; and that the limits of our language are the limits of our cognition. "There may be things hidden behind linguistic phenomena, but they are not of interest to me", says Wittgenstein. It is, however, of interest to those around Quentin Meillassoux, and especially to the originator of the idea of object/subject correlationism himself, Graham Harman. Overly instructive here, as far as the subject is concerned, is his view on correlationism. According to Meillassoux, a significant role is played by the concept of correlation understood as "the idea according to which we only ever have access to the correlation between thinking and being, and never to either term considered apart from the other" (Meillassoux 2008, 5). Hence the current of thought referred to as correlationism.

> Correlationism consists in disqualifying the claim that it is possible to consider the realms of subjectivity and objectivity independently of one another. Not only does it become necessary to insist that we never grasp an object 'in itself', in isolation from its relation to the subject, but it also becomes necessary to maintain that we can never grasp a subject that would not always already be related to an object (Meillassoux 2008, 5).

This is the argumentation in favor of the essential role of the object (although always correlated with the subject) positioned outside language. But let us return to the problem of language, since extra-linguistic reality is not directly available to us.

There is the suggestion of the need to use a semantic definition of truth, more precise than its classical version. According to it, our statements about the past are true when they are relativized by language. This means that a sentence stated as true in one language can be false in another. This is because our statements are value-laden and, therefore, dependent on the cultural context of the phenomenon under study. However, since the cited differences in the so-called level of cognitive advancement of continental and Anglo-American archaeologies are culturally conditioned, the rhetorical question: What do we need methodology for? along with the negative answer contained in it, is, therefore, not an expression of doubt in the sense of methodological orientation of our archaeology, as some statements seem to imply. It is not because the epistemological position (as I have already signaled) is not enough to explain this phenomenon. The cultural factor, as it turns out, plays a significant, if not decisive, role here. Disclosing this problem, substantiated by Stefan Amsterdamski (1983), and thus demonstrating a different opinion than the epistemological viewpoint, is one of the most important, successfully implemented tasks undertaken by the AMU methodologists of archaeology.

Our study of the past is not its reconstruction, as it once was in itself, but its interpretation. We speak of the meaning of past events, graspable in the historical order imposed on them by the researcher, imposed because the revealing the meaning of history proves to be the same failure as the attempt to recall the past. Thus, if the revealing of the meaning of history is not possible, thereby history deprived of the possibility of making sense of past events would be useless. And culture would be useless, if it did not provide this usefulness. This means that the scientific perspective of the past, (pre)history, like Ajdukiewicz's (1985a, 217) scientific perspective of the world, is not determined unambiguously by experiential data, but depends on the choice of these data, the conceptual apparatus, and related linguistic directives. As the conceptual apparatus changes the problems solved using the same data, the same facts, change as well. For the assertions directly based on experience (on facts) are not absolutely dictated by experience alone (by facts); they are also derived from conventions, from the conceptual schemes currently in use (Ajdukiewicz 1985a).

This orientation of questions is related to an apriori determined view of the past; it determines the direction of answers to them, i.e. the direction of interpretation. If our knowledge based on experience is a construct, by the same token, we should assume that our experience is also a construct. It does not transcend the culture in which we function. We know what we discover only because we have experienced it in the reality that surrounds us, not the one being the subject of our study: a past we have never been to. If the objects we discover from the past were not the same as those we have come to know through our senses in the present, they could not be the objects we recognize. This is the sensory experience upon which we base our knowledge, which allows us to perceive objects from the past.

At this point, let me quote Leszek Kołakowski's statement concerning the "'revealing' of meaning" manifested in this teleological, i.e. goal-oriented, historical order imposed by the researcher. A teleologically directed historical order is meant to be an order pertaining, first of all, to the meaning of past events; and secondly to the researcher's imposition of the historical process, provoking a polemic against it from an objectivist perspective. I put the word "revealing" in quotation marks, since it is associated with the researcher's imposition of the historical order. Although archaeological sources (facts) are evidence that potentially falsifies the archaeologist's interpretive hypotheses, as Zalewska (2005, 78ff) and Rączkowski emphasize, they are not imposed by archaeological data but by the researcher's optics, including social and culture theory.

An especially interesting interpretation proposal is offered by Andrzej P. Kowalski (2018, 61–62), who claims that:

> The study of the meaning of empirically graspable cultural realities involves the reconciliation of the emerging knowledge with the existing achievements in this area. Interpretation of culture then becomes a dialog, the establishment of a trans-generational interpretive community, and not just a recapitulation of subjective empirical experiences. Then the recognition of cultural realities will more easily meet the criteria of what is probable or necessary and the requirement of intersubjective controllability of submitted findings. Empirical research retains its validity, but in a perspective of planned interpretation.

In doing so, I will emphasize that both the theory and thus the reality learned through it are subject to constant change. Here is the other of these two perspectives, which assumes that the epistemological perspective is not a sufficient one when reflecting on science. This standpoint, represented by P.K. Feyerabend, T. Kuhn, or S. Amsterdamski, among others, assumes that science and, therefore, archaeology is a sphere of culture. Thus, not only is it the result of evolultion in a non-cultural epistemological perspective, but above all, it is determined by historical and cultural circumstances (Mamzer 2000). The distinction between perspectives: the objectivist perspective, identical to the epistemological orientation; and the culturally oriented interpretive perspective, is not possible without methodological awareness. Therefore, methodology is there to become aware of Josef Mitterer's observation that a description of an object does not refer to the object, but rather starts from the object. The object of a description relates to the description of the object, like a description so far to a description from now on. A new description of the object changes it into a new object of further description(s). The problem, however, is that reaching the object in itself, and thus without description, is not possible. We always access it through description, i.e. we reach objects already deformed by our perception of them from the perspective of the modern world in which we function. Hence Rączkowski's (2009, 16) reiteration of Mitterer's non-dualizing philosophy of change (Mitterer 1996). Only then does the need to reflect on what, how, and why we study become apparent.

Another, more specific answer to the question "What do we need methodology for?" is to realize that common-sense empiricism, which tells us to believe that the data speak for themselves, fails to meet our expectations. It does so, because it is not the facts that speak, but us that do. If so, then, for one thing, our statements about facts are always value-laden, they are never axiologically neutral. Their neutrality can only be viewed, as Hilary Putnam puts it, from the God's Eye point of view. Our statements deform the direct meaning of reality as it was in itself, i.e. objective reality. They cause this reality to be our reality rather than the one that is subject to our study: the objective past existing independently of us. To recall Arthur Schopenhauer, the world (including that of the past) is my representation. This means that "Everything that there is for cognition (the entire world) exists simply as an object in relation to a subject - a 'representation' to a subject. Certainly, this applies to the present as well as to any future..." (Schopenhauer 2011, 31). Secondly, "...one must realize that this reality is gone and there is no real contact with it" as Jerzy Topolski states (1998, 30). One cannot see it, touch it, as Frank Ankersmit would have.

However, we can think, speak, or write about it. The past is, therefore, primarily linguistic in nature. We are thus dealing with the linguistic rather than the actual nature of the past, just as in the case of Richard Rorty's concept we are dealing with the linguistic nature of the world. Facts, as Rorty's "world" or Kmita's "reality beyond us", do not speak. According to Rorty (2009, 22) "The world does not speak. Only we do. The world can, once we have programmed ourselves with a language, cause us to hold beliefs." Following Jan Woleński's assertions leading to a conclusion about the value-oriented nature of images of the past, we make methodological settlements of a postulative nature. Methodological rules are not immutable. They are imposed by historically conditioned ideals of scientific knowledge. In this regard W. v. O. Quine refers to all objects as postulated entities (not just physical objects expressed in common-sense terms). Referring to any reality independent of our speaking or thinking is unwarranted. By this I mean the belief that all that is material (as in the case of archaeological sources) is real, even natural, and therefore objective.

At this point, let me venture a dissenting opinion from that of Danuta Minta-Tworzowska and Michal Pawleta, according to whom the materiality of the past is the strength and asset of archaeology. This kind of statement echoes a peculiar longing for the existence of some extra-cognitive reality of the world, to which, however, the archaeologist would have access thanks to the materiality of the sources at his disposal. It includes a conviction that we can term metaphysical realism as it is perceived by Hartry Field, quoted by Putnam. Meanwhile, it is not possible to materially represent the past without the use of language (Mamzer 2010, 2011). Not only does a nod to realism appear here, but above all to empiricism associated with scientism. This materiality is rather a limitation of archaeology, if only because it marginalizes or even removes from the field of view its cultural sphere, the symbolic sphere, as Minta-Tworzowska mentions in other works. It is significant that a large number of works of both authors of this assertion are works with a decidedly constructivist slant. Because of the restriction of archaeology to the material sphere of culture it is proposed to expand archaeology with research problems hitherto unexplored in it, and thus to expand the term "archaeology" beyond the materiality that has defined it so far. Identifying archaeology with material culture is not enough. On the other hand, the reality of the past is a kind of mythicization of the object of study of archaeology providing the grounds for claiming archaeology to be a natural science. Supporting this metaphysical view of the possibility of reaching the past through materiality is, according to the cited researchers, the possibility of observing changes in the long term. There is a logical inconsistency here. Long duration means immutability, duration devoid of change. So if change means the passage of time, then the absence of change means that time stands still. Thus, the observation of passing time is made possible by the process of change in culture; it becomes synonymous with it. We are dealing here with the use of two types of time by the aforementioned authors: cultural, since the passage of time is a change in culture; and universal Newtonian time.

It follows from the above argument that the methodology of archaeology or its conceptualization is related to the use of metalanguage, as opposed to the object language that an archaeologist usually applies to describe facts. Thus, the prevalence of the belief that there is no need for theoretical thinking in Polish archaeology, tacitly expressed by its absence or reluctance to this type of reflection, is an expression of the lack of awareness that research practice is controlled by its theoretical, value-oriented worldview. "What do you mean?", asks the proponent of empirical archaeology, "What, then, is this past that is the subject of our archaeological study, if it is supposed to be derive from our cultural understandings? (Zybertowicz 1995, 275ff; Kmita 1977; Minta Tworzowska 2000, 530).

Conclusion To answer this question we can use Jerzy Topolski's (1998, 30) words,

> that reality [the past - HM] is gone and there is no real contact with it. (…) the historian [and the archaeologist - HM] does not deal with past reality, but with thinking, speaking, or writing about it, that is, the historian deals with language.

Thus, the archaeologist deals not with the material, but with the linguistic nature of the past, or, given that language is constantly changing, with the linguistic formulation of the past, as Jerzy Kmita would say. Taking into account this cultural studies perspective, we should consider the principle of historism and the subjective 'reconstruction' of past culture. Historism is twofold. While remembering Buchowski's and Zybertowicz's opinions, we should realize

> that the object of study we identify and properly characterize is not something given once and for all, but it changes depending on the cultural context and the historical arrangement of certain features of culture seen from today's perspective. (…) On the other hand, it is an awareness of the evolving ways of conceptualizing the object of study. All in all, then, historism, is a certain dialectical principle of the variability of subject-oriented cultural phenomena and the way of imagining these phenomena within the framework of modern reflection on them (Buchowski 1995, 96).

6 "Vérité à faire": The Manufacturing of the Object of the Study of Archaeology

"Vérité à faire" (Truth to be made) is Danuta Minta-Tworzowska's (1986, 162, 1991, 226) main argument in exposing Pierre Teilhard de Chardin's thesis that "the emergence of an image of the world from the system of sciences is at the same time the design of its evolution" linking the subjective with the objective. According to Minta-Tworzowska (1991, 226):

> The main test of the legitimacy of discovery is for Pierre Teilhard de Chardin the effective construction of the future within the framework of exploratory thinking (…). The image of universal evolution is a projection of cultural change, in which the self-fulfilling factor of "coherence" between the vision of evolution and the real emerging world arises. The contribution of this factor implies at the same time making progress, both in "ordering the vision of the world" and in "indicating greater possibilities in the search for future exploration and construction". Thus, the construction of an image of the structure of the world, around the central element, which is man 'consolidating' in himself the entire system of this world (a spiritually perfecting world) is underway. Evolution thus constitutes a process more in line with its image-design.

This image-design constitutes a prism through which we perceive the past. Let me invoke here Danuta Minta-Tworzowska's view of two worlds: the past world, when it is exclusively about the past, and the "here and now" world, when we deal with the researcher's production of the concept of history. When interpreting the past, we interpret it from the point of view of the present in which we function. The results of such interpretation applied to the study of the past, we again submit to the present, that is, to the cultural community in which we function, thus co-creating our own current culture and the discipline we represent. This is what methodology is for, in order to recognize the fact that access to the object studied and described by the researcher exists only through language. "For what are artifacts without the explanations of the people who use them", says Michał Buchowski (2017, 227). Thus, the nature of the past remains linguistic rather than material. 'Truth to be made' is a consequence of the conclusion of the previous chapter, about the non-existence of an object of study that is not devoid of our participation in its creation. In view of this, I will return for a moment to these two emerging perspectives in establishing and identifying the object of study of archaeology. Two research horizons that determine the final results of the study of archaeology emerge from the above inference: (1) the modern horizon, in which archaeology functions and from which it draws conceptual schemes for interpreting the past; (2) the past horizon, which is the object of archaeological research, viewed, however, from the perspective of the researcher's modern cultural environment. It is a a type of Gadamer's fusion of horizons: past and present, in which the present, however, occupies a dominant position, or a position with the past subordinated to it. This is because interpretation in both cases is made from the perspective of the researcher's cultural environment. In other words, interpretation of the past does not yield objective research results, since we cannot in any way enter the past world, but it is entirely the result of the researcher's gaze by the researcher's present situation. So if we agree with Rorty's (2009, 6) assertion that "The world does not speak. Only we do", then all the more so the material remains of the past world do not speak either. Only we do.

This kind of view is expressed in Minta-Tworzowska's article on archaeological sources in the context of Jerzy Topolski's concept. Minta-Tworzowska's mentions the dualism of the archaeological source, containing, on the one hand, a particle of past reality, and on the other hand, the outcome of the researcher's narration from the cultural standpoint of his or her contemporary world (Minta-Tworzowska 1998, 334). The distinction here has been well-known in the history of historiography, similar to the Hegelian division of history into: *res gestae* and *rerum gestarum*, which are strongly interwoven. This problem is accentuated by M. Kwapiński (2020, 123–124) in his observation of the object co-formed by the subject, which transforms itself under the feedback of its own product.

What then is the past? My answer is that the past is, to a significant extent, a construct from our modern point of view, in other words, a "truth to be made". It is the result of interaction of two worlds: the past and the present, from whose perspective we perceive the past world. It is the "truth to be made" rather than the already existing truth, because the cultural context in which it is made is constantly changing. By the same token, the past, i.e. the subject of archaeological studies, is constantly changing. Therefore, following Topolski's thought, the past is gone just as the future is not yet here. We are only left with the perspective-oriented present in which we are situated. Consequently, any statement about the past is always selective and never a reproduction of it. Therefore, it is not possible to resurrect the past as it once was itself. The meaning we give to the past remains our meaning. It is not possible to reproduce it, because we are unable to give the meaning inherent in the past era we are studying. It is difficult to see past meanings in our ordering procedures or classifications of material remains discovered en masse, such as pottery shards. They are like forms to be filled in through interpretation and imagination.

The ideas of taking care of oneself and working on oneself, taken from ancient Greeks, commands attention here. Paul Veyne (1988, 327) claims these ideas can be successful and illustrates his assertion with the metaphor of Greek columns incorporated into modern buildings, for example, the walls of Christian churches. According to Veyne, this is a "truth to be made", whose modern maker does not rely on tradition anymore. "As failures go, an attempt to recall the past is like trying to grasp the meaning of existence" (Brodsky 1986, 3), unless—as Theodor Lessing (2013) would say—it is a meaning given by ourselves. Self-creation, or in other words co-creation of the environment in which we function, is what P. Veyne (1988) calls a new strategic possibility.

An excellent response to or even justification of Brodsky's or Lessing's assertions is included in Michał Pawleta's (2016) book with the telling title: *Przeszłość we współczesności. Studium metodologiczne archeologicznie kreowanej przeszłości w przestrzeni społecznej* (*The Past in the Present. A Methodological Study of the Archaeologically Created Past in Social Space*). Also interesting in this regard are Arkadiusz Marciniak's works co-written with Michał Pawleta, and devoted to cultural heritage in the modern world (Marciniak et al. 2018). By no means is it about the currently practiced historical reconstructions or archaeology of times we still remember ourselves, as Anna Zalewska (e.g. 2013) seems to suggest. The point is to highlight the archeologists' mindset, guided by Claude Lévi-Strauss's message (1966, 255) that "thought is powerless to extract a scheme of interpretation from events long past". In view of this, we draw on conceptual schemes that arise 'here and now' in the reality in which we ourselves exist. Hence, the past 'reconstructed' on this basis can be described as an 'archaeologically constructed past', which is the subtitle of Pawleta's work. This is because it is not interpreted using terms and concepts from the emergent past we have studied, which we have never been in. We only use the present, in which questions are raised against the past, directing the answers obtained from the analysis of materials we discover. Each such answer is never axiologically neutral or objective, as we would say today. Contemporary thought is used to interpret the past so that we can refer to it as "the past of the present".

The theorization of archaeology is thus tantamount to the production of its object of study, marginalized or absent usually in atheoretical archaeology. This is how I see the methodology of archaeology in the AMU archaeological community. I see the role of methodology in learning about the past in the realization of the role of theory for the study of the past, production of the object of cognition, and co-creation of a cultural environment in which archaeology operates. With his proposal indicating the central role of man, Paul Veyne recalls Michel Foucault's (2000, 293) assertion of "historical ontology of ourselves". It has a practical dimension: not only does it examine historical conditions, but it also shows how to overcome and transcend them. It is related with the new strategic possibility, which is the idea of co-creating the world around us and ourselves (Veyne 1988, 327).

7 Conclusions

The attempted interpretation of the unified thought of Poznań methodologists of archeology is not only the effect of tuning into their research, which is diverse despite their common idea. It is also the effect of identifying with it. This approach is, of course, value-oriented, not only because axiological neutrality is not possible, but above all, because I share the value-laden orientation of this synthesized representation. In other words, it is an interpretation from my own perspective, which is not necessarily shared by the all AMU methodologists of archeology, let alone, each of them individually. This is because each of them follows their own individualized world of values, their own preferences and—as a result—their own choices in their own research practice. In any case, such a unified picture of the Poznań methodologists' thought, as perceived by an outsider, presents itself as follows: Sect. 2 addresses the following question to the Poznań methodologits: 1. **What do we need methodology for?**, in other words: What and how should we study? This question is prompted by the usually reluctant attitude of the archaeological community towards theoretical reflection. This reluctance has its source in the belief that the realness of research results is determined by the facts discovered by the archaeologist. Realism and related empiricism, scientism and naturalism provide the intellectual basis for the belief in the reality of the discovered facts. Section 3: **Archaeology and its sources and the reality of the past** indicates that materiality of sources largely directs archaeology towards natural science. Accordingly, the title of Sect. 4 consists of two phrases separated by the preposition *versus*: **Materiality of the past *versus* the natural-scientific nature of archaeology.** The research of methodologists of archaeology is not so much about research methods, but about the object of study of archaeology. Research methods are a consequence of the specificity of the object of study. If the specificity of the object of study is its materiality, it necessarily draws attention to the methods of natural science.

However, if the first part of the question about the object of study, i.e. "What should we study?", is formulated on the basis of the archaeologist's interpretation of the reality in which he or she functions "here and now", then it has a cultural rather than natural scientific character, especially since the second part of the question, i.e. "How should we study?" is addressed to the community of archaeologists themselves. The present provides archaeology with conceptual schemes that form the basis for interpreting the past. These schemes determine archaeological research results, and the results obtained this way are, in turn, communicated by archaeologists to the contemporary cultural environment, of which it is an emanation, confronting them with social expectations. The value-laden nature of questions raised about the facts uncovered by the archaeologist means that—contrary to expectations—they do not speak for themselves "what it was really like". Doubts arise even in scientifically-oriented archaeological studies. For the meaningfulness of the facts turns out to be the result of cultural arrangements of the questioner and of the source he or she questions. Hence, another problem preoccupying the AMU methodologists of archaeology is Sect. 5. **The concept of the source as a cultural construct**, that is, the linguistic, rather than the material nature of the archaeologically studied past.

Finally, given the ensuing conclusion that the object of our archaeological study does not exist in a readymade undisputed form (as it seems to follow from the exposition of its materiality), another research problem, closely related to the preceding one, arises, which is of particular interest to the methodologists: Sect. 6. **"Vérité à faire": the manufacturing of the object of the study of archaeology.** "Truth to be made", invoking Paul Veyne's views, refers to the manufacturing of the object of study of archaeology. Archaeology not only co-creates the object of its study, but also the surrounding reality in which it functions and itself. In this way, the results of archaeological research provide the material for the formation of modern culture, of which it is also a part. I interpret the research activities of the AMU methodologists of archeology as a culturally oriented program, inspired, among many other things, by the earlier achievements of the Poznań school of methodology.

References

Ajdukiewicz, K. 1985a. Obraz świata i aparatura pojęciowa. In *Język i poznanie, t. I. Wybór pism z lat 1920–1939*, ed. K. Ajdukiewicz, 175–195. Warszawa: PWN.

Ajdukiewicz, K. 1985b. Systemy aksjomatyczne z metodologicznego punktu widzenia. In *Język i poznanie, t. II, Wybór pism z lat 1945–1963*, ed. K. Ajdukiewicz, 332–343. Warszawa: PWN.

Amsterdamski, S. 1983. *Między historią a metodą. Spory o racjonalność nauki*. Warszawa: Państwowy Instytut Wydawniczy.

Anderson, B. 1997. *Wspólnoty wyobrażone. Rozważania o źródłach i rozprzestrzenianiu się nacjonalizmu*. Kraków: Wydawnictwo Znak.

Bintliff, J. 1993. Why Indiana Jones is smarter than the post-procesualist. *Norwegian Archaeological Review* 26 (2):91–100.

Bloch, M. 1949. *Apologie pour l'histoire ou Métier d'historien*. Paris: Librairie Armand Colin.

Brodsky, J. 1986. *Less than one. Selected essays*. New York: Farrar, Straus, Giroux.

Bryant, L. R. 2011. *The democracy of objects*. Ann Arbor: Open Humanities Press.

Buchowski, M. 1995. Kulturoznawstwo a antropologia kulturowa. In: *Perspektywy refleksji kulturoznawczej*, ed. J. Sójka, 91-107. Poznań: Humaniora.

Buchowski, M. 2017. Antropologia kultury a antropologia kulturowa. Kilka refleksji na marginesie książki Grzegorza Godlewskiego. *Filo-Sofija* 17 (36):213–228.

Bugaj, E. 2018. O nowych podejściach w badaniach sztuki greckiej na przykładzie wytwórczości rzeźbiarskiej. In *Archeologia jako humanistyczna interpretacja przeszłości: studia dedykowane Profesorowi Henrykowi Mamzerowi*, ed. A. Posern-Zieliński, J. Sawicka, J. Kabaciński, M. Kara, and K. Zamelska-Monczak, 137–159. Poznań: Wydawnictwo IAE PAN.

Ciesielska, A. 2002. *Elementy teorii społecznej w archeologii. Koncepcje grup, instytucji i struktur społecznych*. Poznań/Kalisz: Uniwersytet im. Adama Mickiewicza, Wydział Pedagogiczno-Artystyczny.

Condorcet, A. N. 1795. *Esquisse d'un tableau historique des progres de l'esprit humain*. Paris: Chez AGASSE.

Deleuze, G., and F. Guattari. 1991. *Qu'est-ce que la philosophie?* Paris: Les éditions de Minuit.

Domańska, E. 2010. Jakiej metodologii potrzebuje współczesna humanistyka? *Teksty Drugie* 1–2:45–55.

Feyerabend, P. K. 1999. *Paul K. Feyerabend: Knowledge, science and relativism*. Cambridge: Cambridge University Press.

Foucault, M. 2000. Czym jest Oświecenie? In *Filozofia, historia, polityka. Wybór pism*, ed. D. Leszczyński and L. Rasiński, 276–293. Warszawa: PWN.

Gabriel, J. 2018. *Teoretyczne aspekty badań nad genezą różnic społecznych w archeologii brytyjskiej XX wieku*. Unpublished PhD thesis, University of Gdańsk.

Gadamer, H.-G. 2004. *Truth and method*. London/New York: Continuum.

Gardin, J.-C. 1998. Cognitive issues and problem of publication in archaeology. In *Theory and practice of archaeological research*, ed. W. Hensel, S. Tabaczyński, and P. Urbańczyk, vol. III, 65–113. Warszawa: Wydawnictwo IAE PAN.

Gellner, E. 1988. *Plough, sword and book. The structure of human history*. London: Collins Harving.

Hacking, I. 1983. *Representing and intervening. Introductory topics in the philosophy of natural science*. Cambridge: Cambridge University Press.

Heidegger, M. 1962. *Being and Time*. Oxford: Blackwell.

Hensel, W. 1971. *Archeologia i prahistoria. Studia i szkice*. Wrocław: Ossolineum.

Ingold, T. 2003. Kultura i postrzeganie środowiska. In *Badanie kultury. Elementy teorii antropologicznej*, ed. M. Kempny and E. Nowicka, 73–88. Warszawa: Wydawnictwo Naukowe PWN.

Johnson, M. 1999. *Archaeological theory. An introduction*. Oxford: Willey-Blackwell.

Johnson, M. 2010. *Archaeological theory. An introduction*. 2nd ed. Oxford: Willey-Blackwell.

Johnson, M. 2013. *Teoria archeologii. Wprowadzenie*. Kraków: Wydawnictwo Uniwersytetu Jagiellońskiego.

Kmita, J. 1971. *Z metodologicznych problemów interpretacji humanistycznej*. Warszawa: Wydawnictwo Naukowe PWN.

Kmita, J. 1977. *Wykłady z logiki i metodologii nauk dla studentów wydziałów humanistycznych*. Warszawa: Wydawnictwo Naukowe PWN.

Kmita, J. 2000. *Wymykanie się uniwersaliom*. Warszawa: Oficyna Naukowa.

Kmita, J. 2015. *Czarnoksięstwa humanistów*. Poznań: Fundacja Instytut im. Jerzego Kmity.

Kmita, J., and A. Pałubicka. 1992. Problem użyteczności pojęcia doświadczenia. In *Poszukiwanie pewności i jego postmodernistyczna dyskwalifikacja*, ed. J. Such, 150–181. Poznań: Wydawnictwo Naukowe Instytutu Filozofii UAM.

Kołakowski, L. 1984. Fabula mundi i nos Kleopatry. In *Czy diabeł może być zbawiony i 27 innych kazań*, ed. L. Kołakowski, 68–82. Londyn: Aneks.

Kołakowski, L. 1989. *The presence of myth*. Chicago/London: The University of Chicago Press.

Kotowa, B. 2001. Archetypy a kultura. In *Skrytość kultury*, ed. K. Zamiara, 35–45. Poznań: Wydawnictwo Fundacji Humaniora.

Kowalski, A. P. 2014. *Antropologia zamierzchłych znaczeń*. Toruń: Polskie Towarzystwo Historyczne.

Kowalski, A. P. 2015. Antropologia kulturowa jako nauka historyczna. In *W krainie metarefleksji. Księga poświęcona profesorowi Czesławowi Robotyckiemu*, ed. J. Barański, M. Golonka-Czajkowska, and A. Niedźwiedź, 143–152. Kraków: Wydawnictwo Uniwersytetu Jagiellońskiego.

Kowalski, A. P. 2018. Znaczniki realiów kulturowych w interpretacji etnologicznej i archeologicznej. In *Archeologia jako humanistyczna interpretacja przeszłości: studia dedykowane Profesorow Henrykowi Mamzerowi*, ed. A. Posern-Zieliński, J. Sawicka, J. Kabaciński, M. Kara, and K. Zamelska-Monczak, 49–63. Poznań: Wydawnictwo IAE PAN.

Kwapiński, M. 2020. *Świadomość mityczna wobec śmierci*. Gdańsk: Muzeum Archeologiczne, Słowo/obraz terytoria.

Lessing, Th. 2013. *Geschichte als Sinngebung des Sinnlosen*. Paderborn: Salzwasser-Verlag Gmbh.

Lévi-Strauss, C. 1966. *The savage mind*. London: Weidenfeld and Nicolson.

Mamzer, H. 2000. Kulturowe konteksty "paradygmatów" w archeologii. In *Archeologia i prahistoria polska w ostatnim półwieczu*, ed. M. Kobusiewicz and S. Kurnatowski, 537–550. Poznań: Wydawnictwo PTPN.

Mamzer, H. 2010. Źródła archeologiczne: Artefakty czy językowe o nich wypowiedzi? *Humanistyka i Przyrodoznawstwo* 16:227–250.

Mamzer, H. 2011. Czy możliwe jest materialne (fizykalne) przedstawianie przeszłości? In *Oblicza przeszłości*, ed. W. Wrzosek, 251–263. Bydgoszcz: Oficyna Wydawnicza Epigram.

Mamzer, H. 2021. *Jak człowiek konstruował świat i siebie samego. Archeologia w procesie humanizacji świata*. Poznań: Instytut Archeologii i Etnologii PAN.

Marciniak, A. 1996. *Archeologia i jej źródła. Materiały faunistyczne w praktyce badawczej archeologii*. Warszawa/Poznań: Wydawnictwo Naukowe PWN.

Marciniak, A. 2005. *Placing animals in the Neolithic: Social zooarchaeology of prehistoric farming communities*. London: UCL Press.

Marciniak, A. 2012. Teoria w archeologii. In *Przeszłość społeczna. Próba konceptualizacji*, ed. S. Tabaczyński, A. Marciniak, D. Cyngot, and A. Zalewska, 84–116. Poznań: Wydawnictwo Poznańskie.

Marciniak, A. 2013. Byki, uczty i kozioł ofiarny. René Girard i jego uczniowie w Çatalhöyük. *Rocznik Antropologii Historii* III (2):169–184.

Marciniak, A., and S. Tabaczyński. 2017. Geneza i rozwój nierówności społecznych w paradygmatach badawczych archeologii. Zarys problematyki. *Przegląd Archeologiczny* 65:45–51.

Marciniak, A., M. Pawleta, and K. Kajda, eds. 2018. *Dziedzictwo we współczesnym świecie. Kultura, natura, człowiek*. Kraków: Universitas.

Meillassoux, Q. 2008. *After finitude. An essay on the necessity of contingency*. New York: Continuum.

Minta-Tworzowska, D. 1980. O metodzie systematyki wytworów kulturowych (na podstawie ceramiki naczyniowej z Kruszy Zamkowej, stan. 3). *Fontes Archaeologici Posnanienses* 31:146–183.

Minta-Tworzowska, D. 1986. *Elementy metodologii prahistorii w historiozofii Teilharda de Chardin*. Poznań: Wydawnictwo Naukowe UAM.

Minta-Tworzowska, D. 1991. Ujęcie nauki w historiozofii P. Teilharda de Chardin. *Studia Metodologiczne* 26:221–236.

Minta-Tworzowska, D. 1994. *Klasyfikacja w archeologii jako sposób wyrażania wyników badań, hipotez oraz teorii archeologicznych*. Poznań: Wydawnictwo Naukowe UAM.

Minta-Tworzowska, D. 1998. Jerzego Topolskiego koncepcja źródeł historycznych a ujęcia źródeł archeologicznych. In *Świat historii. Prace z metodologii historii i historii historiografii dedykowane Jerzemu Topolskiemu z okazji siedemdziesięciolecia urodzin*, ed. W. Wrzosek, 329–340. Poznań: Wydawnictwo Instytutu Historii UAM.

Minta-Tworzowska, D. 2000. Kwestia przełomu metodologicznego w prahistorii i w archeologii polskiej. In *Archeologia i prahistoria Polska w ostatnim półwieczu*, ed. M. Kobusiewicz and S. Kurnatowski, 527–535. Poznań: Wydawnictwo PTPN.

Minta-Tworzowska, D. 2002. Between a community of inspiration and the separatennes od archaeological traditions. In *Archaeologies of Europe. History, methods and theories*, ed. P. F. Biehl, A. Gramsch, and A. Marciniak, 53–64. Münster: Waxman.

Minta-Tworzowska, D. 2011. Człowiek i rzecz w perspektywie archeologicznej (czyli rzecz w perspektywie antropocentrycznej). In *Współczesne oblicza przeszłości*, ed. A. Marciniak, D. Minta-Tworzowska, and M. Pawleta, 39–61. Poznań: Wydawnictwo Poznańskie.

Minta-Tworzowska, D. 2017. Tradycja i inspiracje teoretyczne we współczesnej archeologii. *Filo-Sofija* 17 (36):267–285.

Minta-Tworzowska, D. 2022. Teoretyczne i metodologiczne podstawy archeologicznego dyskursu w Polsce. *Przegląd Archeologiczny* 70:5–40.

Minta-Tworzowska, D., and M. Pawleta. 2013. Teoria i metodologia archeologii na Uniwersytecie w Poznaniu. *Folia Praehistorica Posnaniensia* 18:217–233.

Minta-Tworzowska, D., and W. Rączkowski. 1996. Theoretical traditions in contemporary Polish archaeology. *World Archaeological Bulletin* 8:196–209.

Minta-Tworzowska, D., and W. Rączkowski. 2007. Standardy teoretyczne we współczesnej polskiej archeologii. In *Pół wieku z dziejów archeologii polskiej (1939–1989)*, ed. J. Lech, 219–249. Warszawa: Wydawnictwo Naukowe PWN.

Mitterer, J. 1996. *Tamta strona filozofii. Przeciwko dualistycznej zasadzie poznania*. Warszawa: Oficyna Naukowa.

Nietzsche, F. 1967. *The will to power*. New York: Random House, Inc.

Nietzsche, F. 2019. *On truth and lie in an extra-moral sense*. Oxford: Quadriga.

Ostoja-Zagórski, J. 2015. *Archeologia a historia. Refleksje nad myśleniem historycznym archeologów*. Bydgoszcz: Wydawnictwo Uniwersytetu Kazimierza Wielkiego.

Pałubicka, A. 2018a. Hermeneutyczne i epistemologiczne koncepcje interpretacji. In *Archeologia jako humanistyczna interpretacja przeszłości: studia dedykowane Profesorow Henrykowi Mamzerowi*, ed. A. Posern-Zieliński, J. Sawicka, J. Kabaciński, M. Kara, and K. Zamelska-Monczak, 27–48. Poznań: Wydawnictwo IAE PAN.

Pałubicka, A. 2018b. Ponowoczesność i humanistyka. In *Estetyzacja w archeologii. Obrazowanie w pradziejach i starożytności*, ed. D. Minta-Tworzowska, 71–80. Gdańsk: Muzeum Archeologiczne.

Parczewski, M. 1999. Kilka uwag o krakowskiej szkole archeologii historycznej. *Sprawozdania z posiedzeń komisji naukowych PAU* 43 (2):1–2.

Parczewski, M. 2014. Daleko od przełomu. *Slavia Antiqua* 55:237–253.

Pawleta, M. 2016. *Przeszłość we współczesności. Studium metodologiczne archeologicznie kreowanej przeszłości w przestrzeni społecznej*. Poznań: Wydawnictwo Naukowe UAM.

Platon 1982. Obrona Sokratesa. In *Platon, Uczta, Eutyfron, Obrona Sokratesa, Kriton, Fedon*, 225–303. Warszawa: PWN.

Poleski, J. 2000. Głos w dyskusji. In Archeologia i prahistoria polska w ostatnim półwieczu, ed. M. Kobusiewicz and S. Kurnatowski, 256–257. Poznań: Wydawnictwo PTPN.

Posern-Zieliński, A. 2018. Głos w dyskusji. In *Archeologia jako humanistyczna interpretacja przeszłości: studia dedykowane Profesorow Henrykowi Mamzerowi*, ed. A. Posern-Zieliński, J. Sawicka, J. Kabaciński, M. Kara, and K. Zamelska-Monczak, 324. Poznań: Wydawnictwo IAiE PAN.

Quine, W. V. 1976. *The ways of paradox and other essays*. Cambridge: Harvard University Press.

Rączkowski, W. 1996. Drang nach Westen? Polish archaeology and national identity. In *Nationalism and archaeology in Europe*, ed. M. Díaz-Andreu and T. Champion, 189–217. London: Routledge.

Rączkowski, W. 2002. *Archeologia lotnicza – metoda wobec teorii*. Poznań: Wydawnictwo Naukowe UAM.

Rączkowski, W. 2009. Theory, empiricism and practice: Archaeological discourses in a network of dependency and opposition. *Analecta Archaeologica Resoviensia* 4:7–34.

Renfrew, C., and P. Bahn. 2000. *Archaeology: Theories, methods and practice*. London: Thames & Hudson.

Rorty, R. 2007. *Philosophy as cultural politics*. Cambridge: Cambridge University Press.

Rorty, R. 2009. *Contingency, irony, and solidarity*. Cambridge: Cambridge University Press.

Rydlewski, M. 2016. *Żeby widzieć, trzeba wiedzieć. Kulturowy wymiar percepcji wzrokowej*. Bydgoszcz: Oficyna Wydawnicza "Epigram".

Schopenhauer, A. 2011. *Świat jako wola i przedstawienie*. Warszawa: Wydawnictwo Naukowe PWN.

Stanosz, B. 1991. Wstęp. In *Empiryzm współczesny, ed. B. Stanosz, 5–12. Warszawa: Wydawnictwa Uniwersytetu Warszawskiego.*

Such, J. 2017. Mistrzowie racjonalnego kompromisu. *Filo-Sofia* 17 (36):39–44.

Suchodolski, B. 1957. Wstęp. In *Condorcet. Szkic obrazu postępu ducha ludzkiego* (i–xvi). Warszawa: Państwowe Wydawnictwo Naukowe.

Tabaczyński, S. 2000. Głos w dyskusji. In *Archeologia i prahistoria polska w ostatnim półwieczu*, ed. M. Kobusiewicz and S. Kurnatowski, 259. Poznań: Wydawnictwo PTPN.

Topolski, J. 1978. *Rozumienie historii*. Warszawa: PIW.

Topolski, J. 1998. *Wprowadzenie do historii*. Poznań: Wydawnictwo Poznańskie.

Traba, R. 2009. *Przeszłość w teraźniejszości. Polskie spory o historię na początku XXI wieku*. Poznań: Wydawnictwo Poznańskie.

Veyne, P. 1988. Ostatni Foucault i jego moralność. *Literatura na Świecie* 6 (203):321–329.

Woleński, J. 1996. *W stronę logiki*. Kraków: Aureus.

Żak, J. 1966. Historia pierwotna i jej współczesne metody. *Studia Metodologiczne* 2:67–89.

Zalewska, A. 2005. *Teoria źródła archeologicznego i historycznego we współczesnej refleksji metodologicznej*. Lublin: Wydawnictwo UMCS.

Zalewska, A. 2013. Roadside lessons of historicity. The roles and the material points of refernces to the great war and in shaping historical sensitivity and awareness. *Sensus Historiae* 13:69–85.

Zybertowicz, A. 1995. *Przemoc i poznanie. Studium z nie-klasycznej socjologii wiedzy*. Toruń: Wydawnictwo UMK.

Danuta Minta-Tworzowska and Her Role in Conceptualizing the Poznań School of Archaeology and Advancing the Methodology of Archaeology in Poland

Michał Pawleta

Abstract This chapter examines the influential contributions of Danuta Minta-Tworzowska to the development of the Poznań School of Archaeology. Based on the philosophical and methodological principles of the Poznań School of Methodology, her work established a unique approach to archaeology that has greatly influenced the theoretical and methodological landscape of Polish archaeology for the past three decades. Key themes include archaeological theory and methodology; classification as a reflection of research procedures; the conceptualisation of archaeological sources; and the interpretation of symbolism and prehistoric art. Not only did Minta-Tworzowska adopt ideas from the Poznań School of Methodology, she also transformed them in innovative ways, contributing to the formation of a uniquely Polish archaeological discourse. Her work transcends national frameworks, offering valuable insights into Central European and global archaeological debates.

Keywords Polish archaeology · DanutaMinta-Tworzowska · Poznań School of Archaeology · Archaeological theory and methodology · Classification · Symbolism

1 Introduction

The chapter aims to present the contributions of Danuta Minta-Tworzowska to the development of the Poznań School of Archaeology. Grounded in and inspired by the Poznań School of Methodology (hereinafter referred to as PSM), her approach evolved into an original model of conceptualizing archaeology and markedly impacted the theory and methodology of Polish archaeology. Consequently, it defined how these have been approached in the country for over three last decades.

M. Pawleta (✉)
Faculty of Archaeology, Adam Mickiewicz University in Poznań, Poznań, Poland
e-mail: michal.pawleta@amu.edu.pl

A. Marciniak et al. (eds.), *The Poznań School of Archaeology*, Outstanding Contributions to Archaeology, https://doi.org/10.1007/978-3-032-07848-3_5

In this chapter, the focus will be on main themes and original lines of thought developed by D. Minta-Tworzowska. These include: (1) reflections on the theory and methodology of archaeology, (2) classification as a reflection on research procedures in archaeology, (3) the conceptualization of archaeological sources, and (4) symbols, symbolism and prehistoric art within archaeological research. It is argued that D. Minta-Tworzowska not only drew inspiration from ideas developed within the framework of the PSM but also creatively applied them to the context of Polish archaeology. This resulted in the development of an original approach to the conceptualization of archaeological methodology and significantly contributed to the theoretical grounding of Polish archaeology. Although her contributions have primarily influenced the development of archaeological thought in Poland, the issues she addressed are of broader significance to the discipline. Thus, they shed unique light on the distinct and individual character of Polish archaeology and highlight its substantial contributions to Central European and worldwide archaeological debates.

2 The Poznań School of Methodology and Its Impact Upon Archaeology

PSM was an original philosophical and scientific movement that emerged in the 1960–1970s at Adam Mickiewicz University in Poznań, Poland. The school was notable for its significant contributions to methodology, the philosophy of science, and epistemology. Jerzy Kmita (e.g. 1971, 1973) primarily developed the school, contributing to its interdisciplinary approach and application to the social sciences. Other key figures include Leszek Nowak (e.g. 1980), Anna Pałubicka (1985, 1990), Jerzy Topolski (e.g. 1984), who had a profound impact on methodology of history, and Krzysztof Łastowski (1987), who influenced the discipline of biology.

Generally, the PSM was anti-positivist in character. Researchers associated with this school presented an original research program on the theoretical foundations of the humanities, which was intensively developed in the 1970s. The program took the form of cultural studies (the socio-regulatory theory of culture), the philosophy of science (the idealizational theory of science by L. Nowak), and concepts of historical changes (by J. Topolski). The school proposed humanistic interpretation and functional-genetic explanation (by J. Kmita) as methods for understanding social and historical phenomena. This approach involved examining the functional roles of social practices and their historical origins, providing a more comprehensive understanding of societal changes. The PSM also proposed a redefinition of Marxism, retaining the core principle of historical materialism while expanding the framework to include a more nuanced analysis of cultural and ideological factors, arguing that these elements also play a significant role in shaping society (see more Kotowa 2010; Zamiara 2010; Musiał 2016; Brzechczyn, chapter "Poznań School of Methodology: Institutional History—Research Program—Main Achievements"; Grad, chapter "The Poznań School of Methodology and Its Continuations").

While the PSM was primarily associated with philosophy and the methodology of science, the concepts it developed also influenced Polish archaeology (Kobyliński 1991, 230, 232). In the 1970s, there were efforts within Polish archaeology to move away from its positivist phase and modernize the discipline (Żak and Minta-Tworzowska 1991). Positivism in archaeology was linked to the diffusionist culture-historical paradigm, which did not appeal to Poznań archaeologist Jan Żak (Fig. 1). Consequently, he turned to the dynamically developing PSM led by J. Kmita and L. Nowak, and also influential in history through the contributions of J. Topolski (1984). The works of the representatives of this school deeply intrigued J. Żak. As a result of his methodologically-oriented scientific activity, he established the methodological foundations of the Poznań School of Archaeology and founded the Department of History and Methodology of Prahistory (see more Minta-Tworzowska 2001, 2021b, 2023; Minta-Tworzowska and Rączkowski 2005).

Żak was particularly drawn to the views of J. Kmita, who proposed the concept of humanistic interpretation and functional-genetic explanation. In Żak's scientific works, there is a clear, consistent research approach and a deep appreciation for well-articulated theoretical and methodological foundations. This is most evident in

Fig. 1 Professor Jan Żak in a portrait photo from 1982. (Photo by F. Lachowicz)

his model of Paleolithic communities in the Mediterranean zone (Żak 1975) and in his theoretical articles on settlement studies (Żak 1985), where empirical research was predominant in contemporary archaeology. Żak sought a theory capable of explaining these issues, addressing phenomena in both a synchronic and diachronic perspectives (see more Minta-Tworzowska 2021b, 2023).

In addition to Żak, Stanisław Tabaczyński, a prominent theorist of archaeology in post-war Poland, was also influenced by the ideas proposed by the PSM, which led to the formation of a second theoretical archaeological group centered in Warsaw (Minta-Tworzowska and Rączkowski 1996, 196; 2007, 219). Tabaczyński made significant contributions to the theoretical framework of Polish archaeology by applying PSM principles, thereby enhancing the methodological rigor of the discipline. His engagement with the PSM sparked his interest in Kmita's concept of culture, leading to a period of intense collaboration with Anna Pałubicka, as evidenced in their joint publication (Pałubicka and Tabaczyński 1986). Additionally, Tabaczyński developed an original concept of culture, focusing on processes such as the formation of material remnants from the past, long-term evolutionary processes, ethnogenetic phenomena, and the polysemantization of culture (Tabaczyński 1987). For him, a coherent methodology in archaeological research was crucial for achieving the proposed research objectives.

The ideas, scientific problems, and issues pursued within the Department of History and Methodology of Prahistory were further developed byŻak's colleagues both during his lifetime and particularly after his death. The achievements of archaeologists from Adam Mickiewicz University, noticeable quite early on, can be categorized into two distinct methodological approaches: humanistic and naturalistic orientations (Mamzer 2020, 127). Both perspectives were developed within the framework of the PSM. Following this methodological school, Żak's collaborators at Adam Mickiewicz University, including Danuta Minta-Tworzowska (see below), as well as Włodzimierz Rączkowski (2002) and Arkadiusz Marciniak (2005), furthered these approaches. Danuta Minta-Tworzowska, in particular, has been crucial in developing and refining these methodologies, making substantial contributions to the progress of Polish archaeology's theoretical and methodological framework, which will be examined in more detail below.

3 Danuta Minta-Tworzowska and Her Contribution to the Development of the Poznań School of Archaeology

As mentioned in the introductory chapter (Marciniak, Rączkowski and Pawleta, chapter "The Poznań School of Archaeology: An Outline"), the methodological foundations of Poznań archeology were laid by Jan Żak. A unique mode of thought he initiated has been continuously developed by his close collaborators, mainly Danuta Minta-Tworzowska (Fig. 2). She can be seen as Żak's successor but also made significant contributions of her own to the development of this schools of

Fig. 2 Danuta Minta-Tworzowska

thought. Danuta Minta-Tworzowska studied archaeology at Adam Mickiewicz University in Poznań from 1972 to 1977 under the supervision of Jan Żak. The courses she attended taught her a critical approach to many issues in contemporary archaeology, including the methodology of prehistory, a field that Żak was beginning to develop at Poznań University in the Department of History and Methodology of Prahistory. The ideas of the PSM, which she became acquainted with by attending the classes of its main founder, J. Kmita, also had a great influence on her conceptualization of archaeology.

The scientific career of Danuta Minta-Tworzowska is closely associated with the ideas developed within the PSM. Recognized for her expertise, particularly in archaeological theory and methodology, she has been influential in shaping and developing methodological approaches to the theory and practice of archaeology in Poland. Through numerous publications and academic engagements, she has participated in theoretical debates and discussions, significantly enriching the intellectual and methodological discourse within the discipline. Her work has helped shape the theoretical underpinnings of the Poznań School of Archaeology and its broader impact on archaeological theory and methodology in Poland. This influence has been passed on to numerous students—she has educated entire generations of archaeologists who have significantly broadened the thematic scope of their research.

Alongside the mode of research proposed by the PSM, Minta-Tworzowska has also been a strong advocate for interdisciplinary collaboration in archaeology. She argues that the essence of interdisciplinarity in archaeology lies in the design of research projects. In her view, every stage of the research should involve specialists from related disciplines. However, she believes that a truly interdisciplinary project must meet certain requirements: the research procedure must comply with a single theory and method of inference, and only then can interpretations be proposed based on these foundations. She contends that an insightful consideration of interdisciplinary projects can help researchers overcome the practical limitations of their own disciplines and enable them to undertake projects at the intersection of different fields (Minta-Tworzowska 2006).

Below, I will briefly discuss her contribution to the development of the Poznań School of Archaeology by providing an overview of the main research problems she has undertaken. They include: (Sect. 3.1) the theory and methodology of archaeology, (Sect. 3.2) classification and typology in archaeology as a reflection on research procedures in archaeology, (Sect. 3.3) the conceptualization of archaeological sources, and (Sect. 3.4) symbols and symbolism within archaeological research.

3.1 The Theory and Methodology of Archaeology

Henryk Mamzer(2020, 110–119; see also Mamzer, chapter "An Apologia of Methodology of Archaeology at Adam Mickiewicz University, Poznań") recently and aptly notes that for Poznań methodologists of archaeology, including Danuta Minta-Tworzowska, the central issue is the methodology guiding archaeological research. The theory and methodology of archaeology have consistently been a primary focus of Minta-Tworzowska's scholarly work. Her in-depth studies reveal a clear influence from the PSM, particularly the ideas of J. Kmita and J. Topolski. Following their line of thought, she argues that the theory and methodology of archaeology are closely tied to reflections on paradigms and preferred research procedures within the field. Furthermore, Minta-Tworzowska asserts the primacy of theoretical knowledge over empirical knowledge, emphasizing that theory is crucial for any archaeological research (see also Johnson 2010). She adheres to Topolski's (1983, 184; see also Minta-Tworzowska and Rączkowski 1996, 196) general classification of theoretical approaches used by archaeologists, which includes: (a) general theories that address historical processes, such as evolutionism, structuralism, or processualism; (b) specific theories related to particular systems, classes of facts, phenomena, or fragments of processes; and (c) theoretical concepts that facilitate the conceptualization of complex prehistoric facts. Consequently, archaeologists either apply an existing theory or adapt a specific theory to solve particular problems.

In her works, Minta-Tworzowska has focused on analyzing the concept of theory within the context of archaeology. Her exploration includes its relationship with archaeological practice, its connections to the humanities, its place in the broader intellectual context, and its role in society. Her investigations into theory in archaeology began with a 1986 monograph on Pierre Teilhard de Chardin, an eminent French Jesuit, thinker, historian, geologist, philosopher, and a prehistorian (Minta-Tworzowska 1986). In this work, the influence of the PSM is evident in her attempt to translate Teilhard de Chardin's historiosophy into the methodology of prehistoric research.

Minta-Tworzowska aimed to reconstruct Teilhard de Chardin's concept of prehistory and his methodology of prehistory, based on an original universal historiography. Her focus was on the research procedures it offers and on establishing archaeological foundations. She observes that Teilhard de Chardin transformed all previous static methodological categories into evolutionary ones, as his philosophy of history was centered around the concept of 'universal' evolution. Minta-Tworzowska highlights several methodological contributions of Teilhard de Chardin. These include the universalization and consistent application of the idea of universal evolution, the integration of evolution with biology and geography, the emphasis on the "interior of evolution" (in contrast to the views of G. Childe and L.A. White), a unifying and differentiating approach to pre-phenomenology, strict geneticism rather than a simple succession of stages, and the convergence and multilinearity of human evolution along a central axis.

Minta-Tworzowska views Teilhard de Chardin's proposal as a distinctive model for research procedure in prehistory and a source of inspiration for theorizing about it. His approach integrates methodological proposals from philosophy and empirical sciences, which contributes to its innovative nature. This is evident in his assertion that any research procedure must adhere to a defined methodology. Teilhard de Chardin's approach to prehistory is notably creative, as it incorporates elements of a unique philosophy of history. Minta-Tworzowska highlights two key aspects of his approach. The first is the impossibility of constructing a coherent and universal terminology within any scientific discipline. The second concerns the explanation of prehistoric phenomena. It is not an exaggeration to say that Teilhard de Chardin's historiosophy significantly influenced Minta-Tworzowska's perspective on archaeology. In her subsequent works, she explicitly applied his methodology as a framework for analyzing the approaches adopted by archaeologists, even when these were not clearly articulated.

The initial focus of Minta-Tworzowska's distinctive reflection on the theory and methodology of archaeology was a critique of positivist archaeology (Minta-Tworzowska 1987; Żak and Minta-Tworzowska 1991). In Poland, as in many Central European countries, archaeology long operated within a culture-historical paradigm, which is seen as a pre-theoretical phase. Minta-Tworzowska notes that the fundamental characteristic of the positivist research tradition are methodological naturalism and phenomenalism. Methodological naturalism advocates for the natural interpretation of culture, viewing it as a part od nature. Consequently, cultural regularities and processes are seen as subject to natural laws. Phenomenalism, on the other hand, asserts that valid knowledge is expressed only in perceptual terms. Both perspectives rely on inductionism, which seeks to derive generalizations from observed data. According to this view, the archaeologist's role is limited to the registration and ordering of objective facts from the past. Inductionism, methodological naturalism, and phenomenalism are evident in the research practices of archaeologists, who base scientific statements on inductive generalizations drawn from observation of individual facts. The observation of facts was considered the sole source of valid knowledge. As a result, archaeological research was characterized by a thorough reconstruction of the facts, a focus on establishing chronology, and a belief that only through simple induction could one uncover the truth about the past (Marciniak and Rączkowski 1991, 59).

Aligned with ideas promoted within the PSM, Minta-Tworzowska strongly advocated for an anti-positivist stance in archaeology, aiming to transcend this mode of archaeological practice. She argues that 'raw' archaeological material cannot be considered fully objective due to factors such as selective preservation, the methods used to obtain and document data, and other influences. By rejecting the positivist approach and challenging the belief that facts speak for themselves, she emphasizes the indispensable role of theory in the creation of archaeological knowledge.

An important aspect of the critique of traditional archaeology was the critical reflection on the the concept of archaeological culture. Traditionally understood within a positivistic framework, archaeological culture describes past realities based on material evidence within a defined time-space locality (Minta-Tworzowska 1987, 1996; Minta-Tworzowska and Rączkowski 1996, 207). Minta-Tworzowska acknowledges that archaeological culture is not a real, tangible entity from the past but rather an epistemological construct used by archaeologists to organize and classify source material. According to her, the concept of "archaeological culture" functions solely as a heuristic tool and should not be confused with a living culture. It is crucial to take a firm stance on the issue of ethnicity, emphasizing that entities distinguished by archaeologists in this manner as "archaeological cultures" should not be equated with ethnic groups, as an archaeological culture does not necessarily correspond to an ethnic unit.

The critique of culture-historical archaeology led Minta-Tworzowska not only to challenge its fundamental theses and assumptions but also to reject them outright without falling into cognitive aporia. She argues that positivism, and by extension, evolutionism, neoevolutionism, or diffusionism, do not address the core issues of prehistory. According to her, the solution lies in a fully structured theory that encompasses theoretical theorems extending beyond common experience and popular opinions. This rejection motivated her quest for a more suitable theory and research model for prehistory. Consequently, another crucial aspect of her approach to archaeological theory and methodology is her search for a "paradigm" amidst the diversity of archaeological theories. Central to this pursuit was the conviction, strongly influenced by the PSM, that theory plays a pivotal role in any archaeological procedure. Her critique of positivist archaeology and efforts to advance to a more theoretical stage led her to draw extensively from different orientations in global archaeology, such as processualism, structuralism, Marxism, post-structuralism, and postprocessual archaeology. Additionally, she engaged with theoretically grounded orientations such as symmetrical archaeology, new materialism, and the biography of things (Minta Tworzowska 2011, 2012, 148–151, 2015b, 29–30, 2022).

An important question posed by Minta-Tworzowska related to the methodological breakthrough in prehistory and Polish archaeology (Minta-Tworzowska 2000a). Referring to the ongoing discussion in the humanities about the so-called heuristic breakthroughs in science, she examined whether Marxism, as a theoretical orientation or intellectual tradition, represented a genuine breakthrough in Polish archaeology. She argued that it did not. Marxism in Polish archaeology was perceived as an "enforced methodological revolution" (Tabaczyński 2000). Although some Polish archaeologists attempted to 'domesticate' Marxism, they did so without acknowledging its potential positive aspects as an alternative to the dominant naïve, spontaneous positivism. In general, Marxism as a social theory did not significantly impact Polish archaeology, aside from some creative references in the works of J. Żak or S. Tabaczyński. This limited influence was due to Marxism not being originally conceived for the humanities, which made it difficult to directly adopt and apply in archaeology. The absence of an interpretive bridge to understand Marxist ideas from a humanities perspective further hindered integration. Marxism was foreign to the cognitive tradition of Polish archaeology and merely supplemented the culture-historical model with new concepts, particularly economic categories, allowing traditional archaeology to expand its scope of interests. However, it did not lead to substantive changes in the theory itself (Minta-Tworzowska 2002, 57; see also Barford 1995; Lech 1997; Tabaczyński 2000).

In this context, Minta-Tworzowska's contribution to the debate on whether archaeologists reconstruct or construct images of the past is also significant. This issue emerged from her reflections on the influence of postmodernism in Polish archaeology (Minta-Tworzowska 2000b). Simplified, she identifies two opposing approaches among archaeologists in studying the past, which stem from differing general concepts of culture. The first, objective approach, is based on the conviction that it is possible to reconstruct the prehistoric world. The second, constructivist approach, asserts that only interpretations of the world exist, implying that we construct its image rather than uncover it. Examining the trajectory of archaeology, she points out that traditional archaeology (culture-historical) represents the objectivist approach. Modernist archaeology (processual archaeology) aimed at reconstruction but acknowledged the possibility of different interpretations of the past, thus attempting to merge objectivism with constructivism. In contrast, post-processual archaeology maintains that only interpretations of the past exist and represents the constructivist approach.

Although Minta-Tworzowska supports, albeit with reservations, the constructivist approaches to the past, she argues against the thesis of the incommensurability of successive theories and paradigms in archaeology. Instead, she contends that new theories absorb previous ones, offering better explanations of social reality. Consequently, she advocates for an "interparadigmatic" dialogue in archaeology, referring to the process of communication and exchange between different scientific paradigms (Minta-Tworzowska 1998a, 217; 2000a, 529–530, 2002, 61). This dialogue involves the interaction between distinct theoretical frameworks and methodologies, guiding researchers in their investigations and interpretations of archaeological data. Its essence lies in the idea that theories, models, definitions, and methods should complement each other in the study of prehistoric societies. This mutual supplementation offers archaeologists a view "from the other side of the looking-glass" of our notions of the past, which were once shaped by a single perspective. The goal is to address which paradigm best represents a given part of reality, though it does not imply that paradigms can be fully reconciled. Interparadigmatic dialogue aims to bridge gaps between different scientific perspectives, fostering collaboration and enriching scientific discourse by incorporating diverse viewpoints and approaches. Minta-Tworzowska argues that such dialogue helps prevent stagnation within a single paradigm and encourages continuous development and innovation of the discipline.

Minta-Tworzowska's preferred research approach can be situated within the theoretical framework of processual archaeology combined with post-processual thought. She describes this approach as being "stretched between the modernist dream of scientific truth and the postmodernist, far-reaching reflection of that dream" (Minta-Tworzowska 2021a, 466). This stance reflected a blending of the scientific rigor characteristic of the PSM and the methodological focus of processualism, with an acceptance of the multivocality of archaeological discourse and openness to multiple visions of prehistory emphasized in post-processualism. Such an approach aligns with the neopragmatic approach, which argues that in archaeology, a theory is valuable if it most fully and coherently interprets fragments of past reality (Minta-Tworzowska 2000e, 94; 2017a, 275–276). Applying neopragmatic thought to contemporary archaeology suggests that archaeologists do not merely reflect past reality in their work; rather, they propose statements and theories about it. Since objective truth is elusive, archaeologists should focus on dialogue. This dialogue, occurring between different paradigms, should involve presenting one's own statements and theories about the past and reaching agreements through such exchanges to construct images of the past.

Her reflections on the theoretical standards and traditions in Polish archaeology after World War II (e.g., Minta-Tworzowska 1996, 2002, 2017a, 2022; Minta-Tworzowska and Rączkowski 1996, 2007; Minta-Tworzowska and Pawleta 2013) lead her to conclude that the condition of Polish archaeology is similar to other Central European (continental) archaeologies, which have been shaped by the dominant culture-historical tradition. While traditional archaeology has long been predominant in Poland, there is now a clear trend towards opening up to other theoretical inspirations. Following the fall of the Iron Curtain in 1989, Polish archaeology began to integrate Western theoretical and methodological approaches (Lozny 2002; Marciniak 2011, 182).

While the foundational ideas of Kmita and Topolski remain influential for methodologists of archaeology in Poland, the issues currently addressed by archaeologists, along with new inspirations opening novel fields of research, have significantly broadened. This expansion has involved borrowing or adapting elements from a variety of theories and research approaches across the humanities, social sciences, and natural sciences (Minta-Tworzowska 2017a, 268; 2022). Consequently, archaeology has increasingly engaged with different disciplines, actively seeking answers to complex questions about forms of past social life and culture.

In Minta-Tworzowska's (2002, 2021a, 459) view, while successive research directions have broadened the field of archaeology, the distinctiveness of Polish archaeological methodology has largely diminished. This has created a dilemma "between the community of inspiration and the distinctiveness of archaeological traditions". At the same time, she believes that significant influence on the conceptualization of cognitive problems in archaeology can come from efforts to eliminate the naive empiricism often observed in narrow specializations, the uncritical inclination of archaeologists towards natural sciences, and the dispersion of archaeological knowledge (Minta-Tworzowska 2017a, 282). As a remedy, she argues that only theoretical thinking could overcome this "naivety".

Additionally, Minta-Tworzowska's research extends beyond Polish archaeology to provide a broader assessments of the theoretical landscape in contemporary global archaeology. A central question she raises is the necessity of theory for both the current state and future trajectory of the discipline (Minta-Tworzowska 2015b). She notes a significant ambivalence among archaeologists in recent years: while many considers themselves theoreticians, their focus often shifts merely towards analysis of empirical data frequently lacking solid theoretical foundations underpinning their research. This dichotomy highlights a major challenge in archaeology—balancing the practical aspects of data collection and analysis with the theoretical frameworks essential for interpreting and contextualizing archaeological findings.

Looking ahead, Minta-Tworzowska stresses the critical role of theory in shaping the future of archaeology. She argues that theory is crucial for fostering a unified approach within the discipline, enabling practitioners to establish a common intellectual discourse (Minta-Tworzowska 2015b, 32). Currently, archaeology as a discipline, faces a perceived lack of a cohesive identity, a situation exacerbated by the rapid expansion of other scientific disciplines studying the past. As a result, new fields of interest and theory-driven approaches are gaining prominence within archaeology itself, rather than relying solely on interdisciplinary collaboration. Emerging areas such as the archaeology of the contemporary era, symmetrical archaeology, new materialism, and multi-species archaeology offer novel perspectives on archaeological inquiry amid evolving societal and environmental contexts.

Minta-Tworzowska (2015b, 28) identifies the primary challenge facing European continental archaeology as the need to transcend the "old" paradigm—a rigid framework characterized by strict disciplinary boundaries. This shift is driven by the emergence of new research areas that require innovative theories and methodologies to address contemporary issues in archaeology. Consequently, she emphasizes the increasing need for adapting or creating theoretical frameworks than can accommodate diverse archaeological inquiries and interpretations. Minta-Tworzowska anticipates a growing demand for individual archaeological theories, particularly those rooted in middle-range theory. These theories would bridge overarching theoretical constructs with specific empirical data, offering a nuanced framework for interpreting archaeological evidence and fostering interdisciplinary dialogue.

3.2 Classification as a Reflection on Research Procedures in Archaeology

As mentioned earlier, Minta-Tworzowska primarily focuses on the research procedures in archaeology. Specifically, her methodology centers around the theme of classification, which she views as a central axis of the discipline and a key issue in archaeological discussion, not limited to European archaeology. As discussed earlier, a significant influence on her approach to classification was Teilhard de Chardin's methodological proposal for studying prehistory. Additionally, her approach to theory and methodology of archaeology shows a strong influence of PSM's ideas regarding classification in archaeology. This influence is evident in several assumptions: the conviction that every research procedure represents underlying theories, a focus on the subject-oriented dimension of reality, and an advocacy for a humanistic interpretation that accepts the thesis of the rationality of human actions and cultural norms as fundamental, obligatory activities. These activities refer to behaviors that conform to specific patterns. Moreover, her approach values interdisciplinary research, aiming to create the broadest possible common ground for investigation and reflection in archaeology. Consequently, it encompasses not only archaeology but also cultural anthropology, the methodology of science, the methodology of history, and fields such as logic and statistics.

A crucial aspect of her interest is understanding how archaeological research procedures are influence by particular theories or their guiding role in classification procedures. This issue is central to her most significant book, *Klasyfikacja w archeologii jako sposób wyrażania wyników badan, hipotez oraz teorii archeologicznych* (*Classification in Archaeology as a Means of Expressing Research Results, Hypotheses and Archaeological Theories*), published in 1994. This book served as the foundation for her habilitation degree. She also dedicated subsequent papers to this topic (Minta-Tworzowska 1994b, 1998a, 2000d), presenting refinements and developments of her earlier views on specific issues related to classification in archaeology, its continuity and change, and its role in the archaeological research procedure. Additionally, she explored the role of classification within various schools of thought, such as processual, logicist, and contextual archaeology.

In her book, Minta-Tworzowska examines classification and typology within both global and Polish archaeology, tracing their historical development over the past 150 years. She identifies various meanings of classifications in archaeology and explores its role and place in the archaeological heuristic process. Her goal is to articulate the concept of modern archaeology by drawing upon cultural categories, human actions, and specific theoretical constructs. As she notes (1994a, 8), the development of thought reveals two main lines of reflection: the European continental and the Anglo-Saxon, with their overlap evident only in certain aspects. Within these frameworks, it becomes clear that the method of classification is influenced by research questions, formulated hypotheses, theoretical assumptions, and the adopted research methodology.

Minta-Tworzowska views classification as "a manner of defining concepts, expressing results of studies, hypotheses and archaeological theories, which is a fundamental procedure in archaeology" (Minta-Tworzowska 1998a, 191). Although the meaning of classification has evolved over time, it is used to: (a) present research results, (b) define specific concepts, and (c) express archaeological hypotheses and theories (Minta-Tworzowska 2000d, 121). She argues that classification serves no universal goal: its form and character are always determined by the specific research problem it is designed to address and solve (Minta-Tworzowska 1998a, 192). In her texts, Minta-Tworzowska demonstrates how shifts in conceptual frameworks and paradigms in archaeology—such as culture-historical archaeology, processualism, French logicist archaeology, structuralism, and contextual archaeology—have altered the nature of classification and the goals it expresses. These changes have influenced how concepts are defined, how hypotheses and theories are formulated, and how research results are expressed and presented, particularly in constructing images of the past. As she aptly observes:

> (...) classification constitutes a measure of the direction of changes in conceptualization in archaeology. A change in theoretical orientation involves changes in the concept of the archaeological source as well as shifts in expectations regarding the role, goals, tasks, and character of classification and the grounds for its legitimate application (Minta-Tworzowska 1998a, 194).

Classification in archaeology serves two key purposes: describing and presenting data, and integrating concepts of culture into hypotheses or specific theories. Consequently, Minta-Tworzowska addresses two fundamental issues in archaeology: the classification of archaeological sources and the concept of culture. The former involves using classification to describe and organize archaeological sources and to formulate hypotheses, while the later examines how classification serves as a means for expressing archaeological theories. As she indicates (Minta-Tworzowska 1994a, 31–33), classifications in archaeology fundamentally differ based on whether they adopt an 'internal' or 'external' perspective, reflecting two analytical strategies: 'emic' and 'etic'. The emic approach aims to reconstruct specific patterns or models of human behavior, while the etic approach focuses on external characteristics of archaeological sources, such as the attributes of the producer or user of artifacts and the manifestation of cultural patterns. Minta-Tworzowska views the "emic-etic" strategy as a continuum rather than a dichotomy.

The aim of classification in archaeology is not merely to distinguish between objects, phenomena, or processes; it is to construct images of the past and support arguments based on classification results. Therefore, classification must be considered as a crucial component of theoretical and methodological issues in archaeology. Minta-Tworzowska advocates, though not uncritically, for the systemic approach to classification developed alongside American 'New Archaeology'. This approach allows for perceiving the world as an ordered system and describing it appropriately. It serves both as a heuristic tool and as a representation of reality, functioning as both and ontological and epistemological category (Minta-Tworzowska 1998a, 216). At the same time, she asserts that modern classification in archaeology should be underpinned by its close link to interpretation and explanation procedures and its integration of various approaches, particularly those that apply the concept of culture.

3.3 The Conceptualization of Archaeological Sources

Although archaeologists have been studying and conceptualizing archaeological sources for over a century, this issue remains crucial for contemporary archaeology. The significance of sources in archaeology is paramount, as material culture constitutes the core of archaeological research. Archaeological sources are viewed either as essential tools for reconstructing the past (traditional approach) or for the developing models, explanations, and interpretations of the past (Minta-Tworzowska 2014). Consequently, conclusions and interpretations about prehistory are derived from material culture, guided by the adopted research approach or theory. For example, the model of an archaeological source determines the model of archaeological classification (1998a, 194). Therefore, our approach to archaeological sources is also the result of the theoretical tradition in archaeology (Rączkowski 2011, 15).

Reflection on archaeological sources has provided a unique lens, focusing on the methodological issues faced by methodologists of archaeology from Adam Mickiewicz University in Poznań. Specifically, the conceptualization and critical examination of the concept of sources in archaeology and their role in cognitive processes have been central to Minta-Tworzowska's research interests (1998b, 2000c, 2012, 2014). She treats archaeological sources similarly to historical sources, rejecting the naive belief that these sources inherently contain the truth about the past and that their cumulative accumulation results in a fuller and more objective explanation. In line with her views on classification in archaeology, she firmly rejects the tenets of positivist prehistory. To recall, positivist prehistory assumed that an acceptable model or archaeological research is based on the belief that 'pure' sources or 'raw' facts, exist, and that these facts need only to be collected in large quantities to provide reliable knowledge about the past. It was presumed that the collection of facts is objective and free from theoretical influence (Marciniak and Rączkowski 1991, 61). Minta-Tworzowska thus rejects the notion that archaeological practice is about uncovering a truth 'hidden' in archaeological sources. Her approach is characterized by a humanistic perspective, believing that our engagement with archaeological sources is deeply entwined with our own culture and values.

She begins her reflection on the issue of sources in archaeology with an examination of prehistoric sources as presented by Józef Kostrzewski (1885–1969), the father of Polish prehistoric archaeology (Minta-Tworzowska 1988). This examination includes a critical analysis of his archaeological research methodology. The foundation for this investigation is provided by a modern theory of cultural sources proposed by J. Kmita and J. Giedymin (Giedymin and Kmita 1966).

Minta-Tworzowska studied Kostrzewski's perspective on sources through the lens of positivistic methodology, particularly its Montelius-Kossina version, and the principles adopted by Kostrzewski's Poznań school of prehistory. She reconstructed his concept of sources by examining the definition he provided, the methods for classifying sources, their results, and the historical insights they offer. Kostrzewski focused on archaeological facts, referring to them as 'prehistoric artefacts', which he viewed as traces through which past events could be recognized. According to him, these artefacts were carriers of objective truth. Minta-Tworzowska argues that the methodology of prehistory and the conceptual model employed by J. Kostrzewski align with the school of positivism, characterized by phenomenalism, inductionism, psychologism, and naturalism. Consequently, it can be broadly described as a form of empirical and somewhat commonsense historicism.

In her conceptualization of sources in archaeology, the influence of ideas developed within the PSM is particularly noticeable (e.g., Giedymin and Kmita 1966; Kmita and Pałubicka 1992), especially in her analysis of archaeological sources against the backdrop of the historical source as articulated by Jerzy Topolski (Minta-Tworzowska 1998b, 2000c). She found Topolski's conception of the "dynamic understanding of the historical source", deeply embedded in narrativism and postmodernism, crucially important. Topolski (1984, 324) defines a historical source as "(...) any and all sources of historical cognition (direct and indirect), i.e., all information (including theory and information) on the social past, irrespective of its place, together with its communication (channel of communication)".

According to Topolski, a source is anything that might be useful for constructing a vision of the past and providing history with arguments for this construction. An object becomes a source when it is interrogated in relation to the studied problem. This 'extraction' of information is a constitutive element of the narrativization of the source and the entire interpretive process (Topolski 1996, 338ff). Topolski argues that both the source and historical narrative operate at the same interpretive level. The transition from sources to narrative is not a shift from uncovering a hidden truth within the source to interpretation but rather a movement within the same realm of interpretation (Minta-Tworzowska 2012, 158; 2014, 143). There is no analysis or synthesis; only interpretation exists. However, the category of source is not monolithic. Topolski (1996, 340) distinguishes between direct (persuasive) sources, which aim to convince the receiver of their truth, and indirect (non-persuasive) sources, which merely reflect the past (Topolski 1996, 340). In his research methodology, Topolski prefers argumentation over justification, recognizing that the interpretation of the source is inherently persuasive.

Within Topolski's percpective, one can argue that the content of an archaeological source must be understood through its narrativization, with the narrative structure of the source being crucial for archaeology (Minta-Tworzowska 1998b, 334; 2000c). However, it is essential to remember that the archaeological record represents only a small part of the past world preserved in the earth, while simultaneously serving as a bearer of information about human life. Consequently, Minta-Tworzowska's position aligns with the understanding of sources as texts of culture (*sensu* Hodder 1986), viewing archaeology as an interpretation of the past. In this context, she emphasizes the semiotic nature of material culture, arguing that the reading of material texts must take into account both their content and structure, as well as the accumulated meaning.

In her conceptualization of archaeological sources, Minta-Tworzowska (1998a, b)—following Linda Patrik (1985)—distinguishes between two basic models: the physical model and the textual model. The physical model treats the source as a natural fossil that must be categorized by its generic and typological features and situated within a specific time and space configuration. In contrast, the textual model reintroduces a humanistic dimension to archaeological sources by viewing them as unique 'texts' that convey information to be interpreted by archaeologists (Minta-Tworzowska 1998a, 194; 1998b, 334–335). This model focuses on what the source reveals about its maker or user. This perspective aligns with the approach to historical sources introduced and developed by J. Giedymin and J. Kmita (1966). They divide sources into two categories: indices and signs. An index is considered a relic of a past state of affairs that does not intend to communicate any message and is further divided into humanistic and natural indices. In contrast, a sign is an object intended to convey a specific message according to communication rules accepted within a particular social group.

Minta-Tworzowska (2011, 46; 2021a, 461–462) views an archaeological source as a cultural construct, shaped by archaeologist' decisions in assigning meaning to objects and determining what qualifies as an artifact, ecofact, or source. Thus, it is the archaeologist who effectively creates the source. This perspective challenges the traditional view of the source as a fundamental and objective element in shaping images of the past (Rączkowski 2011, 17). Instead, images of the past result from the researcher's interpretation and are not cognitively neutral. They are based on sources, each of which contains an infinite amount of information, with archaeologists focusing only on aspects relevant to their research questions (Minta-Tworzowska 2014, 142–143). As interpretations of the past evolve, so too does the relationship with its material remains, which are considered sources only insofar as they are connected to specific issues by researchers.

In her later views on archaeological artifacts, Minta-Tworzowska incorporates influences from the methodological turns in the humanities, which have also left their mark on her perspective on this issue. This includes the 'return to things', 'studies on objects', and 'new empiricism' that arose at the end of the 1990s. These approaches broke away from the textual approach to reality (Minta-Tworzowska 2011, 2012, 148–153). She is skeptical of the notion that studies on objects in archaeology present a new problem for the discipline, arguing that archaeology has long been concerned with uncovering the very existence of material sources since the nineteenth century (Minta-Tworzowska 2014, 141). According to her, modern culture has similarly embraced positivistic attempts to align the chronology of human beings with the chronology of objects.

However, Minta-Tworzowska views the 'return to things' in archaeology not as a mere return to recording, describing, and classifying artifacts. These studies extend beyond objects functioning within culture to include elements of nature that are typically excluded from considerations traditionally dominated by culture and society. She argues that a 'return to things' in archaeology should not be seen as a 'return to roots' or to typology and classification. Instead, moves away from models based on chronotopic events or over-generalized systems. Things appear in the context of thought and must be analyzed through the investigation of thinking to generate specific activities in their use. The main idea of the 'return to things' is to prevent archaeology from monopolizing visions of the past, considering that it controls so many of its aspects (Minta-Tworzowska 2011, 57).

3.4 *Symbols, Symbolism and Prehistoric Art Within Archaeological Research*

Another significant aspect of Minta-Tworzowska's research is her focus on symbols and symbolism in archaeological research (Minta-Tworzowska 2000f). Her interests encompass various areas, including the i identity of prehistoric societies, social memory, categories of time and space in archaeology, landscape, and, above all, prehistoric "art". She primarily studies European Paleolithic rock paintings and carvings, such as those from Lascaux, Altamira, and many others, which she views as manifestations of prehistoric art.

Researching prehistoric art poses considerable challenges due to the diversity of research approaches and methodologies employed. While the issues of symbols and symbolism in prehistory have often been raised by archaeologists, this has typically been done in an intuitive manner, lacking a solid theoretical foundation, and usually approached from perspectives of sacrality, cult, religion, or magic (Minta-Tworzowska and Rączkowski 1996, 205). The same can be said about prehistoric art. In contrast to traditional approach, Minta-Tworzowska advocates for a theory-mediated, systematic framework for understanding and interpreting prehistoric art. This approach rejects positivist assumptions that one can read an objective past reality directly from prehistoric ar. Instead, she demonstrates how applying elements of archaeological theory—such as textualism, structuralism, or post-structuralism—can significantly alters interpretations. As she declares: "As a prehistorian, I am interested in those theories and perspectives on art that provide possibilities for interpreting prehistoric art in terms of its meaning" (Minta-Tworzowska 2008, 27), emphasizing the importance of specific theories in understanding prehistoric art.

Consequently, Minta-Tworzowska's research on prehistoric art is inherently interdisciplinary, transcending the boundaries of individual scientific disciplines. While primarily focused on archaeology, her work is complemented by cultural anthropology, art history, and various methodologies from the sciences and history. She also integrates concepts from the PSM developed by figures such as J. Kmita or A. Pałubicka, including notions of the archaic society, magical-mythical thinking, and the socio-regulative theory of culture. In her analyses of prehistoric art, Minta-Tworzowska engages with a range of theoretical perspectives, including Theodor W. Adorno's theories on the sociology of art, John Dewey's category of experience, the perspective of new aesthetics (Minta-Tworzowska 2019), and various phenomenological approaches. For example, she applies the phenomenological approach of Polish phenomenologist Roman Ingarden, whose analysis of cognitive attitudes towards the outer world proves relevant for interpreting prehistoric rock "art" (Minta-Tworzowska 2015a).

In her other works, she addresses the issue of intertextuality in Paleolithic art and the role of color as a key element (Minta-Tworzowska 2017b). A significant concept in this context is the open and interdisciplinary approach to images, derived from the anthropology of images. She emphasizes the importance of understanding images as 'acting' through what she terms 'interimaginal relations' in interpreting these works of art. Her analysis reveals that many Palaeolithic images show interpenetrating traces of other representations, and that cave painting typically employed one or two main colors, rarely using more. She argues that the most significant features of Paleolithic images were movement, color—particularly expressed through contrast of black, red or yellow—and illusion. For prehistoric people, the world of colors was deeply embedded in cultural meanings, with color choices rooted in cultural memory. The 'acting' of images was crucial in the 'mediatisation' of important existential content between the past and present.

An approach to studying prehistoric art from different interpretive angles is not uncommon or genuine, even within Polish archaeology. What makes Minta-Tworzowska's approach distinct and original is its solid theoretical foundation and her effort to define what qualifies as art in prehistory, as well as how archaeologists can approach and interpret it. Her work aims to view prehistoric art as historical sources, advocating for its study as an essential element of the prehistoric world and through the lens of aesthetics (Minta-Tworzowska 2008, 2015a, 2019). She argues prehistoric art can be seen as having a dual character: as an autonomous phenomenon and as a social phenomenon (Minta-Tworzowska 2008, 29). In her view, art responded to the needsand conditions of prehistoric life, reflecting social conditions, traditions, and the psychological processes and temperament of the artist. Art was understood as a form of magic, engaging with the world actively by manipulating the natural forces and the 'spiritual' realm of reality, and was governed by tradition. It also involved repetition, which rendered it ritualistic (Minta-Tworzowska 2008, 33).

Minta-Tworzowska strongly emphasizes the social aspects and role of prehistoric art, building on A. Pałubicka's (1985, 1990) assertion that if we accept the concept of archaic culture as a form of magical culture, then prehistoric "art" should be seen as an integral component of that culture rather than a separate field (Minta-Tworzowska 2015a, 340). This perspective connects to the broader question of whether archaeologists can discuss the symbolic aspect of human life in prehistory, considering the idea of the archaic culture, which posits that distinct spheres did not exist in the world. According to this view, prehistoric culture was magical in nature, with rock paintings and drawings reflecting and mimicking the rhythm of the community's life. Symbols organized perception and ascribed meaning to various aspects of their existence (Minta-Tworzowska 2008, 31).

Through prehistoric art, we gain new insights into the world of past societies, though what we discover is shaped by our goals and intentions. Minta-Tworzowska (2008, 41) argues that

> Reflections on art decisively broaden the scope of archaeological issues to include topics typically absent from its traditional framework. They inspire consideration of both the ways we understand the history and culture of human communities and our own involvement in the culture of which archaeology is a part.

Therefore, in Minta-Tworzowska's view, the primary aim when studying prehistoric art should be an interpretation that leads to understanding its meaning. While no single explanation can fully encompass this meaning, some interpretations are compelling than others. This perspective aligns with the postmodern belief that the past is interpreted no through its own categories, but through the lens of contemporary interests, creating an intellectual construct to explain the significance of prehistoric art (Minta-Tworzowska 2008, 24). This approach also reflects a consistent theme in Minta-Tworzowska's research: every interpretive effort in archaeology reveals a particular research attitude that is theory-driven, where language and culture play a crucial role in constructing images of the world and shaping our understanding of it.

4 Conclusions

The theory and methodology of the Poznań School of Archaeology were profoundly influenced by the concepts developed within the PSM, created by J. Kmita, L. Nowak, and the historian J. Topolski. The PSM's rigorous methodological approach to scientific research, its emphasis on the guiding role of theory in any scientific inquiry, and its advocacy for interdisciplinary collaboration significantly impacted archaeology at Adam Mickiewicz University in Poznań. Archaeologists such as Jan Żak and his students applied these principles to modernize Polish archaeology, moving away from the positivism that had previously dominated the field. This shift gradually refocused archaeological studies from merely describing isolated sources and facts to explaining past processes and obtaining a complex picture of the past. The trajectory of the Poznań School of Archaeology demonstrates that Polish archaeologists were able to promote theoretical concepts that offered an alternative to Anglo-American models, emerging from a unique and original local intellectual tradition (Lozny 2002, 2011, 214–215; Lozny, chapter "Philosophy of Archaeology and the Poznań School of Methodology").

Danuta Minta-Tworzowska played a pivotal role in the development and conceptualization of the Poznań School of Archaeology. Her academic journey was dedicated to the constant search for answers regarding the role of theory and methodology in archaeology. Her approach profoundly influenced her collaborators and students, whose intellectual pursuits reflect their own efforts to explore new paradigms and engage in self-reflection on the role of theory in the discipline. While they maintained some of her general assumptions—such as reflections on archaeological theory, research procedures, archaeological sources, and paradigms in archaeology—they also expanded their focus to include diverse issues, related to theory in archaeology, the history of archaeological thought, heritage management, social functions of archaeology, social memory, public archaeology (Pawleta 2016, 2022), and community archaeology (Kajda, Kobiałka and Marciniak 2022).

The differences between the methodology of J. Żak and the current state of archaeological theory and methodology in Poland should be viewed as a natural progression of subsequent generations their contemporary issues (Minta-Tworzowska 2021b, 164). Despite varying approaches, the core of their work remains a shared commitment to the central role of theory in archaeological practice and the importance of interdisciplinary research. This interdisciplinarity is both theoretical and practical, integrating approaches and methodologies from the humanities, philosophy of science, methodology of history, cultural anthropology, and natural sciences. This integration represents the primary legacy of the PSM in archaeology, which, primarily through Danuta Minta-Tworzowska, has significantly advanced archaeological methodology in Poland over the past four decades.

References

Barford, P. 1995. Marksizm w archeologii polskiej w latach 1945–1975. *Archeologia Polski* 40 (1–2):7–78.

Giedymin, J., and J. Kmita. 1966. *Wykłady z logiki formalnej, teorii komunikacji i metodologii nauk*. Poznań: UAM.

Hodder, I. 1986. *Reading the past*. Cambridge: Cambridge University Press.

Johnson, M. 2010. *Archaeological theory. An introduction*. 2nd ed. Malden/Oxford: Wiley Blackwell.

Kajda, K., D. Kobiałka, and A. Marciniak. 2022. (Eds.) *Archeologia wspólnotowa – poznając przeszłość, nie zapominając o teraźniejszości*. Kraków: Universitas.

Kmita, J. 1971. *Z metodologicznych problemów interpretacji humanistycznej*. Warszawa: Państwowe Wydawnictwo Naukowe.

Kmita, J. 1973. *Wykłady z logiki i metodologii nauk: dla studentów wydziałów humanistycznych*. Warszawa: Państwowe Wydawnictwo Naukowe.

Kmita, J., and A. Pałubicka. 1992. Problem użyteczności pojęcia doświadczenia. In *Poszukiwanie pewności i jego postmodernistyczna dyskwalifikacja*, ed. J. Such, 150–181. Poznań: Wydawnictwo Naukowe Instytutu Filozofii UAM.

Kobyliński, Z. 1991. Theory in Polish archaeology 1960–90: Searching for paradigms. In *Archaeological theory in Europe. The last three decades*, ed. I. Hodder, 223–247. London: Routledge.

Kotowa, B. 2010. Od filozofii humanistyki do filozofii jako dziedziny kultury. Główne kierunki badawcze szkoły poznańskiej od połowy lat siedemdziesiątych. In *Filozofia na Uniwersytecie w Poznaniu*, ed. T. Buksiński, 309–334. Poznań: Wydawnictwo Naukowe IF UAM.

Łastowski, K. 1987. *Rozwój teorii ewolucji: studium metodologiczne*. Poznań: Wydawnictwo Naukowe UAM.

Lech, J. 1997. Małowierni. Spór wokół marksizmu w archeologii polskiej w latach 1945–1975. *Archeologia Polski* 42 (1–2):175–232.

Lozny, L. R. 2002. Far outside, looking in Polish archaeologists and looking-glass self. *Archaeologia Polona* 40:137–148.

Lozny, L. R. 2011. Polish archaeology in retrospective. In *Archaeology in the age of globalization: Local meanings, global interest*, ed. L. Lozny, 195–220. New York: Springer.

Mamzer, H. 2020. "Mowa obrończa" metodologii archeologów Uniwersytetu im. Adama Mickiewicza w Poznaniu. *Folia Praehistorica Posnaniensia* 25:109–148.

Marciniak, A. 2005. *Placing animals in the neolithic: Social zooarchaeology of prehistoric farming communities*. London: UCL Press.

Marciniak, A. 2011. Contemporary Polish archaeology in global context. In *Archaeology in the age of globalization: Local meanings, global interest*, ed. L. Lozny, 179–194. New York: Springer.

Marciniak, A., and W. Rączkowski. 1991. The development of archaeological theory in Poland under conditions of isolation. *World Archaeological Bulletin* 5:57–65.

Minta-Tworzowska, D. 1986. *Elementy metodologii prahistorii w historiozofii P. Teilharda de Chardin. Poznań: Wydawnictwo Naukowe Uniwersytetu im. Adama Mickiewicza.*

Minta-Tworzowska, D. 1987. Kwestia postępu poznawczego w archeologii i prahistorii. *Folia Praehistorica Posnaniensia* 2:5–12.

Minta-Tworzowska, D. 1988. Prehistorical sources in Józef Kostrzewski's view. *Folia Praehistorica Posnaniensia* 3:189–196.

Minta-Tworzowska, D. 1994a. *Klasyfikacja w archeologii jako sposób wyrażania wyników badań, hipotez oraz teorii archeologicznych*. Poznań: Wydawnictwo Naukowe UAM.

Minta-Tworzowska, D. 1994b. Sposoby formułowania hipotez w archeologii i prahistorii. *Folia Praehistorica Posnaniensia* 6:225–255.

Minta-Tworzowska, D. 1996. Culture and society in traditional and processual archaeology. *Archaeologia Polona* 34:59–81.

Minta-Tworzowska, D. 1998a. Continuity and change of classification in archaeology. In *Theory and practice of archaeological research. Dialogue with the data: The archaeology of complex societies and its context in the '90*. vol. III, ed. S. Tabaczyński, 191–223. Warszawa: Institute of Archaeology and Ethnology, Committee of Pre- and Protohistoric Sciences, Polish Academy of Sciences.

Minta-Tworzowska, D. 1998b. Jerzego Topolskiego koncepcja źródła historycznego a ujęcia źródeł archeologicznych. In *Świat historii: prace z metodologii historii i historii historiografii dedykowane Jerzemu Topolskiemu z okazji siedemdziesięciolecia urodzin*, ed. W. Wrzosek, 329–340. Poznań: Instytut Historii UAM.

Minta-Tworzowska, D. 2000a. Kwestia przełomu metodologicznego w archeologii i prahistorii polskiej. In *Archeologia i prahistoria Polska w ostatnim półwieczu*, ed. M. Kobusiewicz and S. Kurnatowski, 527–537. Poznań: Wydawnictwo Poznańskiego Towarzystwa Przyjaciół Nauk.

Minta-Tworzowska, D. 2000b. Archeologiczne rekonstrukcje świata pradziejowego wobec krytyki postmodernistycznej. In *Kultury archeologiczne a rzeczywistość pradziejowa*, ed. S. Tabaczyński and W. Hensel, 185–198. Warszawa: Wydawnictwo Naukowe PWN.

Minta-Tworzowska, D. 2000c. Świat archeologii w świetle źródeł archeologicznych. *Acta Historica et Museologica Universitatis Silesianae Opaviensis* 5:49–59.

Minta-Tworzowska, D. 2000d. Contemporary debates in archaeology: The example of classification – part 1. *Revista De Historia Da Arte E Arqueologia* 3:121–128.

Minta-Tworzowska, D. 2000e. Postmodernizm, myśl postrukturalistyczna a archeologia (zarys problematyki). In *Archeologia w teorii i w praktyce*, ed. A. Buko and P. Urbańczyk, 87–97. Warszawa: IAiE PAN.

Minta-Tworzowska, D. 2000f. Symbole i symbolika w perspektywie badań archeologicznych. In *Kultura symboliczna kręgu pól popielnicowych epoki brązu i wczesnej epoki żelaza w Europie Środkowej*, ed. B. Gediga and D. Piotrowska, 45–54. Warszawa/Wrocław/Biskupin: Polska Akademia Nauk, Oddział we Wrocławiu, Muzeum Archeologiczne.

Minta-Tworzowska, D. 2001. Profesora Jana Żaka poszukiwanie paradygmatu. In *Archeologia. Paradygmat. Pamięć*, ed. D. Minta-Tworzowska and W. Rączkowski, 109–122. Poznań: Wydawnictwo Poznańskie.

Minta-Tworzowska, D. 2002. Between a community of inspiration and the separateness of archaeological traditions. In *Archaeologies of Europe: History, methods and theories*, ed. A. Gramsch, P. Biehl, and A. Marciniak, 53–64. Münster: Waxman.

Minta-Tworzowska, D., and W. Rączkowski. 2005. Overcoming the barriers: Jan Żak's involvement in building contacts with German archaeology. *Archaeologia Polona 43:125–142.*
Minta-Tworzowska, D. 2006. Interdisciplinary research and archaeology. *Archaeologia Polona* 44:41–51.
Minta-Tworzowska, D. 2008. Teoretyczne problemy interpretacji "sztuki" pradziejowej. In *Sztuka pradziejowa i wczesnośredniowieczna jako źródło historyczne*, ed. B. Gediga and W. Piotrowski, 23–47. Wrocław/Biskupin: Polska Akademia Nauk, Oddział we Wrocławiu, Muzeum Archeologiczne.
Minta-Tworzowska, D. 2011. Człowiek i rzecz w perspektywie archeologicznej (czyli rzecz w perspektywie antropocentrycznej). In *Współczesne oblicza przeszłości, ed. A. Marciniak, D. Minta-Tworzowska, M. Pawleta, 39–61. Poznań: Wydawnictwo Poznańskie.*
Minta-Tworzowska, D. 2012. Źródło/Ślad/Artefakt/Rzecz/Przedmiot. In *Przeszłość społeczna. Próba konceptualizacji*, ed. S. Tabaczyński, A. Marciniak, D. Cyngot, and A. Zalewska, 137–161. Poznań: Wydawnictwo Poznańskie.
Minta-Tworzowska, D. 2014. Józefa Kostrzewskiego ujęcie źródeł w relacji do współczesnej archeologii. *Fontes Archaeologici Posnanienses* 50 (1):137–145.
Minta-Tworzowska, D. 2015a. Granice poznania dzieła malarskiego jako dzieła "sztuki" pradziejowej. Na kanwie rozważań Romana Ingardena. *Folia Praehistorica Posnaniensia* 20:341–359.
Minta-Tworzowska, D. 2015b. Czy współczesna i przyszła archeologia będzie "potrzebowała" teorii? *Archeologia Polski* 60:21–37.
Minta-Tworzowska, D. 2017a. Tradycja i inspiracje teoretyczne we współczesnej archeologii. *Filo-Sofija* 17 (1):267–286.
Minta-Tworzowska, D. 2017b. Świat kolorów ludzi paleolitu jako element intertekstualności "obrazów" w jaskiniach. *Folia Praehistorica Posnaniensia* 22:111–142.
Minta-Tworzowska, D. 2019. The program of "new" aesthetics regarding the expectations of archaeologists. *Folia Praehistorica Posnaniensia* 24:149–163.
Minta-Tworzowska, D. 2021a. Are we where we wanted to be? Modernist tendencies versus the postmodern reality of archaeology. Some remarks on the methodology of archaeologists at Adam Mickiewicz University in Poznań. In *Treasures of time. Research of the Faculty of Archaeology of Adam Mickiewicz University in Poznań*, ed. D. Żurkiewicz, 452–473. Poznań: Wydawnictwo Naukowe UAM.
Minta-Tworzowska, D. 2021b. Metodologia archeologii/prahistorii profesora dr. hab. Jana Żaka (1923–1990) po 30. latach. Między modernistycznymi dążeniami a postmodernistyczną rzeczywistością. *Folia Praehistorica Posnaniensia* 26:145–169.
Minta-Tworzowska, D. 2022. Teoretyczne i metodologiczne podstawy archeologicznego dyskursu w Polsce. Zarys problematyki. *Przegląd Archeologiczny* 70:5–40.
Minta-Tworzowska, D. 2023. Nowoczesna archeologia Profesora Jana Żaka (1923–1990). W 100. rocznicę urodzin. *Folia Praehistorica Posnaniensia* 28:231–253.
Minta-Tworzowska, D., and M. Pawleta. 2013. Teoria i metodologia archeologii na Uniwersytecie w Poznaniu – dlaczego właśnie tu? *Folia Praehistorica Posnaniensia* 18:217–233.
Minta-Tworzowska, D., and W. Rączkowski. 1996. Theoretical traditions in contemporary Polish archaeology. *World Archaeological Bulletin* 8:196–209.
Minta-Tworzowska, D., and W. Raczkowski. 2007. Standardy teoretyczne we współczesnej polskiej archeologii. In *Pół wieku z dziejów archeologii polskiej (1939–1989)*, ed. J. Lech, 219–247. Warszawa: PWN.
Musiał, G. 2016. Poznańska szkoła metodologiczna—jej rozwój i znaczenie. *Przegląd filozoficzny — Nowa Seria* 25 (2):553–563.
Nowak, L. 1980. *The structure of idealization. Towards a systematic interpretation of the Marxian idea of science*. Dordrecht/Boston/London: Reidel.
Pałubicka, A. 1985. O trzech historycznych odmianach waloryzacji światopoglądowej. *Studia Metodologiczne* 24:51–76.
Pałubicka, A. 1990. *Kulturowy wymiar ludzkiego świata obiektywnego*. Poznań: Wydawnictwo Naukowe UAM.

Pałubicka, A., and S. Tabaczyński. 1986. Społeczeństwo i kultura jako przedmiot badań archeologicznych. In *Teoria i praktyka badań archeologicznych. Przesłanki metodologiczne*, vol. I, ed. W. Hensel, G. Donato, and S. Tabaczyński, 57–183. Wrocław/Warszawa/Kraków/Łódź: Zakład Narodowy im. Ossolińskich, Wydawnictwo Naukowe IAE PAN.

Patrik, L. E. 1985. Is there an archaeological record? *Advances in Archaeological Method and Theory* 8:27–62.

Pawleta, M. 2016. *Przeszłość we współczesności. Studium metodologiczne archeologicznie kreowanej przeszłości w przestrzeni społecznej*. Poznań: Wydawnictwo Naukowe UAM.

Pawleta, M. 2022. Public archaeology in Poland: Current state of affairs and future directions. *European Journal of Archaeology* 25 (1):103–118.

Rączkowski, W. 2002. *Archeologia lotnicza: metoda wobec teorii*. Poznań: Wydawnictwo Naukowe UAM.

Rączkowski, W. 2011. Theory, empiricism and practice: archaeological discourses in a network of depedency and opposition. *Analecta Archaeologica Resoviensia* 4:7–34.

Tabaczyński, S. 1987. *Archeologia średniowieczna: problemy, źródła, metody, cele badawcze*. Wrocław: Zakład Narodowy im. Ossolińskich.

Tabaczyński, S. 2000. "Wymuszony przewrót metodologiczny" w archeologii polskiej i jego nieoczekiwane skutki. In *Archeologia i prahistoria Polska w ostatnim półwieczu*, ed. M. Kobusiewicz and S. Kurnatowski, 513–525. Poznań: Wydawnictwo Poznańskiego Towarzystwa Przyjaciół Nauk.

Topolski, J. 1983. *Teoria wiedzy historycznej*. Poznań: Wydawnictwo Poznańskie.

Topolski, J. 1984. *Metodologia historii*. Poznań: Państwowe Wydawnictwo Naukowe.

Topolski, J. 1996. *Jak się pisze i rozumie historię. Tajemnice narracji historycznej*. Warszawa: Oficyna Wydawnicza Rytm.

Żak, J. 1975. Próba przedmiotowego modelu kultury społeczeństw młodszego i schyłkowego paleolitu (strefy śródziemnomorskiej). *Fontes Archaeologici Posnanienses* 25:28–59.

Żak, J. 1985. Uwagi metodologiczne nad problemem osadnictwa. *Studia Metodologiczne* 24:77–87.

Żak, J., and D. Minta-Tworzowska. 1991. Problematyka prahistorii pozytywistycznej. *Folia Praehistorica Posnaniensia* 4:9–25.

Zamiara, K. 2010. U początków poznańskiej szkoły metodologicznej. In *Filozofia na Uniwersytecie w Poznaniu*, ed. T. Buksiński, 283–308. Poznań: Wydawnictwo Naukowe IF UAM.

Did/Do Early Traditional Peoples Have a History?

Anna Pałubicka

Abstract The chapter argues that the idea that scientific knowledge without methodological reflection has limited impact on archaeological research. It ultimately focuses solely on describing the archaeological record. Methodology provides a worldview, organises knowledge, offers theoretical explanations and enables comparisons with other perspectives. In this chapter, the author considers the case of different perceptions of time in early traditional societies. A fundamental question arises: how did the concept of past time emerge in European culture? In early traditional societies, time was perceived as cyclical rather than linear. This leads to the next question of how the concept of past time became established in European culture. The text references and discusses the views and arguments of thinkers from antiquity to the present day, organised around the socio-regulative concept of culture. A significant aspect of social development, particularly in communication, is the shift from collective memory to organised historical narratives. Creating coherent historical narratives requires the use of language at an appropriate level of abstraction, enabling the construction of frameworks based on linear time.

Keywords Linear time · Perception of past time · History · Early traditional societies · Communication · Logic of language

Scientific knowledge has limited potential to influence cognitive progress in archaeology in the absence of current methodological reflection. The discipline of archaeology then focuses primarily on the description of archaeological sources acquired in field research. We should remember that methodological reflection first establishes a certain vision of the world and its understanding, creates a framework for organizing empirical knowledge, clarifies it theoretically, provides a perspective for conducting comparative confrontation with materials from other archaeological

A. Pałubicka (✉)
Faculty of Philosophy, Adam Mickiewicz University in Poznań, Poznań, Poland
e-mail: anna.palubicka@amu.edu.pl

A. Marciniak et al. (eds.), *The Poznań School of Archaeology*, Outstanding Contributions to Archaeology, https://doi.org/10.1007/978-3-032-07848-3_6

sites, and outlines the relationship between empirical material and theory. These are just a few examples of the benefits of philosophical and methodological reflection that "rejuvenate" archaeological sources and make archaeology a science like other disciplines.

In the following discussion, I will highlight issues that are understood by archaeologists in an ahistorical way. These concern, for example, the understanding of time among early traditional peoples.[1] The question is how the perception of past time could have been constituted in European culture, as evidenced by the emergence of the concept of past time used in research alongside the commonly functioning present time.

The Kantian characterization of the concept of linear time as an a *priori* form of sensory perception begs the question of why it is time, along with space, that becomes an indispensable condition of scientific cognition. Time and space, as we shall see below, are inseparable from language and writing. By language I mean a set of signs determined by grammatical rules, i.e. not an arbitrary configuration of words, but only expressions suitably put together, ordered according to applicable rules. Thus, not every sign-word sequence or expression is a linguistic sentence. In addition to grammatical rules, language also respects the rules of logic. Grammatical rules are expressed in the language to which they apply, while logical rules serve in an unchanging form the written languages, which are part of symbolic communication, in addition to the use of speech according to grammatical rules. Apart from symbolic communication, performative-imaginative or pictorial communication can be distinguished (L. Lévy-Bruhl, J. Mazurkiewicz, E. Cassirer). The practice of communication based on symbolic language requires operations on a sequence of times, thus distinguishing between past, present and future time.

Figurative communication uses only the present. It was St. Augustine (354–430 A.D.), among others, who formulated a philosophical description of the present. He tried to give the status of real existence only to the present, which was close to the eternal time of Christianity (Czarnowski 1956a, 100). Past and future times exist only in human thought as memory (past time) and expectation (future time). According to St. Augustine:

> What is now clear and unmistakable is that neither things past nor future have any existence, and that it is inaccurate to say, "There are three tenses or times; past, present and future," though it might properly be said, "There are three tenses or times: the present of past things, the present of present things, and the present of future things. These are three realities in the mind, but nowhere else as far as I can see, for the present past things is memory, the present of present things is attention, and the present of future things is expectation. (St. Augustine 1997, 300).

[1] In this article, I will refer to *early traditional peoples* as preliterate communities and all those in which the principal bonds between their members are based on respect for and acceptance of shared collective perceptions and normative-directive beliefs. Mechanical solidarity, in Emile Durkheim's typology, represents the only form of interpersonal ties among community members. In earlier literature on the subject, the term primitive society was used to refer to what I call *early traditional peoples*.

> What, then, is time? If no one asks me, I know; if I want to explain it to someone who asks me, I do not know. I constate with confidence, however, that this is much I do know: if nothing passed away there would be no past time; if there was nothing still on its way there would be no future time; and if nothing existed, there would be no present time.
>
> Now, what about those two times, past and future: in what sense are they real beings, if the past no longer exists and the future does not exist yet? As for present time, if that were always present and never slipped away into the past, it would not be time at all; it would be eternity (St. Augustine 1997, 295–296).

The above statements boldly suggest that past and future times are constructs of human consciousness, not real entities.

1 Past Time and History

According to Stefan Czarnowski, a prominent Polish sociologist influenced by Emile Durkheim's French school of sociology (1956a, 100):

> While we do not need to be convinced that history as we practice it today is the product of only very highly civilized societies, we easily succumb to the illusion that every human group, even the most primitive, gives its tradition the form of this broadly understood "history" as soon as it attains some kind of self-knowledge.

The above passage questions the supra-historical status of the existence of history, i.e. history itself is considered a historical creation that appeared at a certain stage of civilizational development. According to Czarnowski, it is an illusion of contemporaries to ascribe to any human group, especially to early traditional ones, the order of the past according to the principle of linear time, i.e. the consecution and co-occurrence of events. However, such a concept of time is necessary for the practice of history.

History, both in the broader sense and in the narrower sense as understood by Czarnowski, is an orderly sequence of events, things, and human actions according to the principle of succession in time from the designated beginning in the past to the present. Everything that constitutes (creates) the present is the result of a longer or shorter irreversible succession of events, things and people that took place in the past. More generally, the period of time between the establishment of the beginning (somewhere in the past) and the end in the present is represented as history. The scope of the concept of history in the broader sense includes, in addition to scholarly-viewed history (history in the narrower sense), epic poems (regardless of their credibility), genealogical poems, as well as artistic representations of history, provided that they contain a temporal order according to the criteria of consecution and co-occurrence.

Historicism thus underlies the understanding of the term "history" in both its broader and narrower senses. It insists on the irreversible ordering of human actions and events according to temporal succession (consecution and co-occurrence of events), from the established beginning to the present. Examples of artifacts classified as history in the broad sense, which record tradition in a chronologically ordered manner, are lists of rulers or genealogies, which can be found even in societies at a lower level of civilization and meet the condition of historicism, i.e., they are ordered according to the principle of succession. They accompany some early traditional communities in which the sense of duration and passage of time is known. In other words, they appear when the community begins to operationalize the experience of past time.

History in the narrower sense appears at a later stage in the development of European culture, when it becomes accepted in a given community that history is not the result of the operation of a predetermined plan or a destiny assigned to a community or to each of its members individually. Finally, history is not the arena for the fulfillment of a divine purpose, prophecy, or prediction. It is not the result of the intervention of some extra-human forces, but is generated from the relationships between people, things, and events. The study of history in the strict sense, moreover, requires that linear and irreversible events be given narrative form. Narratives, on the other hand, are constructed in a coherent, non-contradictory, and verifiable manner.

For Stefan Czarnowski, scientific (scholarly-viewed) history, formulated according to scientific methods, ensures the acquisition of legitimate, true, and certain knowledge about the past. Scientific knowledge of the past is a narrative organized in time and, I would add, if it is to have a cognitive character, necessarily in space. The distinction between the criterion of spatial order and the criterion of temporal order is crucial. Both orders together form the semantic layer of the language of history, pointing to the existence of states of affairs in time and space (historical facts) independent of individual perceptions of them. They provide experiential access to the world of history as a narrative of events linked by causal relations and, of course, empirical control of constructed narration. Historical research cannot dispense with these prerequisites if it is to produce knowledge. Czarnowski's (1956b) reflections on history suggest that without the formation of history in a given community in the broader sense (*res gestae*), the emergence and functioning of history in the narrower sense, i.e., scientific reflection on history, is impossible.

However, before the emergence of the aforementioned sense of duration, succession, and its irreversible passing, cultural communities used only oral memory. This memory had a collective character in early traditional societies. The emergence of the order of collective memory according to the principles of consecution and co-occurrence of events can be seen as a harbinger of the future constitution of history, first in the broader sense (following Czarnowski).

Jerzy Topolski, in his *Świat bez historii* (*The World Without History*) (1972), while also advocating the historicity of history, did not fully share Czarnowski's concept. Topolski found Czarnowski's notion of history in the broader sense particularly objectionable, and he did not accept the inclusion in historical reflection of all narratives that are merely ordered by linear time but are unreal and, let us say, fictionalized. Topolski (1972, 12) refused to call the history thus identified by Czarnowski history:

> We would easily see the boundary between a historical account and any other (temporally dimensioned) account using the criterion of truth: the former would thus be an account that reproduces past reality in a manner consistent with that reality, that is, a true account. Thus, the difference between a historical account and other types of accounts with a temporal dimension oriented to the past can only be quantitative, not absolute.

For Topolski, history was merely a scientific account of the past. He understood history more or less in the same way that Czarnowski understood the concept of history in the narrower sense. The obvious conclusion is that most tribal historical communities do not have a science-based history. Nor do some contemporary communities have such a history (e.g., India).

In discussing various concepts of the causes of the fall of the Roman Empire, Topolski (1972, 149) raised an interesting hypothesis:

> The Roman Empire was doomed by a lack of history, that is, by a lack of historical thinking, but this type of thinking, in turn, could not have developed further under the circumstances of the time.

Topolski, as a renowned methodologist of history, refers in his explanation to Maurice Halbwachs's determinations of collective memory. According to Topolski, the Roman Empire lacked a sufficient amount of collective content based on the knowledge of the past and the present, which would at the same time unite the various peoples of the Empire. As a result, any collective action within the Roman Empire became impossible.

What was distinctive and innovative in Topolski's argument was his linking of history, historical knowledge, and historical thinking with effective action to modernize the community. In his view, societies that do not evolve and innovate continue to live within the same cultural patterns; they also have no history. He therefore accepted that the transformation of tribal consciousness into national consciousness was impossible without recourse to history and historical thinking.

I would like to leave aside this interesting dispute about the understanding of history in the Polish humanities and turn my attention to the moment of "transition" from collective memory to the ordering of this memory according to the principle of linear time. After all, the intuition of linear time and its practice in a given culture is one of the prerequisites, though not a sufficient one, for the constitution of history in the broader sense.

As Czarnowski and Topolski noted, the transition from collective memory to narrative using a temporal sequence was associated with more significant changes in culture, especially in the sphere of communication. These were transformations with, as I will try to show, momentous consequences for the further fate of European culture. Among other things, the new forms of interpersonal communication in ancient Greek culture gave rise to ancient logic, the most comprehensive form of which is found in the works of Aristotle. A necessary condition for the constitution of logic in this period was the conscious presence of consecution and co-occurrence, indicating the passage of time, followed by the emergence of measures of this passage and, finally, the concept of linear time itself. Without the emergence of a new mode of communication in culture, namely symbolic communication through language and the concept of linear time, it would be impossible to use logic. The rise of logic is the end result of a long process that took place in ancient Greek culture from Thales to Aristotle. An important moment in this process was the emergence of a new attitude toward the world: the attitude of detachment. Around the time of Plato, it was possible to adopt two attitudes towards the world: the attitude of involvement in the world and the attitude of distancing from the world (Arendt 1991, Pałubicka 2013, 43–93). The fully developed attitude of detachment is represented by Plato. The world of ideas is a new, unknown reality, inaccessible through the senses (empirically), which axiologically and objectively dominates the realm of appearances. In the attitude of involvement (of daily concern), expressed in thought in the first person, communication is performative-imaginative.[2]

The attitude of detachment uses third-person language and requires symbolic communication. In this attitude it is necessary to use language, in short, a set of signs organized at least grammatically (and somewhat later logically). By adopting the attitude of detachment, one begins to think conceptually. First-person communication does not require language as defined above. It is closest to Heidegger's notions of speaking and thinking in terms of handedness, or Wittgenstein's notion of language-game. Expressions are tools for acting in the world, not for describing it, which involves communicating images. Pictorial or performative-imaginative communication is historical, as opposed to linguistic, which operates on concepts.

I will consider the transformation of Greek culture and the concomitant emergence of a new conceptual communication using the example of ancient cultures' attitudes toward myth. I am interested in the "transition" from the attitude of experiencing the myth to the attitude of narrating the successive images contained in the myth. There is a shift in attitude: from participating in the world of myth to adopting an attitude of detachment from the mythic reality being experienced. First, however, let me refer to the most familiar classical concept of linear time.

According to Aristotle, linear time is identified by quantitative measurement.[3] Linear time is objectified in motion, and motion expresses time identified by its quantitative measure.[4] The relation of temporal succession, not time itself, is directly present in motion. The passage of time is understood linearly, i.e. it only exists and we have empirical access to it through quantitative measurement. The ability to

[2] Beside symbolic communication, I also recognize performative-imaginative (pictorial) communication (Lévy-Bruhl, Mazurkiewicz, Cassirer). Communication based on symbolic language requires the operation of a temporal sequence—thus distinguishing between past, present and future time - pictorial communication requires only the present time. Expression-based communication undergoes a series of historical transformations before it reaches the level of abstraction required for logical thought.

[3] It was probably Christian monks (Cistercians) in the Middle Ages who first introduced a more accurate way of measuring time by dividing the day according to the activities they performed. The first mechanical clocks were also installed in monasteries (cf. Carr 2010, 33).

[4] Following the adage that "the scale makes the phenomenon" this quantitative measure of duration objectifies linear time in our culture. For me, it gives the category of time the power of objective cultural existence, and for some humanists and popular thinkers, the power of real objective existence.

measure the passage of time objectifies the existence of time and allows us to recognize it as a real and objective state. Time becomes an autonomous ontological entity, although it is difficult to define other than through its measurement. The quantitative measurement of the passage of linear time is what gives the category of time an objective existence in a given culture. For Czarnowski, only linear time is possible and objective, and such a concept of time exists in scholarly-viewed (scientific) history.[5] The conviction that only linear time is true is still maintained in some scientific disciplines, even outside the humanities. Thus, academic history, the entire humanities, our daily life, and the natural sciences would be impossible without a linear understanding of time.

Returning to the discussion of the origins of the development of linear time in European culture, I will continue to refer to the findings of Czarnowski (1956a) and Topolski (1972). Both researchers, in line with the cognitive perspective of Durkheim's school, presumably considering the level of abstraction in thinking necessary to understand the passage of time, argued that the use of objectified linear time[6] in early traditional societies could be nothing more than our illusion. They cited tribal consciousness, i.e. collective memory, as the main obstacle that would have to be appropriately transformed in order for at least the sense of past time to be anchored in it.

Thus, Stefan Czarnowski, following Henri Hubert, accepted the view that linear time was unknown in early traditional cultures and, therefore, did not play the role it is attributed to in our culture. It neither constrained thought in these communities nor set the rhythm of community life. It did not organize memory through relationships of succession, and it was not treated quantitatively. For those communities time was:

> (…) an environment and a specific quality at the same time. (…) Time was presented as a series of cycles that originate in the realization of concrete qualities and return to those qualities to begin a new realization of those qualities, essentially outside of time (Czarnowski 1956a, 101).

Based on the state of anthropological knowledge at the time, Stefan Czarnowski argued for the existence of cyclical time in primitive cultures, in which the perception of the passage of time is subordinated to a renewing cycle. Primitive communities associated duration with a recurring cycle determined by nature, the stars, or the stages of human or community life. Aristotle perhaps aptly characterized this persistence in the present and the state of consciousness in which the intuition of passage necessary for the constitution of past time can be formed:

[5] Incidentally, it is Aristotle who is credited with the philosophical legitimacy that gives linear time an objective status (Aristotle 1983, 42–47).

[6] Let us note that it was a long way for Stefan Czarnowski, as well as for representatives of this school, to come to believe that the category of linear time has the status of a human construct and that the objective status of linear time is limited to one's own culture; this possibility now appears as a result of the development of reflection in the spirit of cultural relativism and cultural-historical constructivism.

> If we do not mark off any change, but the soul seems to remain in one, indivisible, then as a result we think that there was no time, and if when we perceive and mark off [a change] we say that some time has passed, then it is manifest that there is no time apart from change and change. So it is obvious that time is neither change nor apart from change, and since we are looking for what time is, we have to start from this fact and find out what aspect of change it is (Aristotle 1983, 43).

The activities of the members of primitive societies took the form of ritual cultural patterns, the scope of which encompassed the entire tribe in an obligatory manner. Such an existence did not facilitate the observation of transformations necessary for the emergence of an intuition of linear time. The perception of cyclical repetition is an obstacle that must be overcome in order for collective memory to be transformed into history in the broader sense (following Czarnowski's notion).

2 Experience and Narratives of Experience: Present Time and Past Time

The myths of early traditional European and non-European societies represent a different approach to time than the one we are accustomed to. Myths can only be attributed to a cyclical understanding of time. Linear time is fundamental in our culture, and when cyclic time is mentioned, it is only within linear time. In myths we find the opposite situation: within the present time there is a repetition of the cycle. Jan Parandowski's *Mythology* (1992) shows that ancient Greek myths contain the theme of cyclicity and cyclic time, but without references to linear time. For example, in the myth of Demeter and her daughter Persephone, Hades, king of the underworld, abducts Persephone, who lives as his wife in the underworld. Demeter despairs at the loss of her daughter, and finally Zeus becomes merciful and orders his brother Hades to release his wife. Before parting with Persephone, however, Hades gives her some pomegranate seeds to eat. Little does Persephone know that she is now forever bound to Hades (Parandowski 1992, 113):

> From then on, she had to return to her husband for three months every year. When Persephone emerged from the underworld, the world was flooded with all the flowers of spring, so great was Demeter's joy. But equally great was the pain of each parting.

Let us examine the myth of the phoenix rising from the ashes (Parandowski 1992, 97–98):

> The phoenix lived for five hundred years or more. It fed on the resin of balsam plants. When he felt death approaching, he built a nest of fragrant leaves and herbs - myrrh, cinnamon, mat grass - on a palm tree. In this nest he would die burnt by the sun's rays. A new phoenix was born from his bones, and when it grew up, it lifted its nest and placed it in the Temple of the Sun.

Similarly, cyclicity is represented in the myths of Prometheus' liver growing back, or Sisyphus' work.

Mircea Eliade (1964) analyzed myths not only from the European cultural area, focusing on the category of repetition. He emphasized the presence of a pattern contained in myths, reproduced in rituals and sometimes rejuvenated by them. Thus, myths have an objective content in which the cyclicity of nature, the cosmos, the stages of human life, or individual human activities appear, and this cyclicity is constantly cultivated in repeated rituals. The analysis of myths reveals the characteristics of cyclic persistence/existence.

Individual and community life takes place in the present. Whatever we move from the present to the past in our culture is always present, or present in a repeating cycle in primitive cultures. Not only early traditional societies lived in the present. Consider, for example, the annual ritual of the Passion of Christ. According to Christians, it should not be a matter of reaching back to the memory of past reality and recalling the myth in memory of a past event, but rather of invoking and bringing to life the suffering of God, of experiencing it anew, so that it is an emotional experience of the event. The content of the myth, evoked in the present, becomes a present reality. The ritual is an encounter with the real, so it takes place in earnest. In the Passion of Christ, the power is manifested (Mircea Eliade's *kratophany*) and through the participation in the content of the myth, it becomes a reality experienced by Christians. The experience, it must be emphasized, is accompanied by a committed participation in the mythical reality of the present. For participants in mythic reality, the goal is not to imitate a past event, but to evoke its experience. For observers of a ritual taking place before their eyes, in which they do not participate or share the beliefs of the participants (if they did, they would be participants according to the rules of primitive societies), what remains is a mere reenactment, an imitation of tradition. In this situation, the ritual has a performative character. It is a set of actions that brings to life the content of the mythic experience. There is still no space for communication that focuses on the telling of the myth. In what cultural context, then, does the change in attitude toward myth from performative to narrative occur?

Consider a situation, probably experienced by each of us, in which we have a strong negative experience, such as a disaster, loss of property, death of a loved one, loss of social prestige, and so on. If our culture operated only in the present, this experience would continue to exist in the present, would be made present and live in the present. At best, this presence could be cyclical. Remembering this kind of experience involves feelings and emotions, exhausts one's strength, and prolonging this state threatens, in our culture, the loss of mental balance. It is not by chance that a popular proverb says that "time is a great healer", and that Freud suggests that we escape from melancholy into mourning, which ends with time and restores the individual to social life. In a culture where there is a past, we have the possibility of "shifting" this powerful experience to a past time, to a past world, and only remembering the experience. In European culture, as late as the fourth/fifth century AD, the relationship between present time, past time, and future time was thoroughly examined, indicating the axiological supremacy of present time. St. Augustine (1997, 294–306) tries to give the status of real existence only to the present, which is close to the eternal time of Christianity. Past time and future time exist only in the human mind as memory and expectation, respectively.

Remembering, understood as reaching back to the linguistically grounded past, is less emotionally intense than constantly making the present and reliving the images in memory. This "shift" is made possible by constructing one's own narrative about the event and experience, making the experience already past in the story. The narrative is the translation of a performative experience into a story about it, which is communicated outside, for example to one's neighbors. The communicative act, however, is a peculiar repetition of the original experience, because it is carried out with different means of expression, guided by different rules. Being a repetition in a different order (in a spoken narrative), the experience changes its emotional expression, it gains a much lower intensity both for the participants in the communicative interaction and for the storyteller of his/her experience. At the same time, the transformation of the communicative act, which dominates in the experience of the present time, into the past time, and the localization of the present in the past becomes noticeable.

It is important to bear in mind that the transition from an event to a narrative about an event, in addition to the acquisition of emotional distance (besmirching of experiences), takes place through language and in language, through the acquisition of the linguistic awareness of a set of grammatically ordered signs. This is a translation of an activity, a set of events, into a narrative about the activity, in which a change of ontological category occurs: from an activity into a story. Although human activity is intentional (subjectively determined by the motives of action), it adheres to the action, to the consciousness (narrative), which becomes independent of the action or the whole event and relates only to the action. The narrative in question is accompanied by the belief that it is a kind of repetition, by other means and on another level, of the original sensations.

I will now turn to an analysis of the more advanced stage of translation and repetition, emphasizing the preservation of the original in translation. David Hume (1963, 18) made the following distinction between *impression* and *idea*:

> By the term impression, then, I mean all our more lively perceptions, when we hear, or see, or feel, or love, or hate, or desire, or will. And impressions are distinguished from ideas, which are the less lively perceptions, of which we are conscious, when we reflect on any of those sensations or movements above mentioned.

The above definition shows that Hume distinguished between the attitude of participants who feel, love, hear, etc., in the course of their intentional life activity, and the theoretical attitude of observing and reflecting on their impressions. These two attitudes constitute Humean perception. This kind of "strangeness" of Hume's empiricism was also pointed out by Gilles Deleuze in his first work on Hume (Deleuze 1991), and recently recalled by Banasiak (2010, 224) who claims that "repetition changes nothing in the object being repeated, but it does change something in the mind contemplating it.".

Deleuze (1991, 86–88) introduces the notion of difference between experiencing and reflecting. In his own way, Deleuze notes in Hume's conception the two different attitudes to experience that I have already mentioned: direct experiencing and re-experiencing (repetition) in reflecting on the experience. Reflecting on an experience can take the form of a narrative, but it can also, as in Hume's case, take the form of a descriptive, declarative sentence. Hume's thesis of repetition becomes clear when we realize that in perception there is a translation of sensation (source) into consciousness, linguistically articulated in the form of an observational sentence. The transition to the level of language and consciousness does not change the experience, but "transfers" it to another level of abstraction, establishing sensory perception. Sensory perception, expressed in observational descriptive sentences, becomes the primary source of our knowledge as well as its legitimacy. A vivid experience does not transcend the horizon of the present; the transcendence of the present occurs in the linguistic form that places the experience in the past linear time:

> But we note that the movement of self-development and becoming-other is double: the subject transcends itself, but it is also reflected upon. Hume recognized these two dimensions and presented them as the fundamental characteristics of human nature: inference and invention, belief and artifice (Deleuze 1991, 85).

3 The Path from Experience to the Narrative of Experience in Olga Freidenberg's Conception

Two concepts of myth emerge from the discussion so far. In the first, myth is a world, an (imaginary) universe in which the community lives; it is a world in which the members of the community realize their goals. In the second, myth is the narrative of this experienced world, that is, the story of events that belong to myth in the first sense. Olga Freidenberg—a Soviet intellectual who studied the origins of the earliest forms of storytelling, which, according to her, gave rise to literary creativity—emphasized the differences between the two intentionalities of myth. According to Freidenberg (1997, 63).:

> Narration is conceptual myth. But myth was everything: thought, thing, action, being, word; was the sole form of perceiving the world both as a whole and in each separate part. Narration is only a story. And a story "about", and not a myth itself in its subject and object. True, narration retains the entire former inventory of myth in the form of things, living creatures, connections, motifs. But here they have already become characters, scenarios, plot; they are no longer "object" (spatial) and "visible" nature, inseparable from man. Myth which was self-sufficient, became an object of representation; it began to be recreated as something secondary, something created by the subjective, illusory sphere, and from an authentic category it turned into an "image", a "fiction", merely imitating reality.

If we were to use the modern understanding of communication, such as J.L. Austin's concept and other philosophical perspectives, then myth would have two different meanings: performative and descriptive. In the performative sense, myth would express an active stance toward the world, comparable (as indicated earlier) to the late Wittgenstein's notion of the language-game and Heidegger's notion of *speech*, a human stance of engagement with the world. The dual intentionality of myth in a Heideggerian perspective could be characterized as follows. On the one hand, myth would be *the home of being*, and on the other, it would be a description of one's dwelling in the world, i.e., a created narrative of one's being in the world.

Freidenberg further notes that when myth is accorded the form of a narrative, there is a change in the perception and understanding of time in a given culture:

> Narration appears when the past is separated from the present, the world from the other. It becomes an imitation-illusion of that which takes place on earth, everyday real in form, with a special space and a special time, situated "far away", beyond the boundaries of the earth. The past reflects the idea of a removed and frozen time ("Once upon a time, in a far away land…"). As long as there is no sense of duration, there is no narration; I mean as long as there is no sense of changing time, which gives shading to events (Freidenberg 1997, 56).

The shift in approach to myth, from myth-performative to myth-narrative, requires, according to Freidenberg, at least an awareness of duration and transience, hence a distinction between present and past time, and an understanding of the notion of fiction as in fairy tales and legends. This distinction, together with the emergence of narratives, undoubtedly brought communities closer to capturing their collective memory in temporal order. An important stage in the transformation of collective memory into history, in Czarnowski's broader sense, has been reached.

Freidenberg implies that the transition from myth-performative to myth-narrative is accompanied by the emergence of a realm of subjectivity that includes fiction of a subjective nature. This fiction includes "painting" in terms of representations that emerge from the narrative. It is difficult to imagine the emergence of art, the primary forms of literature, without the use of fiction. Without a cultural context that allows for the creation of fiction, art as a cultural field could not have emerged, according to many theorists. This important point is also made by Jerzy Kmita in his essay *Magiczne źródło kultury (The Magical Source of Culture)* (1984, 27–30). He argues that art is created through the repetition in consciousness of the order of metonymic and metaphoric actions that characterize myth-performance. The transition explored by Freidenberg also finds its counterpart, for example, in Hume's repetition mentioned earlier, or in W.V.O. Quine's concept of the indeterminacy of translation.

When the dominant meaning of myth begins to be the retelling, recollection, and representation of past events, the role of verbal communication increases.[7] Prior to the mythic narrative phase, the myth functions in the form of a collective image that is its home and is therefore known to all members of the community. Since understanding and consensus in the process of communication depend on shared meanings—and we are considering a phase of cultural development in which there is no written language yet—communal imaginaries provide the necessary background for communication by offering the necessary communal meanings. Communicative

[7] Narration is linguistic in nature. Language, according to linguists, is derived from speech. Its origins are linked to the imitation of speech. First there appeared (in ancient Greece) a phonetic alphabet (a record of all articulated sounds). Phonetic writing was used to record speech without separating words, in a single string that was read aloud. After the fall of the Roman Empire, language (writing) freed itself from the imitation of oral culture. At the beginning of the second millennium, the rules of word order began to take shape. Syntax became the bearer of meaning (cf. Saenger 1997, 14).

interactions are the place where not sentences (descriptive beliefs) but rather notions (mental images) or their elements, known in a given community, are invoked and manipulated. Verbal articulations serve to evoke these shared images and are subordinated to them. The word in this arrangement is part of the imaginary (*pars pro toto*) and, separated from the image, represents nothing. There is no symbol or symbolic communication.

The transition from the collective imagination to the narrative of the event is thus a variation of the translation of images into expressions-concepts. Olga Freidenberg showed one way of making such a translation. She presented this process by analyzing the oldest literary forms: legends, myths, fables. She also talked about the birth of tragedy and poetry in ancient Greece. In the Polish humanities, the process of symbol formation in culture, from axiological symbolization to semantic symbolization, was studied by Jerzy Kmita (1987, 2005) and Andrzej P. Kowalski (1999).

For a community to embrace tradition in the form of history in the narrow sense (according to Czarnowski), other important cultural transformations must occur in addition to those already mentioned. They concern the dominant world view, the way of thinking and communicating. I will refer to the principle of identity and contradiction, whose adherence to communication is fundamental to the practice of the sciences, including history, and thus to the construction of linguistic creations (theories, narratives) properly ordered by linear time.

4 Opposition, Contradiction, Time, Logic

Philosophy, which originated in ancient Greece, and which for Leszek Kołakowski is a conceptual art, exploits the results of the transition from images to concepts, but also documents this transition. The translation of image into concept illustrates the historical moment in culture when the principle of identity and contradiction is being constituted. Without respecting this principle, any philosophical communication and reflection, which—for the time being—takes place only in language, is not possible.

The principle of identity, whose formulation is attributed to Parmenides, states that *this is being, for it is possible for it to be, while nothing is not.* Every being is identical with itself, say "this man here is identical with himself as a man". It follows from the above sentence that "this observer as a non-human does not exist." Therefore, it is not the case that this observer is a human being and is not a human being at the same time (skipping the law of excluded middle). Kazimierz Ajdukiewicz (1965, 91) articulates the principle of contradiction as follows:

> The principle, which logicians call the principle of contradiction, states that two contradictory propositions, i.e. propositions in which one contradicts the other, cannot be true at the same time. The simplest ontological formulation of this principle is that something can never be and not be at the same time. If we assert something, i.e. that things are so and so, indicated by the letter "p", then the principle of contradiction in its ontological formulation can be expressed by the formula: it is never the case that p and not-p exist at the same time.

The law of contradiction can also be presented as a consequence of the principle of identity. It seems that the derivation of the principle of contradiction from the principle of identity is closer to the historical stages of formation of logical thinking in society. It should be noted that both of the above principles are essential for the formation of coherent narratives, for the emergence of ancient logic and language working with abstract concepts.

It should be remembered that behind the transition from imaginative to conceptual communication there is a significant change in the carriers of messages. Logical principles are fulfilled by propositions, beliefs, or declarative, descriptive, or evaluative statements. They require a language in which the symbol functions. The word is a symbol that refers to something that exists outside the word. In other words, words represent something in the world. The basic component of statements are concepts that apply to (simply put) reality. The assertion of the occurrence of contradiction and identity is profiled by language and its syntactic rules. Language imposes restrictions and dictates requirements, the fulfillment of which determines the content of beliefs. In magical-mythical thinking, which preceded logical thinking, metamorphoses were used, which were imaginary, and all transformations took place on images (Kowalski, 2001). Thus, the basic element of the message is the image. Myths according to Mircea Eliade (1964, 351) include:

> (…) double revelation: 1. on one hand, they reveal the polarity of two divine personalities, stemming from a single principle and destined, in several versions, to reconcile in an eschatological *illud tempus*; 2. and on the other, it makes clear the *coincidentia oppositorum* in the deep nature of the divinities, who are by turns, or simultaneously, benevolent and terrible, creative and destructive, solar and ophidian (that is, manifest and virtual), etc.[8]

Before the rise of philosophy in ancient Greece, gods were characterized as figures full of opposites, neither monistic nor dualistic. The opposites made up the nature of a god, i.e., the opposites existed simultaneously in a god who could show favor, love, kindness; but also hatred, harshness, and evil. A god could eventually metamorphose from evil to good and vice versa. These metamorphoses occur between collective images that are consecutive or concurrent in cyclical motion (Kowalski, 2001).

The opposites emphasized by Heraclitus, e.g., night versus day, war versus peace, love versus hate, as well as his principle of "unity of opposites", are not contradictions in the logical sense of which I have spoken. Let me repeat that the latter occur between propositions (statements). Opposites, on the other hand, occur between two opposing images (imaginations). According to Ajdukiewicz (1965, 91), the principle of contradiction is not refuted by the assertion that the same object can accommodate in itself contrasting features. I think the opposite is also true: the contradiction of propositions does not refute the principle of unity of opposites occurring in images.

[8] French version: (…) double révélation: 1o ils manifestent, d'une part, la polarité de deux personnalités divines, issues d'un seul et même principe et destinées, dans plusieurs versions, à se réconcilier dans un illud tempus eschatologique; 2o d'autre part, la coincidentia oppositorum dans la structure profonde de la divinité, laquelle s'avère tour à tour ou concurremment bienveillante et terrible, créatrice et destructrice, solaire et ophidienne (= manifeste et virtuelle), etc.

Undoubtedly, an important step in the emergence of logical thought was the transition from concepts subordinated to the imaginary (according to the principle of *pars pro toto*) to symbolic concepts (referring to something beyond the symbol).[9] This process took place around the time of the emergence of Plato's philosophy. The philosophy of Heraclitus is one of the first known attempts to reflect on processual and metamorphic thinking by means of still pictorial concepts.

It was even thought that opposites undermine the principle of contradiction. Kazimierz Ajdukiewicz argued against this and showed that the principle of unity of opposites does not violate the law of contradiction. The only propositions that are considered contradictory are those in which, for example, one asserts the existence of a certain property in an object and another asserts the non-existence of the same property in the object. For example, if we describe a god as wise, a contradictory statement only declares that this god is at the same time unwise. The fact that he is not wise does not imply any other property of the god, such as that he is evil or unjust. Good and evil, wisdom and stupidity, are opposites, and the contradictory propositions are "God is wise" and "God is unwise". Therefore, it is impossible to adequately translate the principle of unity of opposites into a pair of contradictory propositions. The non-contradictory narrative of opposites requires the use of the concept of the past tense. It is sufficient to state that in one unit of time (t1) object O manifests property C, and in the next unit of time (t2) it manifests the property S. If we remain only in the present time, it is impossible to give a description of an object having contradictory and opposite properties without violating both principles: identity and contradiction. Thus, it can be concluded that the requirements of constructing narratives in a coherent and non-contradictory manner necessitate the use of language at an appropriate level of abstraction, which makes it possible to create constructions based on linear time. Postmodernists try to deal with the problem of the principle of unity of opposites and the principle of non-contradiction in a different way. They relativize. What for some of us would be a positively evaluated trait of an object or human personality might be negative for another recipient. At the same time, however, we should note that it is necessary to abandon the intersubjective description of an object or a human being in favor of the belief that every description is an evaluation, i.e. it is "cemented" with a chosen value.

With regard to the earliest periods of European culture, the search for a solution to the difficulties outlined can be seen in the emergence of philosophy in ancient Greece, which introduced a dualism of attitudes towards the world: the attitude of detachment (theoretical) and the attitude of involvement. Both attitudes follow different poetics of thinking. The attitude of destachment is based on respect for the laws of contradiction, identity, and excluded middle as well as the logic of language and linear time, and should apply to all sciences, including the humanities. The attitude of involvement, on the other hand, is dominated by the imaginative way of thinking, which uses opposites, but only in pictorial communication, in imagination. Historical thinking in its broadest sense first appears in a culture in which

[9] Expressions subordinated to images, according to the principle of *pars pro toto*, I call pictorial concepts, as opposed to symbolic, abstract concepts.

linear time and logic are constituted. If history, or any other discipline of the humanities, is to be a science, and for the time being it is, it must strive to conform to the above principles of the use of written language, and thus to respect the requirements of logical thought and linear time. The first-person attitude of performative involvement operates in the present.

References

Ajdukiewicz, K. 1965. *Język i poznanie. T. 2. Wybór pism z lat 1945–1963*. Warszawa: Państwowe Wydawnictwo Naukowe.

Arendt, H. 1991. *Myślenie*. Warszawa: Czytelnik.

Aristotle. 1983. *Physics. Books III and IV*. Oxford: Clarendon Press.

Banasiak, B. 2010. U źródeł empiryzmu Gilesa Deleuza – D. Hume'a. *Nowa Krytyka* 24–25:221–241.

Carr, N. 2010. *The shallows: What the internet is doing to our brains*. New York: W.W. Norton & Company.

Cassirer, E. 1995. *Symbol i język*. Poznań: Wydawnictwo Naukowe Instytutu Filozofii UAM.

Czarnowski, S. 1956a. Powstanie i społeczne funkcje historii. In *Dzieła*, vol. V, ed. S. Czarnowski, 99–102. Warszawa: Państwowe Wydawnictwo Naukowe.

Czarnowski, S. 1956b. Historia a historia kultury. In *Dzieła*, vol. V, ed. S. Czarnowski, 84–88. Warszawa: Państwowe Wydawnictwo Naukowe.

Deleuze, G. 1991. *Empirism and subjectivity. An essay on Hume's theory of human nature*. New York: Columbia University Press.

Eliade, M. 1964. *Traité d'Historie des Religiones*. Paris: Payot.

Freidenberg, O. 1997. *Image and concept: Mythopoetic roots of literature*. Amsterdam: Harwood Academic Publishers.

Hume, D. 1963. *Enquiries concerning the human understanding and concerning the principles of morals*. London: Oxford University Press.

Kmita, J. 1984. Magiczne źródło kultury. *Odra* XXIV:24–30.

Kmita, J. 1987. *Kultura jako przedmiot współczesnej refleksji filozoficznej: wykład wygłoszony w Urzędzie Rady Ministrów 1987.05.18*. Warszawa: Polska Akademia Nauk.

Kmita, J. 2005. Symbolizowanie jako relacja aksjologiczna oraz jako relacja semantyczna. In *Aksjologiczne źródła pojęć: zbiór rozpraw*, ed. G. Dominiak, 13–22. Bydgoszcz: Wydawnictwo Epigram.

Kowalski, A. P. 1999. *Symbol w kulturze archaicznej*. Poznań: Wydawnictwo Naukowe Instytutu Filozofii UAM.

Kowalski, A. P. 2001. *Myślenie przedfilozoficzne*. Poznań: Wydawnictwo Fundacji Humaniora.

Lévy-Bruhl, L. 1910. *Les fonctions mentales dans les sociétés inférieures*. Paris: Félix Alcan.

Mazurkiewicz, J. 1958. *Wstęp do psychofizjologii normalnej*. Vol. II. Warszawa: Państwowy Zakład Wydawnictw Lekarskich.

Pałubicka, A. 2013. *Gramatyka kultury europejskiej*. Bydgoszcz: Wydawnictwo Epigram.

Parandowski, J. 1992. *Mitologia. Wierzenia i podania Greków i Rzymian*. Londyn: Puls Publications Ltd.

Saenger, P. 1997. *Space between words. The origins of silent reading*. Palo Alto: Standford University Press.

St. Augustine. 1997. *The confessions*. New York: New City Press.

Topolski, J. 1972. *Świat bez historii*. Warszawa: Wiedza Powszechna.

"Archaeology Is Anthropology or It Is Nothing": But What Kind of Anthropology?

Andrzej P. Kowalski

Abstract The article addresses the significance of ethnological inspirations for archaeological reconstructions of prehistoric cultures. Sixty years later, Lewis Binford's postulate to make archaeology "anthropology of the past" takes on new, unexpected meaning. In the past, archaeology was often seen as a "handicapped" discipline due to its limited use of ethnological data. Meanwhile, anthropology itself has undergone a massive transformation—the study of traditional ("primitive") societies has been marginalised, and modern cultural anthropology now focuses on entirely different issues. Currently, an archaeologist would face a significant challenge in consulting the results of their research with cultural anthropologists. Anthropology is progressively losing its value as a source of cognitively significant solutions for archaeologists. Among ethnological approaches that are interested in the cultural past and offer theoretically mature proposals for archaeologists, structuralism and cultural semiotics remain.

The main thesis of the article is that archaeology has already become the anthropology of primitive societies. Research into the past of these cultures will never disappear—at the very least, it will remain within the scope of archaeologists' interests. They also face the challenge of independently constructing theories of primitive societies. If ethnology were to be of any assistance here, it would only be as a historical discipline, providing an archive of works produced during the classical era of its development.

Keywords Ethnological analogy in archaeology · Paradigms of contemporary ethnology and archaeology · Theory of premodern and traditional societies in archaeology

---

A. P. Kowalski (✉)
Institute of Anthropology, Faculty of History, University of Gdańsk, Gdańsk, Poland
e-mail: andrzej.kowalski@ug.edu.pl

A. Marciniak et al. (eds.), *The Poznań School of Archaeology*, Outstanding Contributions to Archaeology, https://doi.org/10.1007/978-3-032-07848-3_7

The question of whether anthropology is still needed for archaeology becomes clearer when we consider archaeology to be a scientific discipline concerned with the cultural past. I will deliberately leave aside here the tasks fulfilled by field archaeology, museum archaeology, or "supervisory" archaeology serving the purposes of cultural heritage protection. Nor will I address the status of archaeology as it seeks a place among the experimental natural sciences. One might get the impression that archaeology today is increasingly isolated in the study of the culture of so-called primitive societies. In auxiliary sciences of archeology, such as linguistics, religious studies, art studies, sociology, and especially ethnology, there has been a slow but inevitable shift away from their interest in prehistoric cultures. I would like to focus more on the importance of ethnology and cultural and social anthropology in their relationship to archaeology. I do not intend to recapitulate the history of mutual inspiration of these disciplines (Buchowski et al. 2012; Kowalski 2021). I would like to return to only selected aspects of these connections in order to demonstrate some of the consequences of changes within anthropology, especially as it is practiced in Poland, which make it somewhat less promising for archaeologists to make use of current anthropological output.

The pretext for these observations is that it has recently been 60 years since the publication of Lewis Binford's article *Archaeology as Anthropology* (1962), in which the American archaeologist reiterated the remarkable statement: "American archaeology is anthropology or it is nothing" (Willey and Phillips 1958). At the time, Binford's assertion seemed legitimate, since archaeology has yet to develop its own social theory. Although it includes the concept of archaeological culture, archaeology is primarily an analytical and organizing category for collecting relics and artifacts. It turns out that this concept belongs more to the language of museological classifications than to the categories of cultural research. If archaeology lacks the legitimacy of an autonomous theory of culture of primitive societies, then the depiction of the life of such communities must be based mainly on ethnological concepts. The problem is that, on the one hand, archaeologists raise objections to the possibility of using ethnological knowledge in interpreting prehistoric cultures, and, on the other hand, ethnology has the right to address issues according to its own interests, i.e., without considering the needs of archaeology. More than half a century ago, it seemed that the way out of this situation would be the anthropologization of archaeology, for example in the form of ethno-archaeology. Regardless of previous achievements in this field, the situation of archaeology has not changed. Archaeologists who want to learn about prehistoric cultures are still condemned to rely on ethnology alone. Moreover, they must be selective, i.e., they must decide for themselves which version of ethnological research or anthropological interpretation will be most fruitful for them.

The need to anthropologize archaeology boils down to the need to develop archaeological theories of culture or the morphology of past institutions of social life. There are, I believe, two main reasons for this. The first is the long-established intellectual sterility of the practice of merely collecting and presenting archaeological facts. Within the framework of the so-called pre-theoretical model of research conduct, associated with the idea of accumulating empirical data, the knowledge of the cultural past was equated with the number of collected archaeological sources. This quickly proved to be an illusion, especially in the absence of conceptual strategies for transforming data records into a professional interpretation of the state of past cultures (Pałubicka 1971). The other, more momentous reason lies in paradigm shifts within ethnology itself. Let me now discuss the implications of these shifts.

Ethnoarchaeology, as an expression of the need to use ethnology in archaeological interpretation, was an American project. It was developed on a continent with a relatively short history documented by written records, but with an indigenous population that was still numerous. This made it easier for archaeologists to make ethnographic-type observations with a view to applying their findings to archaeology. The development and various applications of ethnoarchaeology in other parts of the world are presented in the book *Contesting Ethnoarchaeologies* (Marciniak and Yalman 2013). In Europe, including Poland, where ethnology has usually had a historical orientation, questions of culture in prehistory were part of the classical strands of ethnology.

The situation has become more complicated for two reasons. First, the split in the original alliance between ethnology and archaeology occurred with the abandonment of the evolutionist definition of the substance of culture. Configurationism had already played an important role in American anthropology. The concept of cultural pattern made it possible to understand culture not as an observable set of accomplishments and material artifacts, but as their organization—a product of a prior ideological order respected by the participants in a given community under study. It was found that certain sets of objects, traces of behavior documented in museums or gleaned from archaeological works, could be variously configured, i.e., mentally stylized, characterized, and felt by past communities according to specific rules that made up that cultural pattern.

In philosophical terms, the cultural pattern is analogous to Kant's *a priori* forms, where it is an *apriorism* that is culturally, historically, but not universally determined. The pattern is therefore a form that shapes the way a community experiences things. This concept makes reference to Herder's *Volksgeist* and Hegelian objective spirit, processed through neo-Kantian thought. They were expressions of a communal way of feeling and valuing the world. The consequences of this way of thinking about culture are far-reaching. It became clear that the collection of objects studied was subject to a certain spiritual structuring—a distinctive axiological tint. For example, the Dionysian configuration described by Ruth Benedict appears as an essential feature of Native American communities, promoting ecstasy, competitiveness, violence, courage, and so on. But this principle cannot be deduced from the mere appearance and arrangement of the objects belonging to these tribes, to be discovered by archaeologists centuries later. And what about the reconstruction of the cultural pattern in the case of people from millennia ago? For we do not know whether the archaeologically documented culture is the remnant of a Dionysian, Apollonian, Faustian, or some other configuration, because this cannot be inferred by archaeologists from material relics alone.

From the point of view of archaeology, the consolidation of an ideological vision of culture in ethnology proved to be a challenge. In Poland, Wojciech Burszta's

(1982) discussion of Ward Goodenough's anthropological concept, according to which culture consists not of material things, but of feelings, knowledge, and views about them cultivated by the participants of the studied community, had a great impact. This concept was in line with the theory of culture proposed by Jerzy Kmita—one of the founders of the "Poznan school of methodology" in the Polish humanities. According to Kmita, culture is a "reality of thought" that implicitly includes socially respected norms and rules for their implementation (Kmita 1983). This notion was close to semiotic and some cognitive accounts of the nature of culture. Archaeologists faced a dilemma that could be reduced to the question of how to access the "mental matrix of meanings" given to artifacts by the socially cultivated minds of past eras. This problem was recognized earlier by Anna Pałubicka, who understood the predicament of archaeologists who wished to treat prehistoric culture as a marker of a non-narrative, social axiology (Pałubicka 1975, 60–62). Understanding culture as an ideational entity created a gap between tangible archaeological objects and their mental representation in the minds of long-extinct peoples.

Second, ethnology, including Polish ethnology, lost interest in the cultural past and cultural images of primitive or traditional communities. In Poland, a manifesto of this reorientation was the article by Michał Buchowski and Wojciech Burszta (1982), in which the deprecation of the "museological" practice of ethnography was reconciled with the promotion of research on contemporary culture (Buchowski and Burszta 1982). This turn initiated a fashion for "ethnography of modernity", which, for obvious reasons, became of no interest to archaeologists. At present, only one university center in Poland offers introductory lectures on archaeology to students majoring in ethnology. Archaeology students do have courses in social sciences, but these are often taught without considering the culture of pre-modern communities. The majority of ethnologists in Poland, who usually refer to themselves as socio-cultural anthropologists, mostly conduct research that is also undertaken by sociologists, cultural scientists, and political scientists. It is difficult to make any accusations here, because in the age of globalization there are basically no more societies of the primitive or traditional type that ethnologists had to deal with when their discipline was born.

Despite the changes that have taken place in ethnology, it is clear that the "demon" of primitive or traditional societies will not yield to an effective exorcism. Prehistoric archaeology will always uncover the remains of such cultures. It will never escape the necessity of interpreting them, it will always be condemned to some kind of interpretation, and even to the development of theories about archaic communities. In light of this, what are the possibilities for archaeologists with ambitions to speak about culture in prehistory? Before attempting to answer this question, I would like to make an important distinction between the theory of archaeological research, i.e. the philosophical reflection on the status and tasks of archaeology as a science of the distant but commonly understood past, and the theoretical reflection on archaeology as a science devoted exclusively to prehistoric culture.

The first of the above mentioned ways of practicing theory in archaeology has been pursued by many outstanding authors in Poland (Danuta Minta-Tworzowska,

Henryk Mamzer, Janusz Ostoja-Zagórski, Stanisław Tabaczyński, Arkadiusz Marciniak, Włodzimierz Rączkowski, Michał Pawleta). I think that the demand for philosophical reflection, i.e. the need to recognize the theoretical component in the interpretation of the prehistoric past, is correct. Not all archaeologists agree. Resistance to accepting a logological view of archaeology stems from liberal ideals that lead to the devaluation of authority, the popularity of post-truth slogans, interpretive voluntarism, and resentment of methodological recommendations that are now considered oppressive. Hence the rejection of those methodologies that identify sources of cognitive value not only in the empirical analyses of researchers, but also in their awareness of theory (Mamzer and Ostoja-Zagórski 2007).

The conditions of cognitive decline have their sources in the changing intellectual culture and are defined by successive breakthroughs in the humanities. Archaeological theory must respond creatively and selectively. As we know, the various turns have implied different understandings of culture and different ways of studying it. A description of this situation can be found in the work of Anna Pałubicka (2013), who notes a shift from attitudes of intellectual distance from the object of study and abandonment of cognitive values in favor of achieving aesthetic experiences caused by "engaging with the world". The linguistic and textual turn presupposed an intellectual factor shaping culture, perceived in it the semantics of local imagery, and subordinated its interpretation to such ontology. Learning these contents or meanings used to require reflection, laborious research, knowledge of exotic or extinct languages, and traditions of research into specific social realities. It was necessary to acquire competencies that prepared one for persistent and arduous intellectual work. But the subsequent turns (iconic, corporate, performative) meant a gradual devaluation of the "ideational" element of culture. After these turns, culture was seen as a set of images, representations, practices, and events without any claim to the veracity of the content shown, leading to concessions from intellectual work to sensory experience.

In a broader sense, the process of de-intellectualization of knowledge after the "performative turn" (Domańska 2007) can be observed. Social reality ceases to be an object of reflection and becomes the scene of various undertakings—a terrain of interventions into the state and course of this reality. Culture has ceased to be a set of "signs of indigenous imagination" and has become a set of practices. These practices cannot be described or "enforced" by the categorizations of language. Rather, it is a matter of re-experiencing them emotionally, of feeling the "past" and the "culture" in ventures that provide some knowledge, but only in passing, that is, at the margins of aesthetic experience. The participatory paradigm, with the slogans of the haptic channel of absorption, reduces culture to a combination of sensations (touch, taste, smell). Objects of direct sensation replace the former objects of discursive knowledge.

The anti-intellectualist tendency indicated here means that archaeology may be slipping from the status of a scientific discipline into a practical form of aesthetics. This is paralleled by the claim of supposedly benevolent inspiration for archaeology from the so-called "new" aesthetics that proclaim the aestheticization of life so widespread today (Zeidler-Janiszewska 1993; Guzowska 2015; Minta-Tworzowska 2019, 158–160). Under such circumstances, the interpretations of archaeologists, deprived of methodological control, are replaced by non-professional narratives circulating in popular culture, resulting from the easy interference of concepts of intellect, imagination, emotion, and overly nobilitated forms of empathy. In the realm of sensations of a somehow felt "past", there are no performances or shows more privileged than others. Their acceptance is measured by the degree of attraction achieved. It was different for narratives that were once judged as logically disciplined (substantiated, strongly factual, intersubjectively agreed statements), or as misguided, idiosyncratic formulations, and therefore scientifically unacceptable.

Much hope was placed in archaeology by the so-called anthropology of things, or "the turn to things". The idea was to rethink the role and function of objects surrounding people in shaping interpersonal relationships, as well as the relationship between people and material objects. There was an elevation of archaeology as a science that deals with material relics, which in some post-humanist perspectives have been given a kind of empowerment in the form of an almost animistic face that appeals to people (Olsen 2013). It is worth noting that certain postulates related to the anthropology of things allude to the aestheticization of life mentioned above. The outdated, supposedly asymmetrical, anthropocentric attitude towards things, dictated by the superstitions of modernist metaphysics, did not make them partners existing on the same level of being with humans, but turned them into objects subjected to scientific research or museum display (Domańska 2008). Restoring dignity to things means that they have been emancipated from the embrace of culture, because they have been freed from the function of axiological signs constraining them. Interestingly, the resulting aesthetic attitude toward things-relics of the "past" is a revival of modernist arrangements. Finally, the emphasis on the textural and sensual qualities of objects leads to a revaluation of archaeology that stops at the viewing of a monument. Thus, through another door, archaeology has a chance to return to the position of phenomenalism (Mamzer 2010, 230–232).

It should be added that the advocates of the archaeological "turn to things" renounce ethnological knowledge about things, especially about the forms of experiencing them in primitive cultures. I believe that the anthropology of things in the archaeological version is still entitled to an emic interpretation, that is, hypothetically taking into account the "point of view" of the participants in the studied culture. It is hard to agree that there is nothing left to be discovered about past cultural perceptions of things, and that archaeology should feel relieved of the obligation to consider achievements in this area in the way that, for example, traditional ethnology does.

As far as images of culture in prehistory are concerned, an expression of their methodological reflection can be found in studies on the application of ethnological analogies in archaeology. Here, too, the prehistoric, general "past" is not always clearly separated from the exclusively cultural past. The renowned Polish ethnologist Aleksander Posern-Zieliński (1987, 1997), to whom archaeology owes many important insights, discussed these issues at a time when ethnology was moving away from the study of primitive communities (Posern-Zieliński and

Ostoja-Zagórski 1977). He made the important claim that analogies should not be used selectively on the basis of superficial similarity of ethnographic and archaeological data, but that archaeologists should consider the place of ethnographic objects in the structure of the "living" culture in their interpretations. Only strong correlations of cultural elements found by ethnologists can be used by archaeologists to infer prehistoric facts. Similar recommendations appeared in an article by Andrzej Prinke (1973, 63), who further suggested that archaeologists interested in ethnographic facts should not resort to ethnological syntheses or "second-hand" studies, but should derive their knowledge as much as possible from the direct reports of ethnologists in the field.

I once described three ways in which archaeologists use ethnology. The first is substitution analogy, which amounts to the common procedure of simply substituting known ethnographic facts for archaeological data in order to interpret the latter. The second, which I described as structural, refers to the conscious choice of a theoretical orientation in anthropology as a component of the *tertium comparationis* in the analysis of archaeological facts. In such a situation, the archaeologist chooses to interpret, respecting the requirements of the chosen research trend, such as phenomenology, structuralism, etc. The third is epistemological analogy, which is the result of recognizing the theoretical, philosophical, or generally non-empirical dimension of the study of the culture of primitive societies (Kowalski 2000). Similar concepts can be found in an article by M. Rajewski (2006), perhaps the last voice in Poland on ethnological analogy in archaeology.

There is also another way of developing archaeological theory, which is to deal only with the cultural past, i.e. culture in prehistory. We already know that ethnology, or contemporary anthropology as it is practiced in Poland, is not useful in this respect. However, it is appropriate to try to answer the question of what ethnology remains useful for archaeologists in view of this. Do ideological conceptions of culture preclude the possibility of proposing reconstructive research? In addition to K. Moszyński's classical and critical evolutionism, a reconstructive attitude was offered by diffusionist cultural-historical ethnology (not to be identified with so-called cultural-historical archaeology). This current developed analytical criteria that facilitate the identification of so-called strong correlations in a group of cultural objects assigned to a particular cultural circle or type of culture. In Poland, an interesting adaptation of reconstructive ethnology was demonstrated by the Polish ethnographer Stanisław Poniatowski. He is the author of important, though now somewhat forgotten, theoretical reflections on the relationship between historical ethnology and archaeology—the so-called parallelism of cultures distinguished by ethnology and archaeology (Poniatowski 1948). He also wrote an interesting study of Polish folk culture in terms of structural identification of its "relic" elements (Poniatowski 1932). He noted that in the archaeological analysis of relics, it is easiest to study their structural (material, technology) and functional (ways of use) aspects, regardless of the level of organization of the social system. Poniatowski (1946) was aware that the ideological aspects, which are responsible for the valorization of products by their original consumers, are the most difficult to interpret.

It seems that, despite the theoretical dilemmas mentioned above, it is mainly anthropological structuralism and cultural semiotics that remain at the disposal of archaeologists in the study of culture as a "reality of thought". These two orientations constitute the most recent and probably the last layer in the stratigraphy of ethnological achievements, which do not study the images of modernity, but the culture of primitive societies. In Poland, as part of creative adaptation of structuralism in the 1970s, we can find examples of the interpretation of non-European cultures and indigenous rural traditions with methods and results useful to archaeologists (Stomma 1981, 132–135, especially on the diachronic dimension of structuralism—Wasilewski 1981). Worth mentioning here are works on the sign system of the traditional world view (Tomicka and Tomicki 1975; Stomma 1986), and especially on the symbolism of rituals (Stomma 1975, 1976; Ciołek et al. 1976). Changes in Polish ethnology at the time included contacts between anthropologists and archaeologists, which resulted in an attempt to establish ethnoarchaeological teaching at the University of Toruń (Kowalewski 2016). This was perhaps the last piece of long-term inspiration with ethnology in archaeology (Kowalski 1991).

In the process of making theoretical proposals, it is not only the reception of literature on the subject that is important, but also various scientific activities. In Poland, opportunities to discuss theoretical issues in archaeology were provided by symposia organized by Bogusław Gediga in Biskupin. Similar events were the Funeralia Legnickie organized by the archaeologist J. Wrzesiński and physical anthropologists. Other activities included the Eidolon Project. Archaic Culture in the Mirror of Words and Things, and subsequent editions of Aesthetics in Archaeology, coordinated by D. Minta-Tworzowska. These were meetings that brought together archaeologists, philosophers and anthropologists. Currently in Poland, only archaeologists in Poznań pursue theoretical archaeology, thanks to which they persistently try to uphold the ideals of *integrated humanities* that prevailed in the Poznań school of methodology in the 1970s (Minta-Tworzowska 2015, 2021; Mamzer 2020).

Unfortunately, there is no permanent agenda for research and exchange of experience in the field of prehistoric anthropology, paleoethnology, or cultural archaeology. Polish archaeology legitimizes successful applications of concepts from religious studies, reflections on art, or analysis of social structures. However, it often does so without recognizing their theoretical basis. As already mentioned, the main reason for this is the change in ethnology, which is no longer an expert discipline on the culture of primitive societies. As a result, a somewhat ironic situation has arisen in which there will soon be a shortage of ethnologists to support archaeologists. Moreover, I dare say that it will be archaeologists who will become the "herdsmen" of cultural issues, the study of which was once the specialty of ethnology. They will also face the challenge of developing their own theories of primitive societies. If ethnology is to be helpful here, it will be only as a historical discipline, providing an archive of works produced during the classical period of its development.

References

Binford, L. R. 1962. Archaeology as anthropology. *American Antiquity* 28 (2): 217–225.

Buchowski, M., and W. Burszta. 1982. Teoretyczny aspekt etnograficznych badań nad kulturą. *Lud* 66:31–44.

Buchowski, M., H. Mamzer, A. Rozwadowski, and A. Posern-Zieliński. 2012. Antropologia a archeologia. In *Przeszłość społeczna. Próba konceptualizacji*, ed. S. Tabaczyński, A. Marciniak, D. Cyngot, and A. Zalewska, 686–720. Poznań: Wydawnictwo Poznańskie.

Burszta, W. 1982. W świecie ideacyjnych kodów - propozycje Warda Goodenough'a. *Lud* 66:45–60.

Ciołek, T. M., J. Olędzki, and A. Zadrożyńska-Barącz. 1976. *Wyrzeczysko: o świętowaniu w Polsce*. Warszawa: Ludowa Spółdzielnia Wydawnicza.

Domańska, E. 2007. “Zwrot performatywny” we współczesnej humanistyce. *Teksty Drugie* 5:48–61.

Domańska, E. 2008. Humanistyka nie-antropocentryczna a studia nad rzeczami. *Kultura Współczesna* 3:9–21.

Guzowska, B. 2015. Estetyzacja – nowa matryca kultury. *Człowiek i Społeczeństwo* 39:57–67.

Kmita, J. 1983. Kultura jako rzeczywistość myślowa. In *Teoria kultury a badania nad zjawiskami artystycznymi*, ed. T. Kostyrko, 15–30. Warszawa: COMUK.

Kowalewski, J. 2016. Antropologia kulturowa i archeologia: etnoarcheologia w Toruniu. *Rocznik Antropologii Historii* 6 (9): 349–387.

Kowalski, A. P. 1991. Z metodologicznych problemów wspólpracy archeologii z etnologią w badaniach symboliki. In *Interpretacje archeologiczne a etnologiczne koncepcje kultury*, ed. K. Przewoźna-Armon, 13–21. Toruń: Wydawnictwo Adam Marszałek.

Kowalski, A. P. 2000. Archeologia kulturoznawcza. Projekt dyskursu “negatywistycznego”. In *Archeologia w teorii i praktyce*, ed. A. Buko and P. Urbańczyk, 115–128. Warszawa: IHKM PAN.

Kowalski, A. P. 2021. Interpretacje archeologiczne w kontekście etnologicznych teorii kultury. In *XXI Sesja Pomorzoznawcza*, T. I., ed. J. Gackowski, K. Adamczak, E. Bokiniec, M. Weinkauf, M. Markiewicz, and D. Bienias, 61–68. Toruń: Wydawnictwo Uniwersytetu Mikołaja Kopernika.

Mamzer, H. 2010. Źródła archeologiczne: artefakty czy językowe o nich wypowiedzi? *Humanistyka i Przyrodoznawstwo* 16:227–250.

Mamzer, H. 2020. “Mowa obrończa” metodologii archeologów Uniwersytetu im. Adama Mickiewicza w Poznaniu. *Folia Praehistorica Posnaniensia* 25:109–148.

Mamzer, H., and J. Ostoja-Zagórski. 2007. Orientacje badawcze w polskiej archeologii. *Nauka* 1:131–148.

Marciniak, A., and N. Yalman, eds. 2013. *Contesting Ethnoarchaeologies. Traditions, theories, prospects*. New York: Springer.

Minta-Tworzowska, D. 2015. Czy współczesna i przyszła archeologia będzie “potrzebowała” teorii? *Archeologia Polski* 60:21–37.

Minta-Tworzowska, D. 2019. The program of “new” aesthetics regarding the expectations of archaeologists. *Folia Praehistorica Posnaniensia* 24:149–163.

Minta-Tworzowska, D. 2021. Are we where we wanted to be? Modernist tendencies versus the postmodern reality of archaeology. Some remarks on the methodology of archaeologists at Adam Mickiewicz University in Poznań. In *Treasures of time. Research of the Faculty of Archaeology of Adam Mickiewicz University in Poznań*, ed. D. Żurkiewicz, 453–473. Poznań: Adam Mickiewicz University.

Olsen, B. 2013. *W obronie rzeczy. Archeologia i ontologia przedmiotów*. Warszawa: Instytut Badań Literackich PAN.

Pałubicka, A. 1971. W sprawie “aprioryzmu” i empiryzmu w archeologii. *Fontes Archaeologici Posnaniensis* 22:252–255.

Pałubicka, A. 1975. O dwóch pojęciach kultury. In *Wartość. Dzieło. Sens. Szkice z filozofii kultury artystycznej*, ed. J. Kmita, 60–72. Warszawa: Książka i Wiedza.

Pałubicka, A. 2013. *Gramatyka kultury europejskiej*. Bydgoszcz: Epigram.

Poniatowski, S. 1932. Etnografja Polski. In *Wiedza o Polsce III*, 191–334. Warszawa: Wydawnictwo "Wiedza o Polsce".

Poniatowski, S. 1946. *Metoda badania genezy wytworów kulturowych w etnologii*. Lublin: Towarzystwo Ludoznawcze.

Poniatowski, S. 1948. Zagadnienie paralelizacji kultur paleolitycznych z kulturami etnologicznymi. *Lud* 38:28–34.

Posern-Zieliński, A. 1987. Etnoarcheologia. In *Słownik etnologiczny. Terminy ogólne*, ed. Z. Staszczak, 82–84. Warszawa/Poznań: Państwowe Wydawnictwo Naukowe.

Posern-Zieliński, A. 1997. Antropologia kulturowa wobec archeologicznej rekonstrukcji społeczeństw pradziejowych. In *Jakiej archeologii potrzebuje współczesna humanistyka?* ed. J. Ostoja-Zagórski, 93–111. Poznań: Wydawnictwo Instytutu Historii UAM.

Posern-Zieliński, A., and J. Ostoja-Zagórski. 1977. Etnologiczna interpretacja i analogie etnograficzne w postepowaniu badawczym archeologii i prahistorii. Uwagi krytyczno-polemiczne. *Slavia Antiqua* 24:39–71.

Prinke, A. 1973. Możliwości porównawczego stosowania danych etnograficznych w archeologii. *Etnografia Polska* 17 (1): 41–66.

Rajewski, M. 2006. Antropologia i archeologia: wybrane problemy współpracy. In *Rola i znaczenie nauk pomocniczych archeologii*, ed. W. Kozak-Zychman and J. Gurba, 81–99. Lublin: Instytut Archeologii UMCS.

Stomma, L. 1975. Rytuał pokładzin w Polsce. Analiza strukturalna. *Etnografia Polska* 19 (1): 83–93.

Stomma, L. 1976. Mit Alkmeny. *Etnografia Polska* 20 (1): 99–110.

Stomma, L. 1981. Metoda strukturalna w etnologii. In *Metody etnologii*, ed. E. Z. Schulz, vol. I, 119–135. Warszawa: Wydawnictwa Uniwersytetu Warszawskiego.

Stomma, L. 1986. *Antropologia kultury wsi polskiej XIX w*. Warszawa: Instytut Wydawniczy PAX.

Tomicka, A., and R. Tomicki. 1975. *Drzewo życia. Ludowa wizja świata i człowieka*. Warszawa: Ludowa Spółdzielnia Wydawnicza.

Wasilewski, J. S. 1981. Semiotyczne badanie symboliki w etnologii. In *Metody etnologii*, ed. E. Z. Schulz, vol. I, 137–157. Warszawa: Wydawnictwo Uniwersytetu Warszawskiego.

Willey, G. R., and P. Phillips. 1958. *Method and theory in archaeology*. Chicago: University of Chicago Press.

Zeidler-Janiszewska, A. 1993. Estetyka jako współczesna "filozofia pierwsza"? *Kultura Współczesna* 2:142–150.

Philosophical Reflections on the Sources of Prehistoric Symbolism

Jacek Woźny

Abstract When examining the philosophical foundations of prehistoric symbolism, we can adopt two analytical paradigms. On one hand, there is "reductive hermeneutics"; on the other, "hermeneutics that establish symbolic meaning". Each of these paradigms is accompanied by a distinct set of conceptual tools. Reductive hermeneutics focus on the semantic aspects of symbols, seeking to demystify their meanings through approaches such as structuralism and historical materialism. These methods analyze symbols both synchronically (within a given time) and historically (across time). Conversely, establishing hermeneutics engage with the non-semantic dimensions of symbols, re-mythologizing their meanings by drawing on methodologies such as phenomenology. This paradigm explores the essence of the sacred in a diachronic, or temporally unfolding, manner. The chapter argues that these two hermeneutic approaches are not mutually exclusive but rather complementary. They work together in the interpretive process—what Danuta Minta-Tworzowska refers to as a "dialogue of paradigms" between past cultures and the present.

Keywords Philosophy · Symbolism · Prehistory · Sources · Interpretation

1 Introduction, Reduction and Constitutive Hermeneutics

When examining the philosophical sources of prehistoric symbolism, we can use two paradigms that help in its analysis. On the one hand, there are reductive hermeneutics and, on the other hand, constitutive hermeneutics as approaches to symbolic meanings (Józefów-Czerwińska and Popielska-Grzybowska 2012, 300–317). These paradigms are accompanied by two groups of concepts. Reductive hermeneutics are

J. Woźny (✉)
Institute of Archaeology and Ancient History, Faculty of History, Kazimierz Wielki University in Bydgoszcz, Bydgoszcz, Poland
e-mail: wozja@ukw.edu.pl

A. Marciniak et al. (eds.), *The Poznań School of Archaeology*, Outstanding Contributions to Archaeology, https://doi.org/10.1007/978-3-032-07848-3_8

associated with the semantic side of symbols, demystifying their meanings using, among others, methods of structuralism and historical materialism, studying symbols synchronously and historically. Establishing hermeneutics, on the other hand, are associated with the non-semantic layer of symbols, which remitizes their meanings using, among others, phenomenology, examining the essence of the sacred in a diachronic way. In the article I will try to justify it, we prove that both hermeneutics are not in conflict with each other, but complement each other during the process of hermeneutical interpretation of symbols or, as Danuta Minta-Tworzowska (2000, 51) stated, during the dialogue of the paradigms of the past culture and the present.

Gilbert Durand (1986, 52–53) described reductive hermeneutics by recalling that:

> our times have become aware of the importance of symbolic images in mental life, thanks to the contributions of psychological pathology and ethnology. (…) But if (…) rediscovered the importance of images and thereby broke (…) the tradition of rejecting and enslaving the imaginary, then these doctrines only discovered the symbolic imagination in order (…) to reduce symbolization to something symbolized, and devoid of secrets.

In this sense, the approach of Jerzy Kmita's Poznań school of cultural studies, rooted in Marxist ideas, has a reductive character. Danuta Piotrowska (2000, 37–38) writes about this, stating that

> symbolic culture in J. Kmita's view includes all manifestations related to the awareness of societies (…). For J. Kmita, the two basic areas of symbolic culture are the culture of practically tangible values, including i.a. language, customs and artistic creativity as well as the worldview-forming area, including, among others, magic and religion. In symbolic and cultural communication, according to J. Kmita, the worldview is very important as the source of assumptions for the semantics of this communication.

According to G. Durand, establishing hermeneutics refer to the imaginary, non-semantic layer of meaning of symbols. This researcher came to the conclusion that a symbol in its most authentic manifestations always refers to some kind of transcendence. Symbolic imagination opens access to this sphere, which is not available to any other forms of cognition. Religious, mystical, poetic symbolism and various mythologies provide convincing evidence that man comes into contact with the transcendent through it. Gilbert Durand, criticized for modernizing "spiritualism", opposed the dominance of rationalism, especially scientism and materialism in all its forms. He attributed great importance to the study of myth in the spirit of phenomenology and hermeneutics (Rowiński 1986, 5–17). He was of the opinion that "there is no break between the rational and the imaginary, for rationalism is only one - among many others - of the particular structures polarizing the field of images" (Durand 1986, 97).

2 Reflections on Mythical Trees of Life and Wisdom

Let's recall the image of a tree in our imagination. It can be spread out, radiating power like a thousand-year-old oak or soaring, directing the gaze to the vastness of the sky like a fir tree. In any case, the plant draws its strength from the Earth, which nourishes the root system. When it dies, the whole tree dies, if it is healthy, only Nature and man decide about the limit of growth. It also happens that a modest ash tree turns into a Cosmic Tree. The world of ancient Greece was supported by the mythical Atlas. In Norse and Sumerian mythology, this role is fulfilled by a giant plant. Germanic Igdrasil—the sacred ash tree—with its top reaching to the sky, its branches embracing the whole world. Its roots penetrate deep into three kingdoms: gods, giants and men. The holy ash draws its strength from Earth and Heaven. At his feet there are miraculous springs of youth and wisdom. And from the sky the dew flows, ensuring eternal life. Igdrasil unites Earth and Sky, human and divine, past and future. At the same time, it is indestructible, able to survive even the twilight of the gods (Ros 1969, 206).

The symbolic imagination has gone beyond the horizon of biology. In addition to the natural form, the cosmic tree can take an inverted form. The Indian Upanishads describe the eternal Aswatth whose roots reach to heaven and whose branches extend above the earth. He is pure cosmic Brahman, the one they call the undead (Radhakrishnan 1959, 183). The motifs of the epiphany of a deity through a tree are often found in mythologies. In addition to the examples mentioned above, there are many others similar to this. Attis is a fir tree, Osiris is a cedar, Greek Artemis was associated with cedar and myrtle (Kopaliński 1990, 72). Above all the images there is a fundamental duality of form. The mythical tree takes a natural or inverted form, irrationally rooted in the heaven.

What could be the reason for the opposing perceptions of this phenomenon? In Indian mythology, there is a male tree inverted and a female tree growing out of the Earth (Jakimowicz-Shah and Jakimowicz 1982, 18). However, will we consider the inherent bisexuality of the human species as an explanation when the symbol refers to the sacred idea of the creation of the world? According to an ancient legend, Plato was to claim that man is an inverted plant, whose roots are directed to the sky and the branches to the earth. Reason, knowledge, cognition have their origin in the world of the gods. Esoteric Jewish science describes the Tree of Life which completely illuminates the Sun as it unfolds from above downwards. Mesopotamian Kiskanu, the abode of the god of fertility and civilized knowledge, rises apex towards the ocean (Eliade 1966, 269). However, the essence of the phenomenon is most fully reflected by the indestructible Aswattha, the male tree of the sky. Its roots are in heaven; its leaves are the hymns of the Vedas; who knows him knows the Veda. Donuts are an object of the senses; beautiful flowers with good and evil, and fruits with pleasure and suffering. Ashwattha's trunk is buddhi (reason), if man splits it with the weapon of metaphysical knowledge and if he rejoices in spirit, it will not return. The mythical, inverted tree marks the horizon of reality accessible to human thought (Eliade 1966, 271).

Let us turn our attention to history. The existence of the writer of these words is derived directly from the parents, their ancestors, etc. By devoting time, strength and resources, we will recreate the genealogy of a number of previous generations. Similarly, we can determine the historical development of an airplane, a ship or even an agricultural plow. Of course, the deeper the chronological penetration, the more questionable its results. With time, guesses and hypotheses overgrow the simple typology of forms (Minta-Tworzowska 1994).

The story of the myth from the beginning presents us with the problem of interpretation. Historical analysis of facts is no longer enough to arrive at their meanings. We are entering the realm of sacral symbolism, where there are no unambiguous answers. The image of the Cross may refer to Christian salvation or the pagan God—the Sun (Forstner 1990, 14). The multiplicity of possible interpretations, however, is not the same as the multiplicity of meanings. The cultural message, in this case the symbol of the Cross, makes sense only in a specific history: prehistoric, early Christian or recent. For a member of the primitive community, its early Christian meaning did not exist, just as the meaning of the German swastika would be alien to an inhabitant of ancient Jerusalem. Only our perspective makes it possible to embrace the existing, already revealed images. Of course, this does not imply the assimilation of all previous meanings of the symbol. Just like for our ancestors, one of his characters is available to us. More importantly, through a certain way of interpretation, we create it ourselves (Mamzer 2003, 156).

3 Reflections on the Interpretation of the Symbolic Meaning

Thus, there is a parallel between the dual form of the Cosmic Tree and the ambiguity of history. The history of objects, technical constructions and, finally, the human species resembles a plant rooted thousands of years ago, climbing towards an inscrutable future. Leaves, flowers and buds are new inventions, improvements and discoveries, serving the further material expansion of man in the world. Their history is based on the knowledge of facts, events chronologically ordered from the oldest to the contemporary. But where can we find the meaning of historical sources, objectively existing in our history? Should we, as in the case of a tool, follow the archaizing form, all the way to the first, oldest human pre-product? It seems that the road thus marked leads to the world of hypotheses that are untestable by any measure or method. The tree of meanings originates from ourselves, not from the so-called objective historical sources (Minta-Tworzowska 2008, 41).

I will try to justify the above thesis on the basis of the philosophy of Edmund Husserl and Paul Ricoeur. The first of them asks in the essay On the Origin of Geometry what sense is and where to look for it. Well, it is just at your fingertips, in the horizon of the surrounding and assimilated present (Husserl 1991, 32). It came here thanks to tradition, just like to the generation of our parents or nineteenth-century ancestors. The present of the twenty-first century has taken it over from earlier eras. Husserl noticed that the problem of meaning permeates all of history, which only acquires specific meanings thanks to it. Their interpretation, however, does not result from objective reconstructions, according to many researchers, of facts and events from the past, aimed solely at the implementation of subjective-rational activities (Kmita 1985, 10). Sensory impressions are temporally and spatially individualized in the world, but not the spiritual form itself, called by Husserl "ideal objectivity". It is realized in an endless process of objectification and idealization, in which many people participate and which lasts for generations. Each question concerning its essence is directed backwards, towards the pre-establishment of meaning, but the epistemological basis of this issue is in our reality (Kmita 1985, 9–10).

The interpretation of the symbolic meaning is the most difficult problem. The hermeneutic approach of Ricoeur refers to some extent to the above-mentioned thoughts of Husserl. Paul Ricoeur describes the symbol as a two-dimensional phenomenon, with a semantic and non-semantic face. The first one is subject to linguistic transcription, enabling the use of structural methods. With reference to a myth, the relationships between individual myths are explained, which simultaneously create a given type of semantic code (Ricoeur 1989, 172).

The non-semantic order is most clearly shown by the symbolism of the sacred. It transforms real space-time into sacred space and sacred time. It may well manifest itself in stones or trees as physical carriers of its power. The human experience of this takes place below the level of language, in the preverbal realm. They are reflected in the bond that exists between myth and ritual. The sacred vision of the world in cosmology is recreated thanks to cultic or linguistic behavior (Eliade 1993, 113–117).

We can now see an analogy between the statements of both philosophers, although we should remember about Ricoeur's critical attitude towards Husserl's phenomenology (Rosner 1991, 203). This connection is very subtle, but also interesting. The linguistic, semantic dimensions of the symbol correspond to the "external historicity", susceptible to factual or structural description. The non-linguistic level of the symbol, closest to the essence of the sacrum, refers to Husserl's "internal historicity", as it is, like it, a sequence of internal, individual experiences. The path leading from objectification to idealization of meaning can therefore be compared with the bond between a myth and a perfectly repeating ritual (Minta-Tworzowska 2000, 46–49).

The problem of interpreting the deep meaning of symbolism, similarly to Husserl's search for pre-establishment, results directly from our existential situation. Ricoeur points out that staying on the sidelines of one's own present, omitting all situations in order to "objectify" research, makes it easier to learn, but makes it impossible to understand the symbolic meanings available to us. Their interpretation is of fundamental importance for modern man, who, according to Ricoeur (1986, 329–330) has lost the sensitivity to feel the symbol of the sacred.

If we recognize the contiguity of the terms knowledge, interpretation and reason, the shape of our cognitive abilities turns out to be similar to the ancient Indian Aswattha, Mesopotamian Kiskan and the Tree of Life of esoteric Jewish science. We are therefore not the crown of the Tree of Interpretation, but its soil. It is only from it that rhizomes of symbolic meaning grow, running through ever more distant horizons of the past present: the Enlightenment, the Renaissance, the Middle Ages, until the dawn of time. As Gilles Deleuze and Felix Guattari (2015, 24–25) write:

> a rhizome connects any point with another, and none of its properties necessarily refers to properties of the same nature (…). The rhizome cannot be reduced to either the One or the many (…). It does not consist of units, but of dimensions, or rather directions of movement. There is neither beginning nor end, but always an environment (…) from which it grows and pours out (…). The rhizome is anti-genealogy (…). The rhizome develops through variation, expansion, conquest, capture, seedling.

The plant metaphor of a rhizome of tangled meanings of culture has an equivalent in the organic expansion of the sacred. According to Ricoeur (1986, 14):

> man reads the sacred above all from the world, from the sky, from the Sun and the Moon, from waters and vegetation (...). This cosmos, in turn, loses its specific boundaries, acquires countless meanings, merges and unites as many departments of the anthropocosmic experience as possible (...). For this kind of reality, to be a symbol is to collect in one node, a node of presence, a mass of meaning intentions (...). Thus, the symbol-thing contains potentially countless meanings.

In this cosmological metaphor by Ricoeur, supported by the concept of rhizome-culture by Gilles Deleuze and Félix Guattari, one can see the ambiguity of interpretation of symbolism. Many of its branches run through one chronological horizon. The deeper we go, the more delicate and entangled they become. A researcher realizing such a situation will not be satisfied with only the right answer, but will pursue new questions.

The conclusion drawn from the above remarks indicates the need to take into account the semantic context. The aim of this operation is not to reconstruct the objective reality of a given epoch and the humanistic coefficient of the societies living in it (Kmita 1985, 40–45), but to cover the widest possible interpretation branches. In order to avoid overly entangled meanings, it is beneficial to choose such a range of analyzed symbols that belong to the category of fundamental images. Thanks to this, we focus our attention on a finite number of possible interpretations, although they refer to even further content (Durand 1986, 105–117).

The course of the hermeneutic interpretation process has a multi-stage character. It consists of the essential understanding of symbolism, its structuring and, finally, the critique of semantic contexts. It is not rectilinear and limited by "ultimate knowledge". Instead, it is a kind of spiral in which we move from initial understanding to explanation, and then from explanation to a new, controlled phase of understanding. The interpretation of a symbolic image is circular, because the assumption that we are dealing with a specific whole is necessary to understand a part of this whole (Rosner 1989, 42–43).

The beginning of the hermeneutical process is the phenomenological understanding of symbols. It consists in reviewing all levels of experience and representation that a given symbol unites. This procedure also allows to place it in a certain whole, of the same kind as it, but much more extensive. The phenomenological method enables a basic understanding of the non-semantic layer of symbolism. Between such basic understanding and critical understanding, there is room for the creation of certain types of mythical messages, due to the presence of a semantic aspect in the symbol (Ricoeur 1989, 72–74).

The phenomena described at the first stage of the hermeneutical process are then included in a broader context, and the analysis seeks to discover the relationships between them. As a result, the multiplicity and variety of symbols from the studied range are ordered and the types of symbolic systems with a similar internal semantic structure are indicated. The way of proceeding that should be chosen in this type of research is indicated by the dialectical thought of Ricoeur. Opposing "figures" should first be found, in order to place intermediate varieties between them, because a symbol acquires meaning for us only when it is distinguished from another—opposite. It cannot occur in isolation, but must be part of a wider system (Ricoeur 1989, 136–139). The aforementioned structural binary pairs such expressions as motion-still, light-dark, evil-good, noise-silence, etc. The question of why a given community valorizes certain activities to the exclusion of others may sometimes have no other explanation than through the binary structure of myths and opposing symbols.

After the phenomenological and structural studies, it is time for a proper critique of the studied symbolism. The essential sense of this activity can be better assessed when it is related to the methodological context. Within historical materialism and structuralism, the rational consciousness of man is distinguished from his mythical thinking. The semantic narrative of the myth and the semantic bonds connecting its individual components are read (Kmita 1985, 11). In turn, the non-semantic layer of the symbol in the phenomenological approach reveals the depth of the sacred, which manifests itself as strength, power, and effectiveness. Regardless of the enigmatic nature of these formulations, they have the value that they arm the researcher against any attempt to reduce mythology to the level of social and economic relations (Durand 1986, 119).

Each of these partial hermeneutics analyzes and establishes the laws governing the subject of its study. In the first one, which considers the semantic aspect of symbolism, relations are discovered between the elements of the mythical story. In the second, examining the non-semantic meanings of the symbol, the inner life of the sacred world is considered. In the whole process of hermeneutical interpretation, however, they do not gain complete autonomy, but constitute its necessary stages, because the balancing dynamism of the symbol inscribes its images into two different worlds: the real and the mythical. Therefore, the materialistic-structural demystification of meanings should be reconciled with their phenomenological remitization. Paul Ricoeur justifies both hermeneutics because, in fact, every symbol is double. As a signifier, it organizes itself historically among determinations and causal chains, then it is a "effect", a symptom (Kmita 1985, 29–34). As a bearer of meaning, however, it is directed towards a certain eschatology, as inalienable as all its embodiments in word or object (Durand 1986, 120).

The connections between phenomenology and structuralism, the remitization and demystification of symbolism, have a further reflex in the diachrony and synchrony of research. The diachronic (temporal, historical) description is contrasted with the synchronic (systemic, structural) approach. In a harmonious hermeneutical process, they too, like remitization and demystification, are fused. The role of diachronic analysis of symbolic sources is fulfilled by phenomenology. The above method is unjustly accused of anti-historicism, as it recognizes the importance of historical sources for discovering the meaning of symbols. Instead, he rejects their naturalistic evolution and purely social determination (Waardenburg 1979, 178). The phenomenological method is based on the results of historical research on religion, it captures the wealth of phenomena and representations that can be combined in formal and intentional terms. The aim of this procedure is to understand religious and symbolic facts, to capture their essence, while refraining from making premature judgments about them (Lanczkowski 1986, 59). Phenomena corresponding to the symbolic epiphanies of the sacred in various periods and cultures are discovered. It is from them that deep, non-semantic content becomes sublime.

In each case, epiphanies and hierophanes document a human choice, a distinction, of what is strong, effective and fertile. The timeliness of this and its structure are matched to the worldview of a particular community, valorized by a specific humanistic coefficient. As Jerzy Kmita (1985, 38) quotes Florian Znaniecki:

> this part of cultural phenomena, objects of humanistic research, their fundamental property that as objects of theoretical reflection they are already objects given to someone in experience, or whose conscious activities, we can call the humanistic coefficient of these phenomena.

Jerzy Kmita (1985, 44) creatively updated Znaniecki's concept, stating that

> the humanistic coefficient of a given phenomenon (…) should be understood as the totality of assumptions of semantics assigning this phenomenon to an appropriate communicative unit as its objective reference.

This statement has a constant value for the study of symbolic culture, however, bearing in mind that the non-semantic layers of the symbol, leading to the essence of the sacred, are equally important in the hermeneutical process (Durand 1986, 118–120).

The course of hermeneutic reasoning eliminates the apparent difference between the two ways of treating sources. The synchronic (structural) analysis is rooted in the phenomenological (diachronic) concept of symbolism. Thanks to this, we enrich the broad, horizontal dimension of depicting culture with the content of individual mythological systems that constitute it. The mapping of individual aspects of symbolism in different frames of reference (cultural and ecological systems) allows for different interpretations in terms of content. The phenomenon of the symbol is actualized through the history of culture, co-creating at the same time specific historical systems and their humanistic coefficients (Kmita 1985, 43).

Here again we encounter two hermeneutics: reductive and constitutive (Durand 1986, 52–95). From the cognitive perspective of Kmita, the reality of the Tree has strictly semantic references. He writes that

> the characteristics of states of affairs, objects, and generally speaking, the world of objective references, are expressed in judgments whose truthfulness is positively decided in advance. For example, it can be assumed that the semantics of the modern Polish language attributes to the expression "tree" as its object reference such a set O that the truth of, among others, the judgments expressed, among others, by the following (sentences of the modern Polish language) concerning any element x of this set is presumed: "x is a plant", "x is rooted in the ground", "x is sometimes deciduous or coniferous", etc. Judgments or assumptions of this type are called (…) assumptions of semantics - respectively cultural communication, or the field of cultural communication, or a given fragment this field, or finally - a given communicative unit, such as. e.g. the above-mentioned assumptions of the semantics of the word "tree" in contemporary Polish (Kmita 1985, 42).

4 Conclusion, Dialogue of Paradigms

Let us return to the mythical, inverted Tree, whose sacrality refers to the non-semantic value of symbols. Mircea Eliade (1966) writes about it that it is a kind of expression of the Cosmos. Both in the natural and inverted form, two opposing spheres are connected: heavenly and earthly. Biological life really rises from Mother Earth, but its destiny and meaning flow from heavenly, divine abodes. Man and his works are at the center of these cosmic processes. The path of a researcher interpreting symbolic meanings is consistent with the gradual complication of their meaning. From the form appropriate to our present, more and more numerous, older branches grow (Deleuze and Guattari 2015, 3–25). While investigating, however, we also move against the development of external historicity, in which the first personification of symbolism took place. Therefore, being in both of these currents—the trunks of the Cosmic Tree, we cannot neglect either of them (Wawrzeniuk and Woźny 2022, 13–14).

This opinion is shared by Danuta Minta-Tworzowska (2000, 51), who states that

> in the field of prehistoric symbolism research, as in any other, we rotate in the plane of two hermeneutical circles: past culture and the present, which consist of past and contemporary cultural meanings (…). It seems that it is not so (…) that there cannot be a fusion of these two circles or horizons of life: the past and the present. (…) this fusion takes place due to the fact that every object, or more generally, the text of culture, does not have its original "arche" meaning and we will never get to know it: once in the sphere of culture, it immediately finds itself in the sphere of translations, hence we translate past forms texts of culture into those that suit us today (…). Therefore, we should return to the problem of symbolism in the aspect of the dialogue of paradigms.

Finally, recalling the image of the mythical Tree of Life, we should reconcile the dialectics of cognitive methods (reductive and constitutive hermeneutics), as well as the sense and physical aspects of symbolism. Because always, after explaining the forgotten meanings, we return to the roots of knowledge, enriching our own vision of the world.

References

Deleuze, G., and F. Guattari. 2015. *Tysiąc plateau*. Warszawa: Fundacja Bęc Zmiana.

Durand, G. 1986. *Wyobraźnia symboliczna*. Warszawa: Wydawnictwo Naukowe PWN.

Eliade, M. 1966. *Traktat o historii religii*. Warszawa: Książka i Wiedza.

Eliade, M. 1993. *Sacrum, mit, historia*. Warszawa: Państwowy Instytut Wydawniczy.

Forstner, D. 1990. *Świat symboliki chrześcijańskiej*. Warszawa: Instytut Wydawniczy PAX.

Husserl, E. 1991. O pochodzeniu geometrii. In *Wokół fundamentalizmu epistemologicznego*, ed. S. Czerniak and J. Rolewski, 9–37. Warszawa: Wydawnictwo IFiS PAN.

Jakimowicz-Shah, M., and A. Jakimowicz. 1982. *Mitologia indyjska*. Warszawa: Wydawnictwa Artystyczne i Filmowe.

Józefów-Czerwińska, B., and J. Popielska-Grzybowska. 2012. Wierzenia – religia. In *Przeszłość społeczna. Próba konceptualizacji*, ed. S. Tabaczyński, A. Marciniak, D. Cyngot, and A. Zalewska, 300–317. Poznań: Wydawnictwo Poznańskie.

Kmita, J. 1985. *Kultura i poznanie*. Warszawa: Państwowe Wydawnictwo Naukowe.
Kopaliński, W. 1990. *Słownik symboli*. Warszawa: Wiedza Powszechna.
Lanczkowski, G. 1986. *Wprowadzenie do religioznawstwa*. Warszawa: Wydawnictwo Księży Werbistów "Verbinum".
Mamzer, H. 2003. Strategie konstruktywistyczne w archeologii. In *Konstruktywizm w humanistyce*, ed. A. Pałubicka and A. P. Kowalski, 149–162. Bydgoszcz: Oficyna Wydawnicza Epigram.
Minta-Tworzowska, D. 1994. *Klasyfikacja w archeologii jako sposób wyrażania wyników badań, hipotez oraz teorii archeologicznych*. Poznań: Wydawnictwo Naukowe UAM.
Minta-Tworzowska, D. 2000. Symbole i symbolika z perspektywy badań archeologicznych. In *Kultura symboliczna kręgu pól popielnicowych epoki brązu i wczesnej epoki żelaza w Europie Środkowej*, ed. B. Gediga and D. Piotrowska, 45–54. Warszawa/Wrocław/Biskupin: Polska Akademia Nauk, Oddział we Wrocławiu, Muzeum Archeologiczne w Biskupinie.
Minta-Tworzowska, D. 2008. Teoretyczne problemy interpretacji "sztuki" pradziejowej. Docieranie do znaczenia sztuki pradziejowej. In *Sztuka pradziejowa i wczesnośredniowieczna jako źródło historyczne*, ed. B. Gediga and D. Piotrowska, 23–48. Wrocław/Biskupin: Polska Akademia Nauk, Oddział we Wrocławiu, Muzeum Archeologiczne w Biskupinie.
Piotrowska, D. 2000. Prof. Zdzisław Rajewski (1907–1974) – badacz Biskupina i jeden z pionierów badań dawnej kultury symbolicznej w Polsce. In *Kultura symboliczna kręgu pól popielnicowych epoki brązu i wczesnej epoki żelaza w Europie Środkowej*, ed. B. Gediga and D. Piotrowska, 17–44. Warszawa/Wrocław/Biskupin: Polska Akademia Nauk, Odział we Wrocławiu, Muzeum Archeologiczne w Biskupinie.
Radhakrishnan, S. 1959. *Filozofia indyjska, t.1*. Warszawa: Instytut Wydawniczy PAX.
Ricoeur, P. 1986. *Symbolika zła*. Warszawa: Instytut Wydawniczy PAX.
Ricoeur, P. 1989. *Język, tekst, interpretacja*. Warszawa: Państwowy Instytut Wydawniczy.
Ros, J. 1969. *Heroje Północy*. Warszawa: Wydawnictwo Iskry.
Rosner, K. 1989. Paul Ricoeur – filozoficzne źródła jego hermeneutyki. In *Język, tekst, interpretacja*, ed. P. Ricoeur, 5–60. Warszawa: Państwowy Instytut Wydawniczy.
Rosner, K. 1991. *Hermeneutyka jako krytyka kultury*. Warszawa: Państwowy Instytut Wydawniczy.
Rowiński, C. 1986. Gilbert Durand i jego "wyobraźnia symboliczna". In *Wyobraźnia symboliczna*, ed. G. Durand, 5–17. Warszawa: Wydawnictwo Naukowe PWN.
Waardenburg, J. 1979. Współczesne kierunki w religioznawstwie holenderskim. *Studia Religioznawcze IFiS PAN* 14:177–205.
Wawrzeniuk, J., and J. Woźny. 2022. Wprowadzenie. In *Interpretacje kultury symbolicznej w badaniach archeologicznych*, ed. J. Wawrzeniuk and J. Woźny, 13–14. Warszawa/Poznań/Zielona Góra: Wydawnictwo Naukowe Uniwersytetu Kardynała Stefana Wyszyńskiego, Stowarzyszenie Naukowe Archeologów Polskich.

Shaping Archaeological Awareness and the Idea of Socializing Archeology

Violetta Julkowska

Abstract Archaeological awareness, alongside historical awareness, is becoming an important aspect of social culture, particularly in relation to the heritage of the past. It refers to a specific type of knowledge about archaeological heritage, typically acquired through education. It is also a form of practice, developed through direct interaction with objects and artifacts in their original context (*in situ*), and is linked to the idea of making archaeology more accessible and engaging to the public.

This chapter explores the potential of the concept of *cultural landscape* as regards historical and archaeological education. The integration of this concept with aerial photography is intended to support the visualization of archaeological artifacts within their spatial context, and to promote their thoughtful protection. The core aim of this approach is to enrich historical education with new content that supports the development of archaeological awareness.

Keywords Archaeological awareness · Historical and archaeological education · Cultural landscape · Aerial photography · Socializing archaeology

1 Introduction

The idea of socializing archeology is a new trend in archaeological research and practice, aiming at meeting archaeologists in the social space with people who inhabit the space studied by archaeologists (Kobiałka et al. 2022). The aim of such a meeting is to go beyond the traditional academic approach to the study of material remains and to learn about the expectations and needs of local communities. This type of practice is conducive to the belief that archaeological cultural heritage is a good that should be protected and one can participate in it. While the inclusion of

V. Julkowska (✉)
Faculty of History, Adam Mickiewicz University in Poznań, Poznań, Poland
e-mail: violetta.jukowska@amu.edu.pl

A. Marciniak et al. (eds.), *The Poznań School of Archaeology*, Outstanding Contributions to Archaeology, https://doi.org/10.1007/978-3-032-07848-3_9

local communities in each of the stages of archaeological research is an important step towards socialized archaeology, the basis for broad participation in archaeological heritage seems to be the development of archaeological awareness as an element of historical education and, at the same time, as cultural competence. Without broader archaeological awareness, this participatory approach to archaeological heritage will be occasional and intermittent. For this reason, the postulate contained in this text is to include important elements of archaeological knowledge in school historical education, which for years has been co-creating the image of the past, but is only marginally present in the wider consciousness as archaeological knowledge. This is a step towards sensitizing students and pupils to the presence of archaeological heritage both in the historical record and in their own environment, and shaping a social attitude of co-responsibility for the protection of this heritage. It can be argued that the conscious perception and interpretation of the surrounding cultural reality, educated as a kind of competence at the level of school history education, provides the basis for a permanent change in the attitude of students to diverse cultural heritage, including archaeological heritage.

2 Educational Potential of Archaeological Aerial Photography

The process of socializing archeology and building archaeological awareness takes place both in the form of independent projects and continuous reflection by archaeologists associated with this trend (Pawleta 2020, 2022; Zalewska 2018, 2021a, b).

In the text, I present my own experiences and concepts, created on the basis of cooperation with archaeologists, inspired by the thought of the Poznań School of Methodology. In 2002, together with Włodzimierz Rączkowski, a specialist in the theory and method of using aerial photos in archaeological research (Rączkowski 2002), we prepared and presented at a conference in Canterbury (University of Kent) an innovative model of educational activities around aerial photos in history lessons entitled: *Let's see the past! The potential of aerial photographs for school history curricula* (Julkowska and Rączkowski 2005). The project focused on including aerial photography in historical education as a cognitively attractive element of visualization and interpretation of archaeological knowledge. The aim of the accompanying project was to draw attention to the presence of archaeological heritage in the immediate vicinity, as well as to the need to protect it with the participation of the local community. According to the age of the students and the level of education, the project assumed the use of photographs to represent objects of cultural archaeological heritage in Wielkopolska. Thanks to Rączkowski's aerial photographs, students were able to observe objects visible on the surface of the earth (e.g. the remains of strongholds, former spatial layouts of cities) and hidden below the surface, but leaving traces that can be interpreted thanks to crop- and soilmarks. The project provided for a comprehensive inclusion of aerial photographs and elements of archaeological knowledge necessary for their "reading" into the practice of teaching history, set the goals and methods of didactic work, and provided professional instruction necessary to conduct classes. It was a pioneer project on a European scale, which in 2002–2009 was presented at many conferences and didactic workshops, both in academic and teaching environments, as well as at lectures and meetings open to the local public and local government authorities. A model study showing the use of aerial photographs in teaching history was illustrated with sample photographic materials and a narrative referring to various towns and areas in Wielkopolska (Kościan, Rydzyna, Nowe Miasto at Warta, Szamotuły, Grodzisk Wielkopolski, Plewiska, archaeological sites in the Obra basin, etc.). This model was also optionally included in the years 2002–2012 in the study program in the field of history with a teaching specialization at the Adam Mickiewicz University in Poznań. For me personally, it was a fundamental stage in developing a didactic method and at the same time actively applying this method, combining the analysis and interpretation of archaeological aerial photos with the dissemination of knowledge about archaeological cultural heritage in the immediate vicinity inhabited by the participants of the meeting. The topic, due to its regional and local approach and the visual attractiveness of the presented aerial photos, invariably enjoyed great interest among students and people present at lectures and presentations. Its culmination was in 2009 an international and interdisciplinary conference devoted to photography as a source of knowledge and research into the past: Photography in the workshop of a historian and history teacher, organized at the Adam Mickiewicz University in Poznań jointly by historians and archaeologists (Julkowska 2012). One of the panels was entirely devoted to the role of aerial photography in research and education.

3 The Category of Cultural Landscape and the Transmission of Archaeological Content

Another opportunity to enrich academic and school history education with the awareness of archaeological heritage was the idea of "reading the cultural landscape", for which aerial photographs, showing the analyzed space in the form of panoramic shots, were an important starting point. The impulse to define the cultural landscape as a conceptual category on the basis of academic didactics was the reflection of geographers and researchers of cultural memory (Traba 2009; Myga-Piątek 2012). The opportunity to exchange experiences arose in 2015 during an interdisciplinary, international conference entitled: *Cultural Landscapes. Actors, methods of construction and narration*, organized in 2014 by the Adam Mickiewicz University in Poznań and the Joint Polish-German Textbook Commission of Historians and Geographers (UNESCO). During the conference study trips to Konin and Gocławice were organized, as well as to Żerków, Ląd and Pyzdry, which was accompanied by the reading of previously prepared source materials and aerial photos.

The cultural landscape and its "reading" is an issue poorly recognized in didactic practice, although present in the form of images and visualizations (aerial photographs, iconographic sources). The definition of the category of cultural landscape on the basis of history didactics, which I proposed together with the didactic model of "reading cultural landscape", takes into account various types of textbook narration (Julkowska 2017, 2020). The proposal to conceptualize this category is a reference to the style of reflection of the Poznań School of Methodology and the references of its representatives to the problems of historical education (Topolski 1983, 1995; Pomorski 1991; Wrzosek 1995, 2010; Pałubicka 2006).

(a) **Defining the Cultural Landscape in the Social Sciences and Humanities**

The concept of "landscape" emerged as an idea and developed during the Renaissance, sensitizing artists and philosophers to the aesthetic quality of the surrounding nature and the contemplative aspect of the landscape (Frydryczak 2013). Since modern times, nature has become a landscape experienced in an emotional way or (according to the classic Kantian idea of looking at a distance) resembled the reception of a work of art. At the turn of the nineteenth and twentieth centuries, photographic shots of the landscape appeared (Belting 2012), and geographers, ethnologists, anthropologists, historians and archaeologists drew attention to the active presence of man in this space, accepting the culturally proved human activity in the space inhabited by him as a constitutive feature of the contemporary definition of "cultural landscape".

The cultural landscape became a separate subject of study when the geographer Artur Hettner defined the cultural landscape in the late 1920s as a harmonious whole of terrain, climate, the plant and animal world, and the human world of culture (Myga-Piątek 2012). In the idea of the Hettnerian landscape, the sense and result of human activity were captured in the natural and social reality, which combined the category of humanistic knowledge with the achievements of natural research. This direction developed after World War II as an American school of cultural landscape, parallel to French human geography, which, growing out of Emil Durkheim's sociology, inspired many researchers. One of its creators, Paul Vidal de la Blaché, claiming that the cultural landscape is the result of history and culture, shifted the emphasis towards the humanistic approach (Vidal de la Blache 1926).

The Vidalian idea was continued in Poland in the works of the representative of the Polish school of landscape, Maria Dobrowolska (1948)—the creator of research methods of cultural landscape. She focused mainly on the historicity and social dimension of the changes taking place in the landscape. Defining the cultural landscape as "a synthesis of the activities of society in its geographical environment", she concluded that

> the cultural landscape best reflects the interdependence and connection of the life of nature and human societies, the intensity and speed of transformations taking place over a long historical development (Dobrowolska 1948, 156).

The change in the perspective of geographical research towards the analysis of the current cultural state of the landscape required taking into account the historical moment of the changes taking place, so that it was possible to capture cultural layers from different historical epochs. Cultural transformations taking place in space were interpreted by humanistic geography as a kind of spatial stratification. The approach of humanistic geography to the cultural landscape is close to the position of cultural history (especially archeology and cultural anthropology), for which the landscape is a form of cultural heritage and a historical place that allows to define the identity of a given community (Barford 1998, 2005; Aston 1999). Historical sciences and geography also have in common the attention to the settlement dimension of landscape research, which introduced further concepts to the discussion, including Heidegger's "surroundings".

(b) **Two Perspectives of the Cultural Landscape and its "Reading"**

Interest in the category of cultural landscape on the borderline of research fields representing the humanities revealed a difference in the understanding of the subject's activity towards the landscape. It is about two concepts with a different scope of meaning: "view" and "surroundings". In the first case (view), the landscape is perceived either from aesthetic positions—as a background of human activities or as a directed object of observation by a distant subject. In the second case (surroundings), the meaning of the landscape changes and becomes a place inhabited by the subject and seen, as it were, from the inside. Thus, new approaches to landscape are aimed at extending its current contemplative and aesthetic approach with a new experience of the activity of a historical human, taking action in the civilizational dimension: economic, social, cultural and symbolic. In this sense, the cultural landscape is close to the understanding of space as a place tamed and managed by man. The neighborhood is a space full of human traces and signs, which are created as a result of the implementation of various life goals and cultural activities undertaken by a given community over the centuries.

Changing the approach also entails changing the aesthetics of the landscape. Traditional aesthetics prefers a distanced and reflective approach to the landscape, which it treats as a work of art. Its "strangeness", resulting from the attitude of distance represented by the subject, is a condition for contemplating the landscape. It is different when the cultural landscape is viewed taking into account the presence and from the perspective of its creators. The observer, who proposes an anthropological and cultural perspective, focuses his attention on the "place" as a space of everyday contacts, introduces an alternative aesthetics of everyday life.

In an attempt to define the features constituting the aesthetics of everyday life, the philosopher Arto Haapala (2014) turns to Heidegger's existential ontology of *Dasein* and reduces the cultural landscape to the dimension of the place where everyday life occurs. Adopting the perspective of the place and everyday life of the subject allows us to focus attention on everyday things and events that are characterized by intimacy. Intimacy and everyday life are the essence of the "place". Thus, Happala proposes a new, existential, anthropological and cultural point of view for studying the cultural landscape, and at the same time shortening the classical cognitive distance. It is the change of research optics that creates a new approach and methods: participant observation and interview—characteristic of historical-cultural research.

Shortening the cognitive distance simultaneously changes the aesthetics and scale of the landscape, abandoning the panoramic landscape in favor of immersing oneself in micro-shots. A change of perspective gives an opportunity to get closer to everyday life, but it requires other sources of information, which include various types of subjective direct and indirect testimonies, i.e. things—including personal items, as well as photographs, memories and oral accounts. Gaining insight thanks to this kind of testimonies makes it possible to enter the cultural landscape, and its "reading" becomes achievable in research and teaching practice.

The clear cultural symbols we see in the landscape are not only a testimony to a given community, but also contain information about its significant identity features. However, all the significant elements of the cultural landscape, everything that is present and has meanings in it, demands interpretation—that is, "reading" the meanings hidden in it. This kind of "reading" space as a sign is proposed by Karl Schlögel (2009), and referring to Kant's ideas, he draws attention to the need for reflective looking, which can be learned. Looking thoughtfully at the surrounding cultural reality engages the intellect in the process of interpreting what is perceived based on historical and cultural knowledge.

4 Didactic and Historical Reflection Around Cultural Landscape

The cultural landscape, defined on the basis of the humanities, fits into the basic conceptual categories of historical science, used by contemporary history teaching, which contribute to the construction of "historical thinking". The development of historical thinking is the result of changing components: the current interpretation of the social function of history, the multi-trend character of historiography, the state of historical awareness of the society and the changing means of didactic influence. In school practice, students' historical thinking is taught through the historical narrative of textbooks, which has a decisive influence on the way the image of the past is constructed. The type of narration determines the specific features of this construction and, as a result, the way of understanding the past (Julkowska 2018). By analogy with the typology of historiography, three types of historical narrative can be distinguished: the first based on the model of classical historiography, the second on the model of modernist historiography and the third on the model of anthropological and cultural historiography. These three dimensions are helpful in interpreting the cultural landscape—both in historiography and in history textbooks.

The historical approach enriches the humanistic reflection on the cultural landscape with a specific constitutive element, which is the "historical human"—a category known in historical research as early as the nineteenth century. "Historical man" is a subject seen in the context of his activity and the space he occupies—in a strictly defined historical time and place. Depending on the methodological basis on which the image of the past is constructed, the "historical human" can be perceived in relation to the cultural landscape in a wide spectrum of roles: as a ruler, host and user, constructor, defender, researcher, priest, artist.

Recognition of space by a historian and an archaeologist takes place from a different perspective than in the case of geography. Space is historically and anthropologically perceived as a stage on which subjects undertake political, economic, social, artistic, religious, scientific and cultural activities. These activities create a network of mutual relations and connections, conditioning further activities or preventing them. From this perspective, historically changing events and processes related to cultural space are studied and observed: including its territorial divisions (politically or economically conditioned), settlement, urbanization, industrialization and colonization processes, as well as the implementation of projects that acquire new spaces or revitalize them. Space is understood in history in a dynamic way, conditioned by human activity. Qualitative changes and the degree of their intensity in historical time are taken into account—different in the case of soil cultivation and farming methods used (e.g. extensive or intensive), different in the case of exploitation activities aimed at obtaining raw materials—and the effects of this activity on the state of the environment (e.g. local ecological disasters).

To summarize this thread: for a historian and an archaeologist, space is the result of the activities of many generations, perceived as a synthesis of overlapping cultural layers, forming a kind of palimpsest (Crawford 1953). For example, the Roman Forum, like many European cities with ancient roots, has preserved the memory of historically shaped space in its layers and architectural arrangement. The cultural landscape is one of the forms of historical and archaeological heritage, and its protection is an expression of historical and archaeological awareness.

The historical perspective relating to the cultural landscape has an inalienable time dimension for history, measured not physically but culturally. This means the historicity of the landscape as a place occupied by man at a particular historical moment. Historical time—both in terms of macrohistory and microhistory—is measured by the lives of successive generations, which leave visible effects of their activities. These accumulating traces and signs of presence can be read and interpreted as determinants of the cultural identity of social groups that live in given places in the long or short term. This palimpsest nature of cultural landscapes is the source of their richness. Taking into account the dynamics of time and space, subject to constant historical changes, are constitutive features of the cultural landscape from a historical perspective.

They are complemented by one more category: "cultural continuity". Some elements of culture retain a surprisingly long durability, others undergo changes, partially or completely obliterating their former meaning. Ancient monuments are an example of the fact that not only individual architectural objects, but also entire urban complexes can survive in a state that allows for research of their cultural meanings. It is similar in the case of medieval monuments: clear traces of former urban and rural layouts have been preserved in the European spatial structure. Their legibility allows you to see a permanent part of the past. Old cultural phenomena rarely appear to our eyes as coherent cultural sequences—they rather occur as single monuments or show their presence in deeper cultural layers. From the historical perspective, the cultural landscape is a mosaic consisting of various—in the chronological and stylistic sense—elements and objects. Here, a significant difference between cultural historians and archaeologists and researchers of living cultures is revealed with all its force—namely, the need to reach the remains, and then the process of analyzing and interpreting historical sources in order to obtain information. Getting to know the old, often culturally alien, communities along with their system of values and evaluations is the starting point for understanding their decisions and examining the effects of their activities in the long term.

An additional domain specific to the history of culture is the study of the cultural landscape in its historical artistic representations—including specific painting, literary, photographic and film representations.

The didactics of history creatively adapts all these dimensions of the past and all the richness and variety of historical sources used in research (written, iconographic and symbolic) for educational purposes. In the field of history didactics, a variant presentation of the cultural landscape takes place—it differs depending on the type of historical narrative proposed by a given textbook. Following Wojciech Wrzosek's (1995) findings, I assumed that, similarly to the types of historiography, also in history textbooks there are at least three models of school historical narrative in history textbooks—which have a fundamental impact on the way of narrative depiction of the past and the representation of the cultural landscape in them (Julkowska 2017, 2020).

In textbooks on classical narrative and event-based history, **the cultural landscape most often appears as a real and symbolic space for the influence of political power**. Examples of addressing this issue are: emphasizing the symbolic political affiliation of a given area to a specific owner; actions of incorporating towns or villages under German law and the accompanying political and economic conditions; the course of political borders defined by treaties, breaking centuries-old local and regional ties; forced migrations of the population irretrievably breaking the cultural continuity of the region. In the classical view of history, the categories of time, space and change are flattened into a political dimension. Their metaphorical meaning is revealed only in the broad context of non-classical approaches.

Non-classical—modernist approaches include those narratives that enrich the current image of the historical world with the social and cultural dimension of space as well as the identity and cultural dimension of time. The new socio-cultural dimension that emerges in this way gives the historical image depth—not illusively as in painting, but metaphorically. Thus, in textbooks on modernist (structural) narrative and processual history, **the cultural landscape appears as a space for socio-economic and cultural activities in the macro dimension**. An example of related issues are all possible attempts at a comprehensive organization of space in the city and in the countryside (settlement, urbanization, industrial processes) together with all the social, economic and cultural consequences of these activities, which in the long term affect the life of groups, communities, local communities. Panoramic aerial photos allow us to capture and visualize the scale of changes taking place in the process of long duration.

The third type of historical narrative is represented by textbooks with an anthropological and cultural approach, often using microhistorical imaging. **The cultural landscape appears in them as culturally empowered, because it is aware of its own identity**. An example of approaching issues in this type of narrative is the presentation of everyday life of a small local community with its problems, which introduces the reader "inside" the cultural landscape as a researcher. The empowerment in this approach is determined by the choice of topics that have not been addressed so far and communities previously absent in the textbook historical narrative, as well as a shortened cognitive, aesthetic and ethical distance. Everything happens in the vicinity of the researched space, thanks to which the classical aesthetics is replaced by the aesthetics of proximity, postulating the active participation of students in the process of learning about the world. The presented three different models of textbook narration are supplemented with new methods, introducing "reading" of aerial photography and cultural landscape to contemporary historical education. One of the methods is direct participation, which most fully expresses the micro-historical approach. Its implementation is possible thanks to the study of the cultural landscape in the field during a historical trip and meetings with testimonies or witnesses of the past. This type of meetings with the past in the present should be accompanied by organizational and methodical consideration as well as substantive preparation for the interpretation of the visited place. In the anthropological, cultural and aesthetic sense, this is a necessary condition for understanding the micro-historical scale of the studied cultural landscape and intimacy as an essential feature of a place whose mental borders we cross both in a metaphorical and literal sense (Julkowska and Moskiewicz 2018).

The methodological tip for conducting in situ research consists of the following steps, leading to "taming the place" and taken into account in conversation with people living in a given place (Happala 2014):

- We look at a place that is foreign from a distance;
- We make it familiar: on a sensory level (we see, hear, touch, smell);
- We recognize the functions and meanings of places (shop, church, school, cemetery, road, market, alley);
- We focus on the aesthetic quality of things and objects;
- We establish an emotional relationship with the place of residence, settlement, we look for examples of attachments because it is the key to understanding the aesthetics of the place;
- We create relationships and build ties that lead to rooting.

The presented phases of "taming the place" are part of historical projects around cultural landscapes, but we also find them in examples of implementation of the strategy of socializing archeology.

Around the category of cultural landscape, I prepared an educational project called Akademia Kształcenia Wyprzedzającego [the Academy of Advance Learning]

(Julkowska and Moskiewicz 2018), which was implemented in 2018–2022 as one of the five modules of the largest educational project in Poland called Cyfrowa Szkoła Wielkopolsk@2020 [Digital School Wielkopolsk@2020]. The project, co-financed by the EU from the ESF funds under the Wielkopolska Regional Operational Program for 2014–2020, covered a total of over 600 schools, including the problem of "cultural landscape", 50 secondary schools got acquainted with and performed their own tasks.

5 Conclusions

The ability to reflect on history built around the cultural landscape should become a component of the historical culture of the history teacher so that the didactic message of history is critical, also in the sense of being aware of the changing visions of the cultural order determined by a given culture.

The perception of the cultural landscape and the perception of cultural heritage in it is the result of methodological reflection and didactic work—often combined with the need for an interdisciplinary approach to the components of this landscape, and thus including archeology in the historical narrative. I believe that the introduction of the category of "cultural landscape" into the teaching of history will allow to obtain a new cognitive and interpretative perspective and to achieve a higher level of education. The new perspective inspires reflection and the search for non-standard methods in the work on educating historical thinking, postulated by the Poznań School of Methodology as an important element of school historical education. The methodological postulate has recently been enriched with new didactic strategies of teaching through emotional involvement and contact with old cultural landscapes, testimonies and witnesses of the past. Their combination gives rise to a reflection on the problem of acceleration in culture, which exposes its participants to the experience of identity change in an unprecedented dimension and scale.

The cited example of aerial photography and the conceptual category of cultural landscape shows that archaeological content can be successfully applied to the teaching of history, provided that it is accompanied by methodological reflection. The image of the past supplemented with the archaeological dimension broadens the horizons of historical thinking and contributes to the development of awareness of the archaeological heritage as an important element of historical culture.

References

Aston, M. 1999. *Interpreting the landscape – Landscape archaeology and local history*. London: Routledge.

Barford, P. M. 1998. Przeszłość Anglii dla teraźniejszości i przyszłości. In *Ochrona dziedzictwa archeologicznego w Europie*, ed. Z. Kobyliński, 143–175. Warszawa: GKZ, SNAP.

Barford, P. M. 2005. Tworzenie krajobrazu: archeologia osadnicza z lotu ptaka? In *Biskupin ...i co dalej. Zdjęcia lotnicze w polskiej archeologii*, ed. J. Nowakowski, A. Prinke, and W. Rączkowski, 379–388. Poznań: Wydawnictwo Ad Rem.

Belting, H. 2012. *Antropologia obrazu: szkice do nauki o obrazie*. Kraków: Universitas.

Crawford, O. G. S. 1953. *Archaeology in the field*. London: Dent.

Dobrowolska, M. 1948. Dynamika krajobrazu kulturowego. *Przegląd Geograficzny* XXI (1): 151–203.

Frydryczak, B. 2013. *Krajobraz. Od estetyki the picturesque do doświadczenia topograficznego*. Poznań: Wydawnictwo Poznańskiego Towarzystwa Przyjaciół Nauk.

Haapala, A. 2014. Estetyka codzienności. In *Krajobraz kulturowy*, ed. B. Frydryczak and M. Ciesielski, 27–42. Poznań: Wydawnictwo Poznańskiego Towarzystwa Przyjaciół Nauk.

Julkowska, V., ed. 2012. *Foto-historia. Fotografia w przedstawianiu przeszłości*. Poznań: Uniwersytet im. A. Mickiewicza w Poznaniu.

Julkowska, V. 2017. Krajobraz kulturowy jako kategoria pojęciowa i perspektywa poznawcza w edukacji historycznej. In *Krajobrazy kulturowe: sposoby konstruowania i narracji*, ed. R. Traba, V. Jukowska, and T. Stryjakiewicz, 46–68. Warszawa: Wydawnictwo Neriton.

Julkowska, V. 2018. Narracje historyczne. In *Historia w przestrzeni publicznej*, ed. J. Wojdon, 57–64. Warszawa: PWN.

Julkowska, V. 2020. Kulturalandschaft als erkenntnisleitende Begriffskategorie der Geschichtsdidaktik. In *Kulturlandschaften in Deutschland und Polen. Akteure und Modi ihrer Konstruktion und Narration*, ed. O. Kühne, T. Strobel, R. Traba, and M. Wiatr, 43–61. Göttingen: Vandenhoeck & Ruprecht Verlage.

Julkowska, V., and K. Moskiewicz. 2018. *Poradnik dydaktyczny dla nauczycieli prowadzących zajęcia pozalekcyjne w ramach projektu CYFROWA SZKOŁA WIELKOPOLSK@2020*. EFS/ Wielkopolski Regionalny Program Operacyjny 2014–2020. Poznań: Akademia Kształcenia Wyprzedzającego.

Julkowska, V., and W. Rączkowski. 2005. Zobaczmy przeszłość Zdjęcia lotnicze w dydaktyce historii. In *Biskupin ...i co dalej. Zdjęcia lotnicze w polskiej archeologii*, ed. J. Nowakowski, A. Prinke, and W. Rączkowski, 353–369. Poznań: Wydawnictwo Ad Rem.

Kobiałka, D., K. Kajda, and A. Marciniak, eds. 2022. *Archeologia wspólnotowa – poznając przeszłość, nie zapominając o teraźniejszości*. Kraków: Universitas.

Myga-Piątek, U. 2012. *Krajobrazy kulturowe. Aspekty ewolucyjne i typologiczne*. Katowice: Uniwersytet Śląski.

Pałubicka, A. 2006. *Myślenie w perspektywie poręczności a pojęciowa konstrukcja świata*. Bydgoszcz: Oficyna Wydawnicza Epigram.

Pawleta, M. 2020. Archeologia odpowiedzialna społecznie. Działania z zakresu public i community archaeology w Polsce. *Folia Praehistorica Posnaniensia* 25:169–190.

Pawleta, M. 2022. Archeologia wspólnotowa jako przejaw zaangażowania dyscypliny w teraźniejszość. In *Archeologia wspólnotowa – poznając przeszłość, nie zapominając o teraźniejszości*, ed. D. Kobiałka, K. Kajda, and A. Marciniak, 25–46. Kraków: Universitas.

Pomorski, J. 1991. *Historyk i metodologia*. Lublin: Wydawnictwo UMCS.

Rączkowski, W. 2002. *Archeologia lotnicza – metoda wobec teorii*. Poznań: Wydawnictwo Naukowe Uniwersytetu im. Adama Mickiewicza.

Schlögel, K. 2009. *W przestrzeni czas czytamy. O historii cywilizacji i geopolityce*. Poznań: Wydawnictwo Poznańskie.

Topolski, J. 1983. *Teoria wiedzy historycznej*. Poznań: Wydawnictwo Poznańskie.

Topolski, J. 1995. *Jak się pisze i rozumie historię. Tajemnice narracji historycznej*. Warszawa: Rytm.

Traba, R. 2009. *Przeszłość w teraźniejszości. Polskie spory o historię na początku XXI wieku*. Poznań: Wydawnictwo Poznańskie.

Vidal de la Blache, P. 1926. *Principles of human geography*. London: Constable Publishers.

Wrzosek, W. 1995. *Historia – kultura – metafora. Powstanie nieklasycznej historiografii*. Wrocław: Wydawnictwo Leopoldinum.

Wrzosek, W. 2010. *O myśleniu historycznym*. Bydgoszcz: Oficyna Wydawnicza Epigram.

Zalewska, A. I. 2018. Prospołeczna, partycypacyjna i "wspólnotowa" archeologia bliskiej przeszłości jako sposób na nadawanie sensu trwaniu (ludzi i rzeczy) oraz jako antidotum na niedostatki wiedzy i trywializację przeszłości. *Studia Humanistyczne AGH* 13 (2): 19–39.

Zalewska, A. I. 2021a. Archeologia prospołeczna i uspołeczniona (public archaeology) z perspektywy polskiej. In *Historia w przestrzeni publicznej*, ed. J. Wojdon, 17–26. Warszawa: PWN.

Zalewska, A. I. 2021b. Pro–social and socialized archaeology (public archaeology) in polish style. In *Public history in Poland*, ed. J. Wojdon, 174–189. London/New York: Routledge.

The Reception of the Poznań School of Archaeology: Other Disciplines, Poland and Beyond

Ethnoarchaeology in Toruń and the Influence of the Poznań School of Methodology

Jacek Kowalewski

Abstract The article presents an overview and intellectual foundations of ethnoarchaeology in Toruń from the 1980s to the early twenty-first century. Its emergence was closely tied to the interdisciplinary collaboration between Polish archaeologists and ethnologists, as well as to inspirations drawn from the Kraków-Warsaw school of ethnology (New Polish Ethnology) and the Poznań School of Methodology founded by Jerzy Kmita. Ethnoarchaeologists based in Toruń—particularly Andrzej P. Kowalski and Andrzej Z. Bokiniec—combined methodological precision with a critical reflection on the cultural foundations of research into prehistoric communities. Special emphasis was placed on studies concerning symbolism, magical thinking, and the cultural conditioning of representations in prehistory.

The article also highlights the role this intellectual tradition played in the establishment and development of the independent Department of Ethnology at Nicolaus Copernicus University in Toruń. It discusses collaboration with intellectual disciples and associates of Kmita, notably Artur Dobosz, as well as research initiatives originating in this milieu, such as the *Eidolon* and *Aesthetics in Archaeology* projects.

Although institutional ethnoarchaeology in Toruń declined due to educational reforms and staff changes in Polish academia, the author underscores its significance as a unique scholarly and cultural experience that left a lasting imprint on the landscape of Polish humanities. The history of Toruń ethnoarchaeology is portrayed as a creative, though concluded, chapter in the wider narrative of collaboration between archaeology and cultural anthropology.

Keywords Ethnoarchaeology · Poznań School of Methodology · Jerzy Kmita · Methodology of the humanities · Magical thinking · Cultural anthropology · Prehistory · New Polish Ethnology

---

J. Kowalewski (✉)
Faculty of Humanities, University of Warmia and Mazury in Olsztyn, Olsztyn, Poland
e-mail: jacek.kowalewski@uwm.edu.pl

A. Marciniak et al. (eds.), *The Poznań School of Archaeology*, Outstanding Contributions to Archaeology, https://doi.org/10.1007/978-3-032-07848-3_10

In 2016, on the occasion of the 40th anniversary of the Institute of Archaeology at the Nicolaus Copernicus University in Toruń (hereinafter referred to as NCU), I published an article in "Rocznik Antropologii Historii" ("Anthropology of History Yearbook") detailing the origins of ethnoarchaeology in Toruń, a specialization in archaeology created by NCU prehistorians in the early 1990s (Kowalewski 2016).[1] The proposal of a new program, which departed from the traditional orientation of archaeological studies, aimed to establish a humanistic path of academic development, offering an alternative to the then-dominant paradigm of natural science orientation in Polish archaeology.

In researching the origins and development of ethnoarchaeology in Toruń, I focused on the key role played by the New Polish Ethnology (hereinafter referred to as NPE) school, a generation of ethnographers institutionalized in the late 1970s and early 1980s who initiated the transformation of their discipline into a modern cultural anthropology (Łuczeczko 2006, 57–78; Buchowski 2012, 46–54). Jolanta Małecka-Kukawka and Stanisław Kukawka (Małecka-Kukawka and Kukawka 2022) complemented the Toruń study program by providing the local contexts and historical conditions that formed the background influencing the directions and dynamics of the ethnoarchaeological project.

It seems that the history of ethnoarchaeology in Toruń is complete, framed in detail, and examined simultaneously from several sides. However, there remains one insufficiently explored, yet extremely important aspect in the history of ethnoarchaeology in Toruń: the relationship of Toruń ethnoarchaeologists with the Poznań School of Methodology, formed in the 1970s (Musiał 2016).

The key event in the history of Toruń ethnoarchaeology was the conference "Methodological Problems of Contemporary Research on Culture and Tradition" organized in May 1986 under the auspices of the Department of Ethnography of the Institute of Archaeology and Ethnography of the NCU, which accompanied an off-site meeting of the Ethnological Sciences Committee of the Polish Academy of Sciences. The initiators of the conference were two Toruń-based ethnologists, Krzysztof Piątkowski and Wojciech J. Burszta, who made a successful attempt to institutionalize the long-standing, privately held social meetings that provided a platform for comparing research views and crystallizing the attitudes of the younger generation of Polish ethnographers. Participants in these discussions included Czesław Robotycki, Wojciech J. Burszta, Michał Buchowski, Krzysztof Piątkowski, Krystyna Piątkowska, Adam Paluch, Zbigniew Libera, Stanisław Węglarz, and Wojciech Łysiak. Czesław Robotycki, slightly older than the others, acted as a generational link between his younger colleagues and the Warsaw-Kraków milieu of the NPE, which was co-founded by Ludwik Stomma, Zbigniew Benedyktowicz, Ryszard Tomicki, and Jerzy S. Wasilewski. The younger colleagues, who were also rebelling against the then current concepts of Polish ethnography, provided solid support for the implementation and development of the new ideas of the NPE

[1] A revised and expanded version of this article was also published in *Konteksty: Polska Sztuka Ludowa* (*Contexts: Polish Folk Art*) magazine (Kowalewski 2021).

manifesto (Benedyktowicz et al. 1980a, b, 1981). In the second half of the 1980s, in the midst of such meetings and discussions, a group of archaeologists specializing in the problems of prehistoric research matured as the future instigators of Toruń ethnoarchaeology. They were Andrzej Z. Bokiniec, Andrzej P. Kowalski, Stanisław Kukawka, and Jolanta Małecka-Kukawka.

In fact, the idea of establishing the Toruń ethnoarchaeology was not only the result of the interest of the main initiators of the project in American ethnoarchaeology. Given the reluctance of the authorities of the NCU Institute of Archaeology at the time, who were not enthusiastic about the activities of young researchers, the use of a relatively exotic name for a new subdiscipline of archaeology in the Polish context was more of a formal ploy than an affirmation of American research traditions. Andrzej Z. Bokiniec, recalling the discussions on the choice of a name for the newly created field, admitted that he preferred the term "ethnography of ancient communities". For institutional reasons, however, this name was rejected, and the term "ethnoarchaeology" was finally adopted (Kowalewski 2016, 358).

The basis for the activities of the Toruń ethnoarchaeologists was the distance from archaeology (borrowed from fellow ethnologists), which focused on the empirical study of evidence obtained through excavation, and rather reluctantly approached cultural and social issues of prehistory. In fact, the original goal of the ethnoarchaeologists was to use the new findings of ethnology in archaeological interpretations. At the same time, an important feature of the Toruń ethnoarchaeologists' approach was pluralism of research, which was clearly related to the epistemological position of the NPE representatives. It was not so much a specific theoretical perspective as the intention to develop the ability to use various tools of the humanities in archaeological analysis that was the overall cohesive force of the nascent research community in Toruń.

Another characteristic of the group was its dissociation from the postmodern ideas that were already being discussed among Polish ethnologists. Their program had a positive character, based on presenting the problems of prehistory in the spirit of ethnological studies. Although free from cognitive naivety, the program had a reconstructive value, based on the conviction that ethnology, as a potential synthetic science, could be useful in archaeological research. However, this did not mean excluding methodological discussions or rejecting paradigmatic pluralism in anthropological approaches to cultural interpretation and in the humanities more broadly.

The above was well represented by the academic teaching program of Toruń ethnoarchaeology, in which theoretical issues, including those of a scientific nature, played an important role. The first specialization curriculum, launched in the academic year 1990/1991, included six lectures and six seminars with a total length of 390 hours, devoted to both theoretical perspectives and subject interpretations. The theoretical courses included "Cognitive Determinants of the Humanities", "History of Anthropological Sciences", "Interpreting Cultures", "Ethnoarchaeology and Selected Theoretical Problems in Archaeology", and "Analytical and Experimental Methods in Archaeology". Meanwhile, the thematic seminars included "Beliefs and Art of Primitive Societies", "Universals of Primitive Cultures", "Ethno-religious and Mythological Comparatism", "Symbolic Culture in Prehistoric Central Europe", "Ethnic Art, and Ethnological Analogies in Archaeology. Examples of Applications" (Kowalewski 2016). The program included a variety of interpretive methods and approaches, particularly relevant to the study of imagery, reflecting the close ties of Toruń ethnoarchaeologists to the NPE community.

The detailed history of ethnoarchaeology in Toruń has been covered in several publications (Kowalewski 2016, 2021; Kowalski 2022; Małecka-Kukawka and Kukawka 2022). However, it is worth noting that the development of the research perspectives of Toruń ethnoarchaeologists took place simultaneously in two separate, but partially overlapping and complementary directions.

The first direction, associated with Andrzej Z. Bokiniec and Stanisław Kukawka, was connected with the research traditions of the older varieties of ethnology, closer to archaeology, with the diffusionist perspective of American ethnological studies, as well as with the cultural-historical school of Friedrich Ratzel, Leo Frobenius, Wilhelm Koppers, George Montandon, or Wilhelm Schmidt. Within this framework, prehistoric cultures subjected to comparative analysis by cartographic methods could be inscribed into various typological schemes based on generalizing socio-cultural models of interpretation. The possibility of combining ethnological concepts of cultural areas or the *Kulturkreis* (Clark Wissler, Otis Tufton Mason, Alfred L. Kroeber) with prehistoric concepts of archaeological cultures was a natural subject of interest for ethnoarchaeologists. More specifically, it included classic ethnoarchaeological questions, such as the relationship between the stylistics of material products and ethnicity, or the conditioning mechanisms and effects of cultural contact (processes of cultural assimilation, adaptation, acculturation, especially in border areas). Important points of reference in terms of archaeology were the works of Vere Gordon Childe on the Neolithic Revolution in Europe, and especially Maria Gimbutas's studies on the cultural consequences of the expansion of the Indo-Europeans in the Bronze Age (Childe 1925, 1926; Gimbutas 1997; Kowalski 2022). Ethnologically based comparative findings on cultural universals sought research support to develop much broader correlations between socio-cultural phenomena and traces of practices as recorded by excavation methods. Based on ethnographic analogies, specific arrangements of preserved material remains would thus be correlated with the hypothetical images of specific social phenomena, e.g., kinship systems. In accordance with the traditions of British and American ethnoarchaeology, Toruń ethnoarchaeologists did not overlook their interest in experimental methods, for example, Jolanta Małecka-Kukawka (2001), is a well-known expert in use-wear analysis.

The use of cultural comparativism in prehistoric research resulted from Andrzej Z. Bokiniec's preference for a strictly systemic view of socio-cultural reality (cf. Kowalski 2022). This was encouraged by references to processual archaeologists, Lewis R. Binford (1962) and David L. Clarke (1968), who posited attempts to reconstruct and understand the functioning of entire socio-cultural systems as the fundamental goal of archaeology. The greatest emphasis was placed on models, supported by system-typological observations from the ethnography of early communities, that depicted global processes of cultural change in prehistory. With a significant degree of generalization in the search for systemic relationships and regularities, even comparative studies of processes in historical epochs, presented from a historiosophical perspective by such eminent scholars as Arnold J. Toynbee, Philip Bagby, Joseph Needham, or Alfred L. Kroeber (in the field of comparative studies of civilizations), proved useful (Kowalski 2022). In the context of the systemic, modernist paradigm of archaeological research, interests in human ecology, along with anthropological population concepts linked to settlement geography (Kowalewski 2016, 316), were also considered important. In the late 1980s and early 1990s, the ecologizing studies of the Poznań physical anthropologists Jan Strzałko, Maciej Hanneberg, and Janusz Piontek (1976, 1980) were highly regarded by the Toruń archaeological community.

The other research direction developed by Andrzej P. Kowalski and Marcin Wiewióra, who was initially associated with ethnoarchaeological circles, drew on inspirations from structural anthropology, semiotics, phenomenology of religion and philosophical hermeneutics,[2] placing issues of cultural symbolism at the core. The topic of the Indo-European community, which is particularly important for prehistorians studying the Neolithic and Early Bronze Age, naturally drew the attention of Toruń ethnoarchaeologists to the works of Russian semioticians of culture of the Tartu-Moscow Semiotic School, Russian ethnolinguistics and folklore studies, popular among NPE scholars (Żyłko 2009). It is no coincidence that in the memoirs and bibliographic entries of the first publications of the Toruń ethnoarchaeologists appear references to Juri Lotman, Boris Uspienski, Vyacheslav Ivanov, Vladimir Toporov, Tamaz Gamkrelidze, Albert Bajburin, Nikita I. Tolstoy, or Oleg Trubachyov. The search for prehistoric meanings led the ethnoarchaeologists to try to adapt in their archaeological interpretations the Indo-European cosmogonic myths reconstructed and analyzed by Russian scholars. This direction was undoubtedly helped by discussions and readings of the works of fellow ethnologists who, among other things, carried out semiotic-structural analyses in the footsteps of the aforementioned Russian scholars (Piątkowski 1981, 1993; Stomma 1976, 1978, 1979, 1986; Tomicka and Tomicki 1975; Wasilewski 1979a, b, 1980, 1987, 1990). In their search for interpretive tools, they also turned to the findings of religious scholars Rudolf Otto, Mircea Eliade, and Gerardus van der Leeuw. The desire to go beyond descriptive and classificatory analyses of excavated material—similar in essence to the calls of the NPE community regarding museum ethnography—prompted us to perceive traces of the anthropologically understood world of cultural symbols and meanings behind prehistoric materiality. This, in turn, required reaching for knowledge outside the source and opening up to the concepts and findings of scholars of anthropologically understood culture. It was important not only to reflect on the

[2] At about the same time similar conclusions—though ultimately "processed" differently—were reached by Ian Hodder, whose works were not easily accessible to Polish archaeologists in the 1980s.

theoretical level, but also to make concrete attempts to analyze strictly empirical materials (Kowalski 1991a, b).[3]

Czesław Robotycki, recalling the aforementioned period of development, expressed surprise at Michał Buchowski's claim of the existence of ethnological "clans" in the Polish academic community of the 1980s (Robotycki 2014, 247). However, in his overview of the recent history of Polish ethnology compiled in the early 2010s, Buchowski aptly emphasized the distinctiveness of the community of Toruń ethnoarchaeologists (Buchowski 2012). In retrospect, Andrzej P. Kowalski also recognized that

> [...] a diversity of perspectives was evident among the new ethnographers. Burszta talked about Kmita, Robotycki spoke more literarily, Wasilewski and Stomma talked about structures. They probably felt the same kind of general sense, but each of them said something slightly different (Kowalewski 2016, 372).

One of the reasons for these "clan" differences was undoubtedly the strong intellectual ties of Wojciech J. Burszta and Michał Buchowski to the Poznań School of Methodology, especially to Jerzy Kmita's cultural studies seminar. This tendency toward the rigor of philosophical, methodological thinking was fundamentally different between the Poznań ethnologists and the NPE Krakow-Warsaw school, which allowed more room for theoretical dissonance (Buchowski 2012, 52–53).

The cultural studies community around the philosopher and cultural theorist Jerzy Kmita based its research on a system of concepts derived from Kmita's social-regulatory theory of culture (Banaszak and Kmita 1991). As Grażyna Musiał (2016, 558) explains:

> Jerzy Kmita emphasized a holistic view of social phenomena and their determinants. Accordingly, he introduced the concept of theoretical history, understood as a science that, unlike history understood in the usual idiographic way, reveals not the role of individual

[3] The early collective research projects of the Toruń ethnoarchaeologists were published in *Interpretacje archeologiczne a etnologiczne koncepcje kultury* (*Archaeological Interpretations vs. Ethnological Concepts of Culture*), edited by Krystyna Przewoźna-Armon (1991), which includes Andrzej P. Kowalski's two articles mentioned earlier, and papers by Krzysztof Piątkowski (*Dylematy współczesnego etnologa—The Modern Ethnologist's Dilemmas*), Andrzej Z. Bokiniec (*Archeologia i etnologia. Uwagi na temat jednego pojęcia - Archaeology and Ethnology. Notes on a Single Concept*), and Stanisław Kukawka and Jolanta Małecka-Kukawka (*Neolit - subneolit. Z problematyki kontaktu interkulturowego w świetle badań chełmińskiej grupy kultury pucharów lejkowatych—The Neolithic—the Subneolithic. On Issues of Intercultural Contact in the Light of Studies of the Chelmno Group of the Funnelbeaker Culture*). Another edited collection, *Szkice prahistoryczne. Źródła—metody—interpretacje* (*Prehistoric Sketches. Sources—Methods—Interpretations*) (Kukawka 1998) a festschrift in honor of Prof. Krystyna Przewoźna-Armon, contains in its first part a number of interesting ethnoarchaeological papers, e.g., Andrzej P. Kowalski's *Dar, import i obcość. Szkic z prahistorycznej etnologii kontaktu kulturowego* (*Gift, Import and Foreignness. A Sketch on the Prehistoric Ethnology of Cultural Contact*); Marta Śliwa's *Koncepcja bricolage'u w strukturalistycznej interpretacji sztuki pierwotnej u C. Lévi-Straussa* (*The Concept of Bricolage in C. Lévi-Strauss's Structuralist Interpretation of Primitive Art*), Jacek Woźny's *Wstęp do hermeneutyki przestrzeni w archeologii* (*Introduction to the Hermeneutics of Space in Archaeology*); and Andrzej Z. Bokiniec's *Ukryty czynnik. Esej etno-archeologiczny" (The Hidden Factor. An Ethno-Archaeological Essay).*

> events and individual causes in the historical process, but general regularities and recurrent conditions [...].

This strategy, generally similar to the Toruń ethnoarchaeologists' attempts to use synthetic anthropological models and generalizations, created a field for historically oriented cultural research not only for ethnologists, who at that time were gradually losing their interest in model-based synthesis, especially from a historical perspective, but also for prehistorians.

The Poznań School of Methodology focused mainly on the methodology of science, taking as its foundation the premise of the cultural and historical nature of science. For example, Michał Buchowski, while characterizing the Poznań ethnological "clan" that emerged under the leadership of Jerzy Kmita, argued that within the framework of social-regulatory (culturally oriented) anthropology, the main concern was critical theoretical reflection on the cognitive status of anthropology. Roughly speaking, it was a matter of subjecting the ethnological concepts that arose in the course of the development of this discipline to a cultural-historical deconstruction (Buchowski 2012, 53–54). Undoubtedly, such problems, which nota bene fit well with the postmodern tendencies of the younger generation of Polish ethnologists and cultural anthropologists, were the subject of the doctoral dissertations of Michał Buchowski (1986) and Wojciech J. Burszta (1986). The study of the cultural determinants of the historically changing theoretical paradigms and research practices of various anthropological orientations provided the Toruń ethnoarchaeologists, especially Andrzej P. Kowalski, with an excellent testing ground for working with various theories and research perspectives of the humanities. The metalinguistic understanding and comparison on a common ground of the premises of different research approaches was an extremely valuable skill that was lacking in many academic communities, which took the methodological and philosophical precision of constructed interpretations all too lightly.

However, from the point of view of the needs of the Toruń ethnoarchaeologists, equally important, if not more so, were the thematic insights derived, so to speak, in parallel by the Poznań cultural scientists and anthropologists from the social-regulatory cultural theory. Particularly relevant in this regard were Anna Pałubicka's reinterpretations of evolutionist and structuralist concepts of magical thinking, combined with Jerzy Kmita's (1987a, b) findings on the logical-historical relations between axiological and semantic symbolization and the characterization of mythological thinking. For academic ethnology, the subject of magical culture was still an important area of research at that time, as evidenced by the studies of magic, ritual, and magical speech conducted by Poznań ethnologists in the late 1980s and early 1990s in the spirit of the Poznań School of Methodology (Buchowski 1986; Burszta 1986, 65ff; Buchowski and Burszta 1992, 15ff). The growing pressure of postmodernism, which undermined the status and value of subject knowledge, and the growing interest of Polish ethnologists in the perspective of contemporary anthropology, eventually led to the abandonment of the subject. In the atmosphere of the all-consuming fever of deconstructionism, the existing cognitive goals of ethnology lost their importance. The declining interest in traditional culture, for the study of which the competence of historical disciplines became increasingly necessary, and the clear turn of Polish ethnology towards the study of contemporary culture, while abandoning the optics of the historical depth of cultural phenomena, led to the gradual disappearance of the professional ties between ethnologists and archaeologists. Paradoxically, the Toruń ethnoarchaeologists attempted to fill the resulting cognitive void in Polish ethnology. It is not surprising, therefore, that they formed the essential core of a group that, at the turn of the twentieth century, attempted to reestablish the university center of ethnology in Toruń (Kowalewski 2016, 2021).

The reinterpretation of anthropological concepts of magic by cultural scientists in Poznań became firmly established in the mid-1980s thanks to Anna Pałubicka (1984, 1985). On the basis of the social-regulatory theory of culture, using the concept of culture as a reality of thought, Pałubicka presented a model account of three types of cultural-humanistic structures responsible for three possible ways of valorizing social practices. The first, historically the oldest, was referred to as magical valorization, the second as religious valorization, and the third as modern European (or, interchangeably, metaphysical-philosophical) valorization. In the model approach, each valorization determined the systems of cultural convictions appropriate to each variety. As Pałubicka (1985, 56) explained:

> In a cultural system dominated by magical valorization, there is actually only one type of cultural convictions: normative and simultaneous technical-utilitarian and symbolic-worldview convictions [...] The worldview beliefs of magical valorization do not form a separate system that can be deductively systematized, but can only be abstracted from the realized direct-practical values around which they are chaotically concentrated.

Anna Pałubicka's and Jerzy Kmita's concepts provided the theoretical framework for the construction of a general model of historical transformations of Western culture, which took place in a triadic order of successive or dominant cultural types in a given epoch: magical, religious, and modern European. Within this model it was possible to consider more detailed cultural issues, verified by independent types of historical records, for example, linguistic, mythological, religious, folkloristic, ethnographic as well as archaeological. For the Toruń ethnoarchaeologists, the findings on the syncretism of magical-type culture, first applied by Wojciech J. Burszta and Michał Buchowski in their reflections on anthropological concepts of magic, magical speech, and ritual in primitive societies, were key (Buchowski 1986; Burszta 1986, 65 ff; Buchowski and Burszta 1992, 15ff). Through Burszta's and Buchowski's mediation, the ideas of Poznań cultural studies permeated the emerging milieu of Toruń ethnoarchaeologists.

In 1992, with the help of Wojciech J. Burszta, Andrzej P. Kowalski contacted Anna Pałubicka and, under her supervision, began work on his dissertation "Methodological Problems of Researching the Symbolism of Death Rituals in Early Traditional Peoples". The study was designed to achieve two parallel goals. The first, in accordance with the perspective of the Poznań School of Methodology, focused on epistemological and methodological issues, constituting a critique, mainly directed at archaeology, of the cognitive determinants of the ways of defining, analyzing, and understanding aspects of symbolic funerary practices in early traditional societies. The other goal was to offer an object-oriented, reconstructive interpretation of the issue of early traditional death rituals, based on theoretical models derived from the socio-regulatory conception of culture. The empirical basis for his findings was material from the field of Indo-European traditions of poetics, rhetoric, mythology, Indo-European linguistics, as well as ethnology and archaeology. His most notable research achievement was the development of cultural characteristics of metamorphic thinking. According to Burszta, magical metamorphosis was the earliest practiced, pre-metaphoric (pre-symbolic) form of human thought/action, corresponding to the earliest variety of human culture, which, according to Anna Pałubicka, is called a magical type culture.

The concept of "culture without symbols" was introduced both in Andrzej P. Kowalski's doctoral dissertation and in his article "Metamorphoses as a manifestation of the oldest magically practiced form of thinking" (Kowalski 1994).

An extended interpretation of the idea of magical metamorphosis was also offered a few years later in the study *Symbol in Archaic Culture* (Kowalski 1999). The claim that symbolic thinking was essentially a late cultural innovation also became the subject of debate in the Poznań cultural studies community. This was documented in a book edited by Jerzy Kmita *Does Metamorphosis Compensate for the Lack of a Symbol?* (Kmita 2001), which includes extensive analyses by Andrzej P. Kowalski ("On the Imagination of Metamorphosis and the Magical Experience in Early Traditional Culture"), Artur Dobosz ("On Magical Metamorphoses, the Fusion of Words with the World and Jacques Derrida's Concept of Language"), Grzegorz A. Dominiak ("The Enigma of Metamorphosis"), and Dorota Angutek ("What Do the Concept of Magical Metamorphosis and the Theory of the 'Rite of Passage' Challenge Each Other in Their Systems?") (Angutek 2001; Dobosz 2001; Dominiak 2001; Kowalski 2001a).[4]

Andrzej P. Kowalski also discussed the oldest forms of human thought, reinforcing the central thesis of the cultural origins of the mind, in *Myślenie przedfilozoficzne. Studia z filozofii kultury i historii idei* (*Pre-Philosophical Thinking. Studies in the Philosophy of Culture and History of Ideas*) (2001b). Starting from a philosophical perspective, this time he focused on the prehistory of philosophical thought in the context of cultural archaeology.[5] As he argued, the subject of pre-philosophical thought referred to

> [...] socially shared beliefs, and especially to a way of thinking, within the framework of which a philosophical attitude could not spontaneously arise, but whose characteristics would be indirectly reflected in the sphere of later philosophical ideas (Kowalski 2001b, 6).

[4] Artur Dobosz worked in the Department of Ethnology at NCU, which was co-founded by the Toruń ethnoarchaeologists at the turn of the twentieth century.

[5] The project of cultural-anthropological archaeology as an independent research method serving cultural studies was proposed by Andrzej P. Kowalski as early as 2000 (Kowalski 2000). "Cultural studies anthropology" was also discussed by Michał Buchowski (2012). Both projects were directly influenced by Jerzy Kmita's cultural studies seminar.

This kind of research drew on "the findings of historical ethnology, historical anthropology, sociology and, above all, the philosophy of culture" (Kowalski 2001b, 6). It was no coincidence that, in constructing the characteristics of magical thinking, the findings of Lucien Lévy-Bruhl, Ernst Cassirer, Edmund Husserl, or Martin Heidegger were used, i.e., the ideas of philosophers who attempted to outline the source image of human culture (Kowalski 2001b, 6). Kowalski's analysis focused on the evolution of cognition, determined by the limits of prehistoric humans' capacity for self-determination, reflecting on the question of being, life, and death. In addition to the complementary demonstration of the idea of metamorphosis as the basis of the most ancient "metaphysics", Kowalski also showed the practical application of the method of cultural archaeology he created, which aims to identify and describe the formation of layers of cultural tradition that constitute the metaphysical basis of modern thinking about the world (cf. Kowalski 2000). The author's intention is that the method should support the acquisition of a genealogical "stratigraphic" image of key cultural elements, the historical assembly of which led to the emergence and ultimate objectification of a culture based on metaphor and symbol (Kowalski 2001b, 9).[6]

The development of the research ideas described above coincided with the ethnoarchaeologists' quest for institutional independence. A favorable moment was the possibility of separating an independent Department of Ethnology from the NCU Institute of Archaeology and Ethnography. The leading Toruń ethnoarchaeologists Andrzej Z. Bokiniec and Andrzej P. Kowalski, together with ethnologists Krzysztof Piątkowski and Wojciech Olszewski, were instrumental in creating an independent research and teaching unit. In its first years, the newly established ethnological center in Toruń was characterized by a clear focus on historically oriented ethnology and anthropology (Piątkowski 2011), with a strong dominance of the anthropology of early traditional communities. The influence of the Poznań cultural studies circle was strengthened by the fact that Artur Dobosz, a cultural philosopher and student of Jerzy Kmita, joined the Department of Ethnology. Dobosz (2001, 2002, 2013) conducted in-depth research on magical thinking and magical metamorphosis. The lecturers of the newly established department were soon joined by young Ph.D. graduates of Toruń archaeology, who had acquired their cultural-anthropological inclinations under the influence of the ethnoarchaeological milieu.[7]

[6] After leaving the Department of Ethnology and the end of ethnoarchaeology in Toruń, Andrzej P. Kowalski continued his research and published some of his most renowned works: *Mit a piękno: z badań nad pochodzeniem sztuki* (*Myth and Beauty: Studies on the Origins of Art*) (Kowalski 2013); *Antropologia zamierzchłych znaczeń* (*The Anthropology of Ancient Meanings*) (Kowalski 2014), and *Kultura indoeuropejska. Antropologia wspólnot prehistorycznych* (*Indo-European Culture. Anthropology of Prehistoric Communities*) (Kowalski 2016). Kowalski's other research interests are presented in *Antropologia zamierzchłych znaczeń*, where he deals with methods of researching primitive culture, reflection on the function of signs in the cultural-pragmatic dimension, ethno-aesthetics and the origins of art, culture as a reality of thought from the perspective of cultural pragmatics, and the status of language and its meaning in the context of researching archaic communities.

[7] Joanna Wawrzeniuk and Jacek Kowalewski.

The two directions of Toruń ethnoarchaeology had a significant impact on the research of its students. Their M.A. theses, Ph.D. dissertations, and publications were marked by influences from the NPE community and from Poznań cultural studies, especially as Andrzej P. Kowalski's research developed. Master's theses often dealt with prehistoric symbolism, Indo-European culture, methodological issues with an anthropological and cultural focus, as well as more traditional ethnoarchaeological topics. The influence of the ethnoarchaeological milieu of the 1990s was also evident in the topics of the Ph.D. dissertations of the archaeology graduates. The dissertations of the first generation of Toruń ethnology graduates were also influenced by research on magical metamorphosis.[8]

It is also worth mentioning that after leaving the Nicolaus Copernicus University in Toruń, Andrzej P. Kowalski was the supervisor of two doctoral dissertations, which were based on his previous research. These were "Sztuki plastyczne starożytnej Grecji w świetle myśli wybranych przedstawicieli niemieckiej filozofii życia oraz refleksji kulturoznawczej" ("The Plastic Arts of Ancient Greece in the Light of the Thought of in the Light of the Thought of Selected Representatives of German Philosophy of Life and Cultural Reflections") by Arkadiusz Bednarczuk (dissertation defense in 2010) and "Sacrum w wyobrażeniach pogańskich Prusów. Próba interpretacji na pograniczu historii i etnologii religii" ("The Sacrum in the Images of the Pagan Prussians. An Attempt at Interpretation at the Crossroads of History and Ethnology of Religion") by Paweł Kawiński (dissertation defense in 2012).[9]

At the turn of the twentieth century, Andrzej P. Kowalski also co-founded research teams centered around the "Eidolon" and "Estetyka w archeologii" ("Aesthetics in Archaeology") projects. A key role in these projects was played by scholars with intellectual ties to the Poznań School of Methodology. The history of the research activities of the scholars of the Eidolon project has already been discussed by Marian Kwapiński (2013). The results of the other project, led by Andrzej P. Kowalski in cooperation with Bogusław Gediga and Danuta Minta-Tworzowska, were published in four books: *Estetyka w archeologii* (*Aesthetics in Archeology*) (Gediga and Kowalski 2003), *Estetyka w archeologii: kopie i naśladownictwa* (*Aesthetics in Archaeology: Copies and Imitations*) (Kwapiński 2006), *Estetyka w archeologii: antropomorfizacje w pradziejach i starożytności* (*Aesthetics in Archaeology: Anthropomorphizations in Prehistory and Antiquity*) (Bugaj and Kowalski 2010), and *Estetyka w archeologii: obrazowanie w pradziejach i starożytności* (*Aesthetics in Archaeology: Imagery in Prehistory and Antiquity*) (Minta-Tworzowska 2018).

[8] Both dissertations were published in extended book formats: Ł. Dominiak, *Partycypacja i dyskurs: mentalność pierwotna w badaniach Luciena Lévy-Bruhla* (*Participation and Discourse: Primeval Mentality in Lucien Lévy-Bruhl's Research*—2015); M. Rydlewski, *Żeby widzieć, trzeba wiedzieć. Kulturowy wymiar percepcji wzrokowej* (*To See, One Must Know. Cultural Dimensions of Visual Perception*—2016).

[9] Paweł Kawiński's dissertation was published in 2018 under the same title in the series "Monumenta Literaria Prussiae" by Oficyna Wydawnicza Towarzystwa Naukowego Pruthenia in Olsztyn.

The institutional end of ethnoarchaeology at the NCU Institute of Archaeology was brought about by changes in the organization of higher education in Poland, resulting from the adoption and implementation of the Bologna Process in 1999. The ethnoarchaeology curriculum was partially incorporated into the elective specializations available to archaeology students under the supervision of Stanisław Kukawka and Jolanta Małecka-Kukawka. The program survived in this form for several more years, although it never regained its former dynamism and vitality.

On the other hand, the traditions of ethnoarchaeological research were continued in the early 2000s in the NCU's Department of Ethnology. However, even there, as a result of changes in the research profile of the department forced by the departmental authorities, the faculty that made up the ethnoarchaeological community eventually dispersed. Andrzej P. Kowalski accepted a position at the University of Gdańsk, Andrzej Z. Bokiniec remained at the Institute of Archaeology in Toruń, and Artur Dobosz returned to the Poznań University of Technology. Most of the younger faculty members, who had been raised in the spirit of Toruń ethnoarchaeology, did not maintain ties with the NCU in Toruń. They found employment in distant research centers in Wrocław, Warsaw, and Olsztyn. Some of them resigned from their academic positions.

The almost 20-year social and intellectual project of ethnoarchaeology in Toruń has come to an end. In the words of Czesław Robotycki—only the myth remains.

References

Angutek, D. 2001. Co wzajemnie w swoich systemach kwestionują koncepcja metamorfoz magicznych i teoria "rytuałów przejścia"? In *Czy metamorfoza magiczna rekompensuje brak symbolu?* ed. J. Kmita, 177–180. Poznań: Wydawnictwo Naukowe Instytutu Filozofii UAM.

Banaszak, G., and J. Kmita. 1991. *Społeczno-regulacyjna teoria kultury*. Warszawa: Instytut Kultury.

Benedyktowicz, Z., C. Robotycki, L. Stomma, R. Tomicki, and J. S. Wasilewski. 1980a. Antropologia kultury w Polsce – dziedzictwo, pojęcia, inspiracje, cz. I. *Polska Sztuka Ludowa* 34 (1):47–60.

Benedyktowicz, Z., C. Robotycki, L. Stomma, R. Tomicki, and J. S. Wasilewski. 1980b. Antropologia kultury w Polsce – dziedzictwo, pojęcia, inspiracje, cz. II. *Polska Sztuka Ludowa* 34 (1):117–125.

Benedyktowicz, Z., C. Robotycki, L. Stomma, and J. S. Wasilewski. 1981. Antropologia kultury w Polsce – dziedzictwo, pojęcia, inspiracje (materiały do słownika, cz. III). *Polska Sztuka Ludowa* 35 (1):58–60.

Binford, L. 1962. Archaeology as anthropology. *American Antiquity* 28 (2):217–225.

Buchowski, M. 1986. *Magia. Jej funkcje i struktura*. Poznań: Wydawnictwo Naukowe Uniwersytetu im. Adama Mickiewicza.

Buchowski, M. 2012. *Etnologia Polska: historie i powinowactwa*. Poznań: Wydawnictwo Nauka i Innowacje.

Buchowski, M., and W. J. Burszta. 1992. *O założeniach interpretacji antropologicznej*. Warszawa: Wydawnictwo Naukowe PWN.

Bugaj, E., and A. P. Kowalski 2010. (Eds.), *Estetyka w archeologii: antropomorfizacje w pradziejach i starożytności*. Poznań: Wydawnictwo Poznańskie.

Burszta, W. J. 1986. *Język a kultura w myśli etnologicznej*. Wrocław: Wydawnictwo Polskiego Towarzystwa Ludoznawczego.

Childe, G. V. 1925. *The down of European civilisation*. London/New York: K. Paul, Trench, Trubner & Co., A.A. Knopf.

Childe, G. V. 1926. *The Aryans a study of Indo-European origins*. London/New York: K. Paul, Trench, Trubner & Co., A.A. Knopf.

Clarke, D. 1968. *Analytical archaeology*. London: Methuen.

Dobosz, A. 2001. O metamorfozach magicznych, łączeniu się słów ze światem oraz Jacques'a Derridy koncepcji języka. In *Czy metamorfoza magiczna rekompensuje brak symbolu?* ed. J. Kmita, 95–156. Poznań: Wydawnictwo Naukowe Instytutu Filozofii UAM.

Dobosz, A. 2002. *Tożsamość metamorficzna a komunikacja językowa*. Poznań: Wydawnictwo Naukowe Instytutu Filozofii UAM.

Dobosz, A. 2013. *Myślenie magiczno-mityczna a schizofrenia*. Bydgoszcz: Oficyna Wydawnicza Epigram.

Dominiak, G. 2001. Zagadka metamorfozy. In *Czy metamorfoza magiczna rekompensuje brak symbolu?* ed. J. Kmita, 157–176. Poznań: Wydawnictwo Naukowe Instytutu Filozofii UAM.

Dominiak, Ł. 2015. *Partycypacja i dyskurs: mentalność pierwotna w badaniach Luciena Lévy-Bruhla*. Kraków: Zakład Wydawniczy Nomos.

Gediga, B., and A. P. Kowalski 2003. (Eds.), *Estetyka w archeologii*. Gdańsk: Muzeum Archeologiczne.

Gimbutas, M. 1997. *The Kurgan culture and the Indo-Europeanization of Europe: Selected articles from 1952 to 1993*. Washington: Institute for the Study of Man.

Kawiński, P. 2018. *Sacrum w wyobrażeniach pogańskich Prusów. Próba interpretacji na pograniczu historii i etnologii religii*. Olsztyn: Oficyna Wydawnicza Towarzystwa Naukowego Pruthenia.

Kmita, J. 1987a. Myślenie mitologiczne w świetle strukturalistycznej opozycji metonimii i metafory. In *Symbol i poznanie*, ed. T. Kostyrko, 139–156. Warszawa: Wydawnictwo Naukowe PWN.

Kmita, J. 1987b. Symbolizowanie jako relacja aksjologiczna oraz jako relacja semantyczna. In *Symbol i poznanie*, ed. T. Kostyrko, 188–198. Warszawa: Wydawnictwo Naukowe PWN.

Kmita, J., ed. 2001. *Czy metamorfoza magiczna rekompensuje brak symbolu?* Poznań: Wydawnictwo Naukowe Instytutu Filozofii UAM.

Kowalewski, J. 2016. Antropologia kulturowa i archeologia: etnoarcheologia w Toruniu. *Rocznik Antropologii Historii* 6:349–387.

Kowalewski, J. 2021. Siła wpływu – o nieznanej historii etnoarcheologii w Toruniu. *Konteksty. Polska Sztuka Ludowa* 75 (1–2):484–494.

Kowalski, A. P. 1991a. Z metodologicznych problemów współpracy archeologii z etnologią. In *Interpretacje archeologiczne a etnologiczne koncepcje kultury*, ed. K. Przewoźna-Armon, 13–21. Toruń: Wydawnictwo Uniwersytetu Mikołaja Kopernika.

Kowalski, A. P. 1991b. Próba analizy kontekstowej symboli wczesnego neolitu Europy Środkowe. In *Interpretacje archeologiczne a etnologiczne koncepcje kultury*, ed. K. Przewoźna-Armon, 31–41. Toruń: Wydawnictwo Uniwersytetu Mikołaja Kopernika.

Kowalski, A. P. 1994. Metamorfozy jako przejaw najstarszej magicznie praktykowanej formy myślenia. In *Szkice z filozofii kultury*, ed. A. Pałubicka, 11–21. Poznań: Wydawnictwo Naukowe Instytutu Filozofii UAM.

Kowalski, A. P. 1999. *Symbol w kulturze archaicznej*. Poznań: Wydawnictwo Naukowe Instytutu Filozofii UAM.

Kowalski, A. P. 2000. Archeologia kulturoznawcza. Projekt dyskursu "negatywistycznego". In *Archeologia w teorii i w praktyce*, ed. A. Buko and P. Urbańczyk, 115–128. Warszawa: Instytut Archeologii i Etnologii Polskiej Akademii Nauk.

Kowalski, A. P. 2001a. O wyobrażeniu metamorfozy i o doświadczeniu magicznym w kulturze wczesnotradycyjnej. In *Czy metamorfoza magiczna rekompensuje brak symbolu?* ed. J. Kmita, 19–93. Poznań: Wydawnictwo Naukowe Instytutu Filozofii UAM.

Kowalski, A. P. 2001b. *Myślenie przedfilozoficzne. Studia z filozofii kultury i historii idei*. Poznań: Wydawnictwo Fundacji Humaniora.

Kowalski, A. P. 2013. *Mit a piękno: z badań nad pochodzeniem sztuki*. Bydgoszcz: Epigram.

Kowalski, A. P. 2014. *Antropologia zamierzchłych znaczeń*. Toruń: Polskie Towarzystwo Historyczne.

Kowalski, A. P. 2016. *Kultura indoeuropejska. Antropologia wspólnot prehistorycznych*. Gdańsk: Wydawnictwo Uniwersytetu Gdańskiego.

Kowalski, A. P. 2022. Etnologia i antropologia historyczna Andrzeja Z. Bokińca. *Rocznik Antropologii Historii* 11–12 (14):27–37.

Kukawka, S. 1998. *Szkice prahistoryczne. Źródła – metody – interpretacje*. Toruń: Wydawnictwo Naukowej UMK.

Kwapiński, M. 2006. (Ed.), *Estetyka w archeologii: kopie i naśladownictwa*. Gdańsk: Muzeum Archeologiczne.

Kwapiński, M. 2013. ΕΙΔΩΛΟΝ – genealogia symbolu. *Rocznik Antropologii Historii* 2 (5):352–356.

Łuczeczko, P. 2006. *Zrozumieć własną kulturę. Antropologia współczesności w Polsce*. Kraków: Zakład Wydawniczy Nomos.

Małecka-Kukawka, J. 2001. *Między formą a funkcją: traseologia neolitycznych zabytków krzemiennych z ziemi chełmińskiej*. Toruń: Wydawnictwo Naukowe Uniwersytetu Mikołaja Kopernika.

Małecka-Kukawka, J. 2022. Antropologia kulturowa i archeologia: etnoarcheologia w Toruniu. *Rocznik Antropologii Historii* 11–12 (14):159–173.

Minta-Tworzowska, D. 2018. (Ed.), *Estetyka w archeologii: obrazowanie w pradziejach i starożytności*. Gdańsk: Muzeum Archeologiczne.

Musiał, G. 2016. Poznańska Szkoła Metodologiczna – jej rozwój i znaczenie. *Przegląd Filozoficzny – Nowa Seria* 25 (2):553–563.

Pałubicka, A. 1984. *Przedteoretyczne postaci historyzmu*. Warszawa/Poznań: Wydawnictwo Naukowe PWN.

Pałubicka, A. 1985. O trzech historycznych odmianach waloryzacji światopoglądowej. *Studia Metodologiczne* 24:51–76.

Piątkowski, K. 1981. Niewerbalne komunikowanie się w obrzędzie weselnym w polskiej kulturze ludowej XIX wieku. Podstawy analizy pragmatycznej. *Lud* 65:33–54.

Piątkowski, K. 1993. *Semiotyczne badania nad kulturą w etnologii. Studium metodologiczne*. Toruń: Wydawnictwo Uniwersytetu Mikołaja Kopernika.

Piątkowski, K. 2011. *Mit – historia – pamięć. Kulturowe konteksty antropologii/etnologii*. Łódź: Dom Wydawniczy Księży Młyn.

Przewoźna-Armon, K. 1991. (Ed.), *Interpretacje archeologiczne a etnologiczne koncepcje kultury*. Toruń: Wydawnictwo Naukowe UMK.

Robotycki, C. 2014. Moja etnografia – spojrzenie z oddali. *Konteksty. Polska Sztuka Ludowa* 48 (1):236–247.

Rydlewski, M. 2016. *"Żeby widzieć, trzeba wiedzieć". Kulturowy wymiar percepcji wzrokowej*. Bydgoszcz: Epigram.

Stomma, L. 1976. Mit Alkmeny. *Etnografia Polska* 20 (1):99–110.

Stomma, L. 1978. Magia dzisiaj: rozważania o metodzie. *Polska Sztuka Ludowa* 32 (3–4):131–142.

Stomma, L. 1979. Magia dzisiaj: zasady analizy. *Polska Sztuka Ludowa* 33 (4):195–204.

Stomma, L. 1986. *Antropologia kulturowa wsi polskiej XIX wieku*. Warszawa: Instytut Wydawniczy PAX.

Strzałko, J., M. Henneberg, and J. Piontek. 1976. *Wstęp do ekologii populacyjnej człowieka*. Poznań: Wydawnictwo Naukowe Uniwersytetu im. Adama Mickiewicza.

Strzałko, J., M. Henneberg, and J. Piontek. 1980. *Populacje ludzkie jako systemy biologiczne*. Warszawa: Wydawnictwo Naukowe PWN.

Tomicka, J., and R. Tomicki. 1975. *Drzewo życia: ludowa wizja świata i człowieka*. Warszawa: Ludowa Spółdzielnia Wydawnicza.

Wasilewski, J. S. 1979a. ...Po śmierci wędrować. Szkice z zakresu etnologii świata znaczeń (I). *Teksty* 3:97–120.

Wasilewski, J. S. 1979b. ...Po śmierci wędrować. Szkice z zakresu etnologii świata znaczeń (II). *Teksty* 4:58–84.

Wasilewski, J. S. 1980. Podarować-znaleźć-zgubić-zbłądzić. Niektóre kategorie języka symbolicznego związane z opozycją życie-śmierć. *Etnografia Polska* 24 (1):283–303.

Wasilewski, J. S. 1987. Tabu, zakaz magiczny, nieczystość, część I. *Etnografia Polska* 31 (1):19–41.

Wasilewski, J. S. 1990. Tabu, zakaz magiczny, nieczystość, część II. *Etnografia Polska* 34 (1–2):7–45.

Żyłko, B. 2009. *Semiotyka kultury: szkoła tartusko-moskiewska*. Gdańsk: Wydawnictwo Słowo/Obraz Terytoria.

Philosophy of Archaeology and the Poznań School of Methodology

Ludomir R. Lozny

Abstract In the late 1960s and 1970s, Poznań philosophers and historians developed a cognition theory rooted in Marxism but with expanded epistemology. This critical and reflective thought rejected the simplified functional and rigid deductive-nomological explanations but remained structural, maintaining a solid foundation in its principles.

Comparing the methodological proposition of the Poznań School of Methodology with the Western thoughts of the time, I focus on two features prominent in the School's curriculum: historical epistemology, also known in its French and Anglophone variants, and non-Marxist historical materialism. I conclude that a shift towards interpretive approaches, engagement with postmodern critiques, reflexivity, and interdisciplinary perspectives have marked the evolution of the Poznań School's historical epistemology and showcased the School's adaptability and relevance to contemporary academic discourse. The School proposed a unique in Eastern Europe Marxist-inspired epistemology, which is fodder for philosophical analysis and understanding. Importantly, it ushered in a new way of viewing historical processes, particularly relevant for archaeological reasoning. The School's intellectual impact, presented at times of thawing contacts between communist Poland and the West, contrasted with Anglo-American thoughts on researching historical processes and remained within the reach of the progressive European philosophy of the time, adding to the understanding of the political ecology of communist Poland and the Eastern Bloc in general.

In 1997, while on a Fulbright fellowship in Poland, I visited Poznań, and Prof. Minta-Tworzowska gave me her newly published book *Klasyfikacja w archeologii*, with the following author's note: "To Mr Ludomir Lozny with wishes of success in studying archaeological theories." It motivated me to spend the next 25 years trying.

L. R. Lozny (✉)
College of Science, Long Island University Brooklyn, Brooklyn, NY, USA

A. Marciniak et al. (eds.), *The Poznań School of Archaeology*, Outstanding Contributions to Archaeology, https://doi.org/10.1007/978-3-032-07848-3_11

Keywords Archaeology · Philosophy · Epistemology · Historical materialism · Methodology · Theory · Marxism · Non-Marxist historical materialism · Poznań School of Methodology

1 Introduction

In the late 1960s and 1970s, a group of Marxist philosophers and historians from Adam Mickiewicz University, Poland, formed the Poznań School of Methodology. Its notable representatives, Jerzy Kmita, Jerzy Topolski, and Leszek Nowak, theorized on Marxist-based historical epistemology, championed by French philosophers Jean Cavaillès, Gaston Bachelard, Georges Canguilhem, and Michel Foucault (Lecourt 1975; Rheinberger 2005), and its application to study historical processes. Historical epistemology has been discussed chiefly by historians of science and neglected by philosophers, especially in the United States.

The School's members also theorized about non-Marxist historical materialism (Nowak 1983; Brzechczyn 2022), an umbrella term that covers approaches to considering economic conditions behind social change but not in strictly deterministic Marxist terms. It logically continues the School's principles, explicitly linked to Leszek Nowak's philosophy (Borbone 2011)

Before discussing the School's intellectual contribution, I want to argue that assessing the process of acquiring archaeological knowledge raises a bigger question regarding the nature of archaeology, which falls in the philosophy of science (Palubicka & Tabaczynski 1986; Fuller & Collier 2004; Gingras 2010; Humphreys 2016). It involves deliberations on the dichotomic view of archaeology. Thus, traditionally, scholars asked whether archaeology is a humanistic, historical field that relies on positivist empiricism or a scientific discipline ruled by logical empiricism. Seeing archaeology as a historical discipline or a social and behavioral field causes various approaches to confirming and explaining past events and processes. It is very refreshing to see alternative explanations of the past. Still, we must uphold tests of the adequacy of those approaches to the past that are more satisfying, which may also mean more explanatory than others.

Thus, in a larger scheme, the topics discussed here underscore the philosophy of archaeology influenced by the Poznań School's position on cognitive processes (Kmita 1991). Some may wonder whether we need philosophy in archaeology at all (Krieger 2006). To many, archaeology is still primarily about collecting and cataloging artifacts. It is time to abandon such an approach rooted in the nineteenth-century taxonomist paradigm. The philosophy of archaeology enriches the intellectual outlook on the field. It serves as a critical lens through which we can explore various aspects of our discipline, such as the history of theories, epistemology, ethics of archaeological research and practice, interpretations of artifacts and their symbolism, the relationship between philosophy, archaeology, and science in general, and the impact of philosophical perspectives on methodology (McQuillan 2010). Archaeology is no longer a vocation to collect artifacts from the past or "undisciplined" scientific discipline. It matured. Methodologically, current archaeology is multi-cross-disciplinary and occasionally interdisciplinary (see contributions in Lozny 2011; see Minta-Tworzowska 2022 for a review of methodological and theoretical fundations of archaeology in Poland). However, because its cognitive processes involve humanistic and natural sciences ideas, a challenging-to-follow methodological mix, methodological chaos could easily be confused with creative ferment.

This essay has two sections and addresses three major themes. In the first section, I discuss epistemology and philosophy of archaeology; in the second, I relate these topics to the Poznań School of Methodology. Focusing specifically on historical epistemology (Kmita 1988; Daston 1994; Hyder 2003; Feest & Sturm 2011) and non-Marxist historical materialism, I describe how the Poznań School challenged positivist assumptions and provided alternative perspectives on historical knowledge-making (Topolski 1983, 1984, 1996), especially historical materialism central to Marxist philosophy. Finally, I compare the School's Marxist-inspired epistemological approach with similar approaches within Marxist philosophy. Examining the similarities and differences between these frameworks aims to comprehensively understand their unique contributions to epistemological discourse (Kmita 1996). The three themes that intersect throughout the text are the significance of philosophical inquiry, the epistemological foundation of archaeology, and the role of interpretation in archaeology (Trigger 1978). By exploring these topics, I aim to deepen our understanding of how knowledge has been shaped throughout history and how it continues to impact present-day understanding of the past.

One of the questions lingering throughout the text is: Can researchers achieve objectivity in analyzing archaeological data, or does subjectivity inevitably influence interpretations? It raises follow-up questions about biases, personal perspectives, and cultural influences that shape our understanding of the past. Thus, I briefly discuss how historical contexts influence archaeological analysis and acknowledge that researchers operate within specific temporal periods with unique social, political, and economic conditions that can impact their interpretations. Comprehending these contexts allows for a more nuanced understanding of cultural dynamics throughout time.

By critically examining these aspects from a Poznań School's perspective, I aim to shed light on the strengths and weaknesses of its methodology, contributing to a broader understanding of archaeology and its potential for future research.

2 Part One: Contextual Background: Philosophy of Archaeology, Historical Epistemology, and Non-Marxist Historical Materialism

I briefly outline three critical components: philosophy of archaeology, historical epistemology, and non-Marxist historical materialism present in the Poznań School of Methodology.

2.1 *Philosophy and Archaeology*

The relationship between philosophy and archaeology is complex and not always realized, especially by those who uncritically follow certain lines of thought. It impacts our approach to archaeological investigations. Examining this relationship helps to bridge gaps between these two disciplines while highlighting their shared goal of seeking knowledge about humanity. Exploring philosophical perspectives' influence on archaeological methodologies provides valuable insights into how different theoretical frameworks shape research design and data interpretation. Understanding these influences enhances our ability to evaluate archaeological findings within a broader context critically.

Historical Exploration of Archaeological Theory

In his famed essay "The Golden Marshalltown: A Parable for the Archeology of the 1980s" (1982), Kent Flannery stated that there is no archaeological theory; we borrow theories from others, primarily sociologists, economists, political scientists, and philosophers, and apply them uncritically to often inadequate datasets. I agree, but I will use "archaeological theory" as shorthand. Archaeological theories have undergone a fascinating historical exploration, leaving scholars and researchers captivated by their complexities and enigmatic nature. As the field matured, archaeologists confronted a fundamental question: What is the purpose of studying material remains? Such intellectual curiosity led to the development of cultural-historical theory, which aimed to reconstruct past societies through their material remains. Scholars argued that artifacts could provide insights into social organization, economic systems, and beliefs. This perspective brought attention to broader societal contexts and challenged previous assumptions about archaeology's role as merely an ancillary discipline.

In the mid-twentieth century, another paradigmatic shift occurred with the rise of processual archaeology. Influenced by developments in other social sciences, such as anthropology, geography, and sociology, processualists sought to explain cultural change through rigorous scientific methods. They emphasized systematic data collection and analysis using statistical techniques, aiming for objective interpretations rather than subjective speculations. However, as with any intellectual pursuit, criticisms arose regarding processualism's limitations in addressing individual agency and symbolic meanings associated with material culture (Hodder 1987). This realization prompted post-processual archaeologists to challenge prevailing orthodoxies and advocate for more interpretive approaches. Post-processualists argued that archaeology should embrace subjectivity, recognizing that the researcher's background and biases inevitably influence interpretations. They emphasized the importance of understanding past societies through their unique cultural perspectives, challenging the notion of a universal scientific truth.

The historical exploration of archaeological theory reveals a rich tapestry woven from diverse perspectives and intellectual debates. This journey highlights how archaeology has matured into a multifaceted discipline encompassing scientific methodologies and subjective interpretations. As we continue to unearth discoveries and grapple with complex questions about the past, scholars in this field must engage with these historical explorations and remain open to novel theoretical frameworks that challenge established paradigms.

Relationship Between Philosophy and Archaeology

The intricate relationship between philosophy, science, and archaeology has long been a subject of profound contemplation. Philosophy and archaeology converge in a symphony of intellectual inquiry, each playing a distinct yet harmonious role in unraveling the mysteries of our past. I argue that philosophy, with its profound introspection and dialectical reasoning, provides the foundation for scientific exploration and archaeological investigations. It dares us to question the assumptions underlying our understanding of history and demands a rigorous examination of the epistemological frameworks we employ.

Science is the conduit through which philosophy's inquiries are transformed into empirical reality. Guided by logical positivism and grounded in empirical evidence, scientific methods provide archaeologists with tools to explore past societies systematically. From excavation techniques to absolute dating, science equips archaeologists with the means to uncover artifacts and contextualize them within broader historical narratives. Yet, science is full of philosophical implications; it grapples with questions surrounding objectivity versus subjectivity, causality versus correlation, and the nature of truth itself (Psillos & Curd 2014).

Archaeology is the tangible manifestation of this intricate dance between philosophy and science. Like all social sciences, it stands at the intersection of science and philosophy, translating philosophical musings into concrete discoveries while grappling with their inherent complexities. Archaeologists immerse themselves in theoretical debates about interpreting material remains and practical fieldwork that unearths them from soil layers. They navigate through fragmented reality to construct meaningful narratives about human existence through time—a task requiring scientific rigor and philosophical reflection (Salmon 1982; Wylie 2002, 2017; Chapman & Wylie 2016).

Philosophy infuses archaeology with critical lenses for interpretation—an endeavor that extends beyond mere data collection or cataloging artifacts. Thus, epistemological questions arise: How do we know what we know about the past? What biases might influence our interpretations? By examining these foundational assumptions, philosophers challenge archaeologists to reflect on their methodologies critically—a process necessary for advancing knowledge while minimizing potential biases.

Likewise, archaeological findings often spark philosophical contemplations about the nature of human existence, the complexities of culture, and our place in the world. They invite us to ponder questions that transcend material remains: What can we learn from past societies? How do these discoveries shape our understanding of who we are today? By intertwining philosophy with archaeology, we embark on a journey that transcends disciplinary boundaries and enriches our comprehension of both past and present.

The above ponderings suggest that the relationship between philosophy and archaeology is a profound symbiosis that elevates our understanding of history. Philosophy lays the groundwork for scientific investigations, while science provides the empirical tools necessary for archaeological discoveries. Archaeology bridges these disciplines by translating philosophical inquiries into tangible insights while inviting philosophical contemplation about our place in the world. As we probe deeper into the mysteries of our past, let us embrace this interdisciplinary union that amplifies our intellectual endeavors and challenges us to question everything we think we know.

Impact of Philosophical Perspectives on Archaeological Methodologies

Archaeologists must critically examine their practices and assumptions by delving into philosophy and enhancing their research's rigor and validity. Here are some examples of persuasive ideas other than the positivistic rigor of natural sciences. One such influential philosophical perspective is postmodernism, which challenges the notion of objective truth and emphasizes the importance of multiple interpretations. Postmodernism encourages archaeologists to recognize that their cultural biases and preconceived notions shape their understanding of the past. This realization prompts researchers to adopt a more reflexive approach, constantly questioning and reevaluating their assumptions to gain a more nuanced understanding of the archaeological record.

Another significant philosophical perspective is phenomenology, which emphasizes the human experience's role in shaping our understanding of reality. When applied to archaeology, this perspective highlights the importance of considering how past human experiences have shaped material remains. By examining artifacts and the context in which they were found, archaeologists can gain insights into aspects such as social structures, religious beliefs, and daily practices. Phenomenology encourages researchers to move beyond a purely scientific/empirical approach and embrace a more holistic understanding that considers human emotions, intentions, and motivations.

In addition to postmodernism and phenomenology, hermeneutics plays a crucial role in shaping archaeological methodologies. Hermeneutics focuses on interpretation and meaning-making processes within different cultural contexts. Archaeological findings are fragmentary or ambiguous, requiring researchers to engage in interpretation to reconstruct and understand past societies accurately. Adopting a hermeneutic approach enables archaeologists to navigate these interpretive challenges by employing various methods such as contextual analysis, comparative studies with ethnographic data or historical documents, and engaging with local communities with intricate knowledge about their cultural heritage.

Furthermore, Marxist-inspired critical theory has profoundly impacted archaeological methodologies by emphasizing societal power dynamics. Critical theorists argue that archaeological interpretations can perpetuate existing power structures and marginalize certain groups. By critically examining the biases inherent in their research practices, archaeologists can strive for more inclusive and equitable approaches that give voice to marginalized communities.

The philosophy of archaeology also sheds light on the complex relationship between empirical evidence and interpretation in archaeological research. While empirical data play a crucial role in validating or rejecting hypotheses, this philosophy argues that they do not possess inherent meanings independent from human perception. Rather than assuming a direct correspondence between objects found at an archaeological site and historical events or social practices, it emphasizes that meaning arises from contextual analysis within specific theoretical frameworks. Moreover, philosophy challenges traditional notions of objectivity by highlighting how personal experiences, cultural backgrounds, and societal contexts influence interpretations. It contends that multiple valid interpretations can emerge from the same empirical data set, reflecting archaeologists' diverse perspectives and biases. This recognition of subjectivity encourages dialogue and collaboration among researchers, fostering a more reflexive and self-aware archaeological practice.

Historical Epistemology

Historical epistemology is a multidisciplinary field investigating the origins, evolution, and critique of knowledge production throughout history (Pena-Guzman 2020). Although primarily of interest to historians of science, it has a philosophical context in exploring how societies have understood and constructed knowledge, shedding light on how cultural, social, and historical contexts have shaped ideas.

The study of historical epistemology traces its roots back to the Enlightenment when European philosophers began questioning traditional, religion-based sources of knowledge. Since then, it has evolved into a rich scholarship body encompassing diverse disciplines such as philosophy, history, sociology, anthropology, and science studies. Historical epistemology offers valuable insights into the diversity of human understanding by investigating how different cultures have impacted the interpretation of reality and generated knowledge systems over time.

However, historical epistemology is not without its critics. It fails to adequately address issues related to race or gender bias in knowledge production processes. Moreover, empirical methods within historical epistemology face challenges due to the scarcity or unreliability of primary sources from earlier periods. The reliance on textual evidence hinders a comprehensive understanding of past contexts and limits researchers' ability to reconstruct accurate narratives about the development of ideas. Understanding the implications for contemporary knowledge production is another crucial aspect of historical epistemology. By critically examining the historical construction of knowledge, we can gain insights into the biases and limitations inherent in our current systems. This awareness allows questioning and challenging dominant paradigms, fostering a more inclusive and nuanced approach to knowledge production.

Origins and Evolution of Historical Epistemology

The origins and evolution of historical epistemology go back to the works of philosophers and historians who sought to unravel the intricate relationship between knowledge, time, and human experience. A fundamental aspect of understanding the development of historical epistemology lies in comprehending its philosophical underpinnings, which have undergone a transformative journey over centuries. At its core, historical epistemology investigates the construction of knowledge within specific contexts. Immanuel Kant and Johann Gottfried Herder recognized the importance of historical perspectives in shaping our understanding of truth, the idea advocated by Karl Marx. These early pioneers laid the groundwork for subsequent scholars to delve deeper into the interconnectedness between history and epistemology.

In the twentieth century, historical epistemology became a distinct field of study. The contributions of Gaston Bachelard (1938) and Michel Foucault (1969, 1972) propelled this intellectual endeavor forward by challenging traditional notions of objectivity and highlighting the role played by history in shaping knowledge systems. Bachelard's concept of "epistemological rupture" emphasized how scientific progress often necessitates a break from established paradigms, underscoring the dynamic nature of knowledge production across time. Furthermore, Foucault's notion of "episteme" introduced a novel perspective on historical analysis by examining how power structures influence knowledge and who has access to it. This critical approach brought attention to previously overlooked voices within history while emphasizing that knowledge is not solely determined by rationality (Pearce & Wolenski 1988) or empirical evidence but rather emerges through complex social processes.

As the methodological approach to historical epistemology progressed, scholars began exploring interdisciplinary avenues that enriched its theoretical framework. The integration of anthropology allowed for a deeper examination of cultural factors that shape knowledge production throughout different periods.

Presently, the advent of digital technologies has revolutionized the field of historical epistemology by providing novel tools for data analysis and visualization. The availability of vast archives and databases has opened new possibilities for researchers to trace the evolution of knowledge systems across time in unprecedented ways. This interdisciplinary approach, coupled with technological advancements, has broadened the scope of historical epistemology and enhanced our understanding of the construction of knowledge within specific temporal contexts.

A continuous reevaluation of established philosophical frameworks has marked the origins and evolution of historical epistemology. From Kant to Bachelard, Foucault to contemporary scholars, this field has undergone significant transformations that reflect an ongoing quest to unravel the intricate relationship between history, knowledge, and human experience. By embracing interdisciplinary approaches and leveraging technological advancements, historical epistemology continues to shed light on how different societies have come to know what they know.

Key Thinkers and Their Contributions

Ideas from Michel Foucault's (Rabinow et al., 1983) examination of power dynamics shaping scientific discourses to Thomas Kuhn's (1996) paradigm shifts in scientific revolutions or Ludwik Fleck's concept of thought collectives influencing scientific discovery have revolutionized our perception of how facts are established.

Michel Foucault's "discourse analysis" examines how language shapes and governs our perception of reality. By reviewing historical texts, he revealed how power structures operate through discursive practices, influencing what can be known and who gets to define knowledge. Thomas Kuhn challenged traditional views about scientific progress by introducing the notion of paradigm shifts. Kuhn argued that scientific disciplines are not characterized by gradual accumulation of knowledge but rather undergo radical transformations when new ones replace old paradigms. This idea sparked debates about the nature of scientific revolutions and fundamentally changed our understanding of how science progresses. In addition to Foucault and Kuhn, Donna Haraway has made significant contributions to historical epistemology through her work on situated knowledge. Haraway argues that all forms of knowledge are situated within specific social contexts and influenced by power relations. She emphasizes acknowledging multiple perspectives and voices to challenge dominant narratives perpetuating inequality and exclusion. Sandra Harding's standpoint theory is another crucial contribution to this field. Harding contends that knowledge is always shaped by the social position from which it emerges; therefore, marginalized groups possess unique insights that challenge dominant epistemic frameworks. Her work highlights the need for diverse voices in shaping our understanding of the world. Lastly, Bruno Latour's actor-network theory has influenced historical epistemology by challenging traditional distinctions between human and non-human actors in knowledge production. Latour argues that objects and technologies actively shape knowledge, often mediating between humans and the world. Actor-network theory encourages us to consider the complex interaction network contributing to knowledge formation, transcending the traditional focus on human agency. These thinkers have significantly shaped historical epistemology by challenging conventional ideas about the production of knowledge, its dissemination, and legitimization. Their contributions have paved the way for critical examinations of power dynamics, social contexts, diverse perspectives, and non-human actors in shaping our understanding of the world.

Critiques of the Scope of Historical Epistemology

Critiques of the scope of historical epistemology have been a subject of debate. One fundamental critique revolves around the narrow focus of historical epistemology, prioritizing well-established disciplines and academic traditions over marginalized or overlooked forms of knowledge production. Doing so perpetuates a biased narrative that favors dominant perspectives while neglecting alternative ways of knowing.

Another critique targets the tendency of historical epistemology to overlook social and cultural contexts in favor of abstract analyses by emphasizing conceptual frameworks and intellectual genealogies, which neglect crucial societal factors influencing knowledge production. They contend that understanding how social structures shape what is considered valid or valuable knowledge is essential for a holistic comprehension of epistemic practices throughout history.

Does historical epistemology adequately address power dynamics and ideology in knowledge construction processes? By focusing solely on tracing ideas across time, this approach fails to fully address how various forms of oppression have shaped what counts as legitimate knowledge in different periods. The exclusionary nature of specific disciplines or dominant paradigms may perpetuate biases rooted in classism, racism, sexism, or other oppressive ideologies. In response to these critiques, I propose expanding the scope of historical epistemology by incorporating interdisciplinary approaches into its framework. By integrating insights from sociology, anthropology, and critical theory, historical epistemology can better examine the interplay between knowledge and power structures. Such an approach would enable a more nuanced understanding of the social, political, and cultural dimensions that shape knowledge production processes.

Challenges to Empirical Methods in Historical Epistemology

Empirical methods in historical epistemology encounter challenges that impede acquiring accurate knowledge about the past. The first formidable obstacle is the scarcity and unreliability of available historical sources. Documents, artifacts, and testimonies often suffer significant gaps, biases, or deliberate distortions. Consequently, historians need help sifting through these imperfect pieces of evidence to construct a coherent narrative approximating the truth. Interpreting historical sources necessitates grappling with the complexities of context and perspective. Each historical event unfolds within a unique social, cultural, political, and economic web that shapes its meaning and significance. However, discerning these intricate connections from fragmentary evidence can be an intellectual labyrinth for historians and archaeologists. Their interpretations are inherently subjective as they attempt to reconstruct past realities through culturally conditioned lenses.

Another challenge arises from the limitations imposed by temporal distance. As time separates us from historical events, our understanding becomes unavoidably colored by present knowledge and beliefs. Anachronistic interpretations often arise when we apply contemporary concepts to past phenomena. Such distortions hinder our ability to grasp how people at a particular time understood their world and made sense of their experiences.

Historical epistemology recognizes that individuals' subjective perspectives are crucial in shaping knowledge production throughout history. However, capturing these subjective elements poses significant difficulties for empirical approaches reliant on quantifiable data or observable phenomena. Empirical methods need help to account for intangible aspects of human experience that elude objective measurement. They prioritize broad patterns over individual experiences or exceptional cases due to their reliance on statistical analysis and generalization. While this approach is valuable in identifying overarching trends or regularities in historical development, it risks overlooking unique insights offered by marginalized voices or unconventional narratives that defy established norms. Finally, empirical methods confront inherent limitations when addressing questions that transcend materiality or causality alone. Historical epistemology embraces the complexity of human thought, belief systems, and values as central to understanding historical developments. However, empirical methods primarily focus on observable phenomena and causal relationships, often paying attention to the intricate interplay between ideas, emotions, and social dynamics that shape knowledge production. Overcoming these challenges requires historians to exercise caution in their interpretations while embracing interdisciplinary approaches incorporating diverse methodologies to approach historical inquiry holistically.

Contemporary Knowledge Production

Contemporary knowledge production is profoundly impacted by the insights and methodologies of historical epistemology, yielding significant implications that challenge conventional modes of inquiry. This approach unveils the contingent nature of truth claims and disrupts the notion of universal, ahistorical knowledge by investigating the historical contexts of knowledge production. Such an understanding engenders a rupture within traditional disciplinary boundaries, fostering interdisciplinary collaborations that enrich scholarly endeavors. One implication lies in the deconstruction of binary oppositions between science and humanities. Historical epistemology illuminates how social and cultural factors shape scientific knowledge by moving beyond positivist notions that posit science as an objective pursuit detached from human subjectivity. This realization challenges contemporary knowledge production by emphasizing the entanglement of scientific practices with historical contingencies, thereby enabling a more nuanced understanding of scientific claims. Integrating historical perspectives prompts scientists to reflect critically on their methods and assumptions, encouraging reflexivity within their fields.

Because historical epistemology also sheds light on the role of power structures in shaping knowledge production processes, it reveals how dominant narratives serve interests or reinforce existing power dynamics. Contemporary scholars question taken-for-granted assumptions and explore alternative ways of producing knowledge that challenge oppressive systems by disrupting hegemonic discourses and fostering critical engagement with established knowledge. The implications for contemporary knowledge production extend beyond individual disciplines to encompass broader societal dimensions. Contemporary historical epistemology emphasizes the situatedness of knowledge within specific historical moments, urging scholars to recognize how different forms of knowing emerge within diverse socio-cultural contexts. This recognition engenders a more inclusive approach to scholarship that acknowledges multiple ways of knowing and values diverse perspectives.

Furthermore, contemporary historical epistemology underscores the importance of reflexivity in research practices by unveiling hidden biases embedded within scholarly work. It compels researchers to evaluate their positionality critically and acknowledge how their social, cultural, and historical contexts shape their understanding of knowledge. This self-reflexivity enhances the rigor and validity of contemporary knowledge production by fostering awareness of potential blind spots or biases.

Non-Marxist Historical Materialism

Non-Marxist historical materialism is a critical framework that analyzes historical development and societal change through the lens of material conditions while departing from traditional, deterministic Marxist interpretations. The key principles of non-Marxist historical materialism emphasize the significance of economic factors in driving historical progress but also acknowledge the influence of non-economic factors such as cultural practices, political structures, and individual agency. Unlike traditional Marxist approaches that prioritize class struggle as the primary driver of history, non-Marxist historical materialism recognizes multiple sources of social change. It thus emphasizes the complex interplay between various forces.

One crucial aspect to consider when critiquing Marx's interpretation is his deterministic view that economic factors control all other aspects of society. This reductionist approach neglects the role of culture, ideology, and individual agency in shaping historical events. Proponents of non-Marxist historical materialism offer alternative perspectives that challenge Marx's emphasis on economic determinism. These perspectives highlight how cultural practices, ideas, and social institutions can independently affect historical development. Examining the role of culture and ideology in shaping history reveals their significant impact on societal change. Cultural beliefs shape people's motivations and actions while influencing political ideologies, further molding social structures. Understanding these dynamics is crucial for comprehending how societies evolve.

Nevertheless, critically evaluating this theory's applicability and limitations is essential. While non-Marxist historical materialism offers valuable insights into understanding societal development beyond strict economic determinism, it has its restrictions. Analyzing these limitations will help us understand where this theory falls short or requires further refinement.

Key Principles of Non-Marxist Historical Materialism

Non-Marxist historical materialism encompasses a set of fundamental principles that provide a distinct perspective on analyzing history, society, and culture. At its core, this theoretical framework emphasizes the significance of material conditions in shaping human existence and development. Contrary to Marxist historical materialism, non-Marxist theorists emphasize the multidimensionality and complexity of historical processes, rejecting the reductionist tendencies inherent in traditional Marxist approaches. One fundamental principle of non-Marxist historical materialism is acknowledging the interplay between various factors in shaping history. Rather than reducing history to an economic base or class struggle alone, theorists recognize that multiple social, cultural, political, and ideological forces operate simultaneously within a given historical context. This recognition expands our understanding by illuminating how different power relations intertwine with economic factors to shape societal dynamics. Non-Marxist historical materialism highlights the importance of agency in shaping history. While acknowledging the influence of structural factors such as economic conditions or institutional arrangements, this perspective rejects determinism and recognizes that individuals possess the agency to act upon their circumstances. By considering agency as an integral part of historical analysis, theorists shed light on how individual actions can challenge or reinforce existing structures and contribute to transformative social change.

Another critical principle is its emphasis on contingency and unpredictability. Unlike deterministic approaches that suggest a linear progression or predetermined outcomes based solely on economic forces or class struggle, this perspective recognizes that history unfolds through complex interactions among various actors and unforeseen events. By acknowledging contingency, non-Marxist theorists embrace uncertainty as an inherent characteristic of human affairs rather than seeking to impose a rigid teleological framework onto historical processes.

Furthermore, non-Marxist historical materialism underscores the importance of cultural production in shaping societies throughout history. It recognizes that ideas do not merely reflect economic conditions but actively participate in constructing socio-cultural realities. Theorists explore how cultural practices, discourses, and symbolic systems contribute to the reproduction or transformation of social structures. This emphasis on culture enriches our understanding of history by highlighting the role of language, art, religion, and other cultural forms in shaping human experiences.

Criticisms of Marx's Interpretation of Historical Materialism

Marx's interpretation of historical materialism has garnered significant criticism from non-Marxist scholars, who argue that it oversimplifies the complexities of historical development and fails to account for other essential factors shaping societies. One of the primary criticisms is that Marx's focus on economic forces as the driving factor in history neglects the influence of political, cultural, and ideological dynamics. While Marx acknowledges the role of class struggle in shaping societies, his reductionist approach overlooks the intricate interplay between various social forces. Non-Marxist scholars contend that economic determinism cannot explain historical change. They argue that political institutions and power structures are crucial in shaping societies. For instance, in pre-industrial civilizations such as ancient Egypt or feudal Europe, political elites held significant sway over societal progress and development. The decisions made by rulers and their ability to consolidate power were vital factors in determining the trajectory of these societies. Marx's narrow focus on economic class struggle overlooks the importance of other social divisions such as gender, ethnicity, and religion. These aspects contribute significantly to social stratification and uniquely impact historical processes. For example, feminist scholars highlight how gender dynamics shape societal power relations and influence socioeconomic structures.

Moreover, non-Marxist critics contend that cultural factors play a substantial role in shaping historical development. Culture encompasses shared values, beliefs, customs, arts, and intellectual achievements within a society or civilization. It influences human behavior and shapes societal norms, which can profoundly affect political systems or economic practices. In his history analysis, Marx undermines our understanding of how ideas shape material conditions by disregarding these cultural dimensions.

Another criticism centers around Marx's teleological view of history—his belief that history inevitably moves towards communism through a series of predetermined stages. Critics argue that such a deterministic perspective fails to acknowledge contingency—that historical outcomes are influenced by chance events or individual agency. Historical processes are complex and multifaceted and cannot be reduced to a linear progression towards a predetermined end. Non-Marxist scholars have offered substantial criticisms of Marx's interpretation of historical materialism. In sum, Marx's analysis oversimplifies the complexity of historical development by focusing primarily on economic factors and class struggle while neglecting political, cultural, and other social dynamics. Furthermore, his teleological view of history fails to account for contingency and individual agency. Incorporating these overlooked factors into archaeological analysis is crucial to gaining a more comprehensive understanding of history.

Alternative Perspectives Within Non-Marxist Historical Materialism

Within non-Marxist historical materialism, alternative perspectives challenge conventional notions and present a more nuanced understanding of socio-historical dynamics. These perspectives offer fresh insights into the complex interplay between material conditions and human agency, fostering a more profound comprehension of historical processes. One such alternative perspective is the traditional cultural-material approach, which posits that culture plays a pivotal role in shaping historical developments alongside economic factors. Contrary to traditional Marxist accounts prioritizing economic determinism, the cultural-material approach emphasizes the reciprocal relationship between culture and material conditions. It recognizes that culture influences and is influenced by economic structures, acknowledging their inseparable nature. This perspective contends that historical transformations cannot be fully comprehended without considering how cultural practices and beliefs interact with material circumstances. By recognizing culture as an active force within history, this approach challenges deterministic narratives that reduce social change solely to economic forces.

Another alternative perspective within non-Marxist historical materialism is the gendered-material analysis, which illuminates how gender relations intersect with material conditions to shape historical outcomes. This perspective highlights how patriarchy and gender hierarchies are intertwined with economic systems, exerting a profound influence on social dynamics throughout history. By foregrounding gender as a critical analytical category, this approach uncovers overlooked dimensions of power relations and sheds light on women's experiences in different socioeconomic contexts.

Postcolonial approaches within non-Marxist historical materialism contest Eurocentric biases inherent in traditional Marxist frameworks. These perspectives expose how colonial encounters shaped economic relationships, cultural identities, and knowledge production. They emphasize the agency of subaltern groups in resisting colonial domination while analyzing how imperialism altered local economies and disrupted indigenous ways of life.

Furthermore, ecological approaches highlight the intricate connection between environmental factors and human societies throughout history. These perspectives underscore the impact of ecological constraints on socioeconomic systems while investigating how human activities have shaped ecosystems over time. Integrating environmental considerations into historical analysis challenges anthropocentric perspectives and emphasizes the reciprocal relationship between humans and nature.

The Role of Culture and Ideology in Shaping History

Culture and ideology have played a pivotal role in shaping history, fundamentally influencing the trajectory of societies and civilizations. The intricate interplay between culture and ideology has been instrumental in determining historical events and their outcomes. This multifaceted relationship between culture, ideology, and history has been a subject of great interest among scholars studying non-Marxist historical materialism. One cannot undermine the influence of culture in shaping history. Cultural norms and traditions have molded societies' behaviors and interactions with one another, driving social change or reinforcing existing power structures. For instance, in ancient Greece during the fifth century BCE, cultural values such as democracy and individual freedom laid the foundation for Athenian society to flourish into what is now considered one of history's greatest civilizations. These cultural ideals shaped political institutions and influenced artistic expressions like theater and philosophy. Similarly, during Europe's Age of Enlightenment in the eighteenth century CE, intellectual movements centered around reason and scientific inquiry led to profound shifts in societal thinking about politics, religion, and human rights.

Ideology also plays a critical role in shaping history by providing frameworks through which individuals interpret their worldviews. Based on their nature, ideological systems can catalyze change or reinforce existing power dynamics. For example, nationalist ideologies emerged during the nineteenth century as powerful forces that fueled struggles for independence across various regions worldwide. These ideologies united people under common beliefs regarding national identity and self-determination while challenging colonial powers' dominance.

Ideologies have often acted as drivers for economic systems shaping societies' development. Capitalism's emergence during Europe's Industrial Revolution fundamentally transformed economic structures globally by promoting free markets based on private ownership of resources. This ideological shift had profound implications for labor, trade, and wealth distribution, resulting in unprecedented social and economic changes. The complex interaction between culture and ideology becomes particularly evident when examining historical events involving clashes of civilizations or ideological conflicts. The intricate relationship between culture, ideology, and history is a fascinating subject of study within the context of non-Marxist historical materialism.

Evaluating the Applicability and Limitations of Non-Marxist Historical Materialism

To comprehensively assess the applicability and limitations of non-Marxist historical materialism, it is imperative to delve into its intricacies and analyze its merits within a broader academic framework. This theory presents a novel perspective on historical development, emphasizing the role of material conditions in shaping societal structures without subscribing to Marxist ideals. However, while it offers valuable insights into the dynamics of social change, it is not immune to criticism and faces certain limitations that must be acknowledged. One significant strength of non-Marxist historical materialism is its ability to provide a nuanced understanding of how material conditions interact with cultural, political, and ideological factors in shaping societies over time. By focusing on the interplay between economic forces and other societal dimensions, this theory offers a comprehensive lens through which historians can examine historical processes. For instance, when analyzing the rise of industrial capitalism in Europe during the nineteenth century, non-Marxist historical materialism allows scholars to appreciate how technological advancements and changing modes of production influenced economic systems, social hierarchies, and cultural norms.

However, it is crucial to recognize that this theory has certain limitations. One key concern pertains to its potential reductionism—an oversimplification or neglect of complex factors. Non-Marxist historical materialism prioritizes economic determinants over other aspects, such as ideology or individual agency. Consequently, there is a risk of overlooking the multifaceted nature of historical events and underestimating their complexity. For example, when examining revolutions or independence movements in various regions across the globe, solely focusing on economic factors may fail to capture the full range of motivations behind these transformative events.

Another limitation relates to potential cultural biases inherent within non-Marxist historical materialism. The theories developed under this framework have primarily emerged from Western academia and are often rooted in Eurocentric perspectives. As a result, there is a risk of overlooking or downplaying the significance of non-Western cultures and histories in shaping global dynamics. For instance, when studying the economic development of Asian societies, exclusive reliance on non-Marxist historical materialism may inadvertently perpetuate Orientalist views that fail to recognize the agency and contributions of these societies. Evaluating the applicability and limitations of non-Marxist historical materialism necessitates a meticulous examination of its strengths and weaknesses within a broader academic discourse. While this theory offers valuable insights into the interplay between material conditions and societal structures, it has shortcomings. Its potential reductionism and cultural biases must be acknowledged to ensure a comprehensive understanding of historical processes. By critically assessing this theory's merits and limitations, scholars can strive towards more nuanced interpretations incorporating diverse perspectives while analyzing the complexities of social change throughout history.

3 Part Two: The Poznań School of Methodology: Criticisms, Relevance, and Influence

3.1 Philosophy of Archaeology of the Poznań School of Methodology

I will show that the Poznań School of Methodology has significantly contributed to our understanding of the epistemological foundations, interpretation practices, objectivity, subjectivity in research and historical contexts, challenges, and limitations within archaeology.

Epistemological Foundations of Archaeology

The philosophy of archaeology of the Poznań School of Methodology and the epistemological foundation of archaeology offers a thought-provoking perspective on nature and the acquisition of archaeological knowledge. The School advocated for an approach that embraces empirical evidence and theoretical frameworks to uncover the past. It acknowledged the subjective nature inherent in archaeological research by challenging traditional positivist views that perceive knowledge as objective and independent from human interpretation. Recognizing theory's pivotal role highlights how human subjectivity intertwines with knowledge production in archaeology.

The archaeological evidence is inherently fragmentary and incomplete. Instead of perceiving these gaps as obstacles to overcome or dismiss, the Poznań School encourages archaeologists to embrace them as opportunities for creative thinking. Acknowledging what is missing or uncertain in their data prompted researchers to develop alternative hypotheses and critically reflect on their assumptions and biases.

Role of Interpretation in Archaeological Practice

The role of interpretation in archaeological practice is multifaceted and intricate and requires careful consideration. Interpretation has its challenges and complexities. It involves deciphering the meaning behind archaeological evidence and constructing narratives about past human behavior and cultural practices. The Poznań School recognized that this task is inherently subjective and influenced by various factors such as personal biases, theoretical frameworks, and social contexts. This recognition highlights the need for archaeologists to engage with their interpretations critically and be aware of how these may shape their understanding of the past.

One key aspect the Poznań School emphasized is the importance of reflexivity in archaeological interpretation. Reflexivity encourages archaeologists to ponder their positionality and acknowledge how it may influence their interpretations. By doing so, researchers can strive for a more nuanced understanding of archaeological data and avoid perpetuating biased or misleading narratives about the past. Furthermore, interpretation in archaeological practice is not solely concerned with individual artifacts or features but extends to broader questions about cultural change and continuity over time. The Poznań School stressed that interpretations should be grounded in a comprehensive understanding of historical contexts, cultural dynamics, and social processes. Such an integrative approach enables archaeologists to move beyond mere description towards more meaningful explanations about past societies. However, interpreting complex archaeological evidence presents considerable challenges due to fragmentation, preservation biases, or incomplete datasets. When constructing interpretations, these limitations require careful consideration to ensure they relate to robust evidence rather than speculative assumptions. Moreover, interpretation in archaeology does not end with individual excavations or research projects but extends into broader conversations within the discipline.

Objectivity and Subjectivity in Archaeological Research

Objectivity and subjectivity are fundamental concepts in archaeological research, shaping how scholars approach the study of the past. Within the philosophy of archaeology of the Poznań School of Methodology, these notions intertwine, leading to a complex understanding of how knowledge is constructed and interpreted. In this context, objectivity refers to the idea that archaeological research should aim for impartiality and to uncover an unadulterated truth about past human societies. However, subjectivity recognizes that researchers bring their perspectives, biases, and preconceptions to their work. Due to inherent limitations, the School argued that complete objectivity is unattainable in archaeological research.

Consequently, archaeologists must rely on subjective interpretations when analyzing available evidence. This subjectivity arises from individual backgrounds, personal experiences, and cultural influences that shape a researcher's worldview. Moreover, subjectivity plays a crucial role in determining which aspects of the past are considered significant for investigation.

The Poznań School emphasized that archaeological research is not isolated but remains within specific socio-cultural contexts. These contexts significantly impact how archaeologists frame research questions and select sites for excavation. For instance, a society might prioritize studying monumental architecture while neglecting everyday domestic artifacts, as was common in colonial Egyptian archaeology or studies of the Roman influence in French archaeology. Such biases can lead to incomplete understandings of past societies without alternative perspectives. The School advocated for a hermeneutic approach that acknowledges that meanings are inherently subjective and vary across different cultural contexts. This perspective rejects any notion of an objective truth waiting to be uncovered but emphasizes multiple valid interpretations based on diverse viewpoints. Within this framework, archaeologists must navigate between acknowledging their subjective positionality and striving for objectivity in their research. The relationship between objectivity and subjectivity becomes even more intricate when considering interpretation and analysis methods employed by archaeologists.

The Poznań School suggested that researchers embrace reflexivity, acknowledging and critically examining their biases and assumptions throughout the research process. By doing so, archaeologists can minimize the impact of subjectivity on their interpretations to approach a more objective understanding of past societies. By embracing reflexivity and considering multiple valid interpretations, archaeologists can navigate the complexities of objectivity and subjectivity to construct a nuanced understanding of past human societies.

The Influence of Historical Contexts on Archaeological Analysis

Examining archaeological artifacts and sites necessitates a comprehensive understanding of their historical contexts, as these contexts profoundly influence the analysis. The philosophy of archaeology of the Poznań School recognizes this fundamental relationship between history and interpretation. This philosophy offers valuable insights into historical contexts and provides crucial background information that helps archaeologists interpret artifacts accurately.

Finally, the Poznań School recognized that different historical periods necessitate distinct analytical approaches due to varying social norms and material culture practices. By considering these specificities, archaeologists gain a deeper understanding of how people lived during different eras. Historical contexts illuminate broader societal dynamics that shape archaeological landscapes. Political upheavals or socioeconomic systems can significantly impact material remains found at sites. Understanding such dynamics allows researchers to explore how power relations affected various aspects of society—from technological advancements to artistic expressions.

Challenges and Limitations in the Poznań School's Methodology

However, it is crucial to acknowledge the challenges and limitations that arise from this approach. One such challenge lies in its heavy reliance on empirical data. While this commitment to empirical data is commendable, it can be problematic due to archaeological materials' inherent biases and limitations. These materials are fragmentary, incomplete, or subject to various interpretations, resulting in potential inaccuracies in reconstructing the past.

Furthermore, another limitation lies in the approach to interpretation (Kuokkanen 1988). This methodology incorporates elements of a positivist perspective, mixing objectivity and scientific rigor with subjective interpretations. While objectivity is undoubtedly valuable within archaeology, it is essential to recognize alternative interpretive frameworks that may offer new insights into the past.

We must also consider the temporal and regional limitations. The Poznań School's methodology primarily focuses on Central European archaeology during specific historical periods, neglecting other regions or periods outside this scope. This narrow focus limits its applicability and relevance beyond these specific contexts, potentially overlooking critical insights that archaeologists might gain from studying different regions or periods.

Moreover, one key challenge facing the School's methodology is its limited engagement with alternative theoretical debates within archaeology. While it heavily relies on empirical data collection and interpretation techniques, there appears to be a relative absence of engagement with broader theoretical frameworks that have emerged within archaeology over recent decades. By not integrating these theoretical advancements into their methodology, they may miss opportunities for innovative approaches and deeper understandings of human societies in the past.

Lastly, the methodology's heavy emphasis on reconstructing past societies based on material remains may inadvertently overshadow other important aspects of archaeological inquiry. By focusing on artifacts and structures, there is a risk of neglecting the social, cultural, and ideological dimensions that shaped ancient societies. A more holistic approach considering a broader range of evidence, such as oral traditions or ethnographic analogies, could provide a more comprehensive understanding of past civilizations.

3.2 *Poznań School and Historical Epistemology*

The School focused on criticism of the positivist understanding of history that emphasizes the dominance of empirical evidence and objective analysis and delved deeper into the complexities of historical research, questioning the limits of the positivist framework. This methodology gradually shifted towards a more interpretive and hermeneutical approach, recognizing that historical knowledge does not simply derive from objective facts but also subjective interpretations. Historians are not neutral observers but active participants in constructing historical narratives. This realization prompted scholars to explore how various social, cultural, and political factors influence the production and interpretation of historical knowledge.

Another significant development in the School's historical epistemology is its engagement with what is presently labeled as postmodern critiques, i.e., challenging traditional notions of objectivity and truth and arguing that all knowledge is inherently situated within specific socio-cultural contexts. In response to these critiques, scholars within the Poznań School sought to incorporate postmodern insights into their methodology while maintaining a commitment to rigorous scholarship. This integration between postmodernism and traditional historiography resulted in an increased emphasis on reflexivity within the School's approach. Historians were encouraged to reflect critically on their biases, assumptions, and preconceptions throughout their research. Acknowledging their positionality as historians, they aimed to produce more nuanced and self-aware interpretations of historical events.

Furthermore, incorporating interdisciplinary perspectives into historical research is a notable aspect of this evolution. The School recognized that history cannot be studied in isolation from other disciplines such as sociology, anthropology, or philosophy. By drawing insights from these diverse fields, scholars could enrich their understanding of past events and develop more comprehensive explanations for historical phenomena.

Critique of Poznań School's Methodology Principles

While the Poznań School of Methodology has contributed to the study of historical epistemology, it is not immune to critique. The emphasis on historicity, questioning of universal truths, focus on subjectivity, and lack of clarity regarding application have all been identified as areas of concern. Addressing these critiques would strengthen the methodology principles associated with this School and enhance our understanding of how knowledge is constructed.

One fundamental criticism pertains to the School's heavy reliance on historicity as a guiding principle. While acknowledging the importance of historical context is crucial in studying epistemological developments, an exclusive focus on historicity may lead to neglecting other equally significant factors. By privileged historical circumstances, this approach risks oversimplifying complex (entangled is the fancy word used these days) epistemological phenomena and disregarding the role played by other factors, such as social dynamics or individual agency.

Rejecting universal truths has raised concerns. While it is certainly important to recognize that knowledge is socially constructed and influenced by various contextual factors, completely dismissing universal truths may undermine our ability to make meaningful claims about human understanding across time and space. By denying any possibility of objective knowledge or overarching theories, this approach limits our capacity to engage in fruitful comparative analysis or identify common patterns within different epistemic frameworks.

Influence of Social Factors on Poznań School's Epistemology

The historical epistemology of the Poznań School is deeply intertwined with various social factors that have significantly influenced its development. One is the broader intellectual foundation (especially the legacy of Tadeusz Kotarbiński, Kazimierz Ajdukiewicz, Stanisław and Maria Ossowski, (see Wybraniec-Skardowska & Garrido 2018) dated to the early twentieth century, marked by significant socio-political changes and intellectual debates. The tumultuous socio-political landscape of this era, characterized by World War I and the aftermath of political revolutions, profoundly impacted the epistemological orientation adopted by members of the Poznań School who claimed affinity to this tradition. The disruptive nature of these events spurred a reevaluation of traditional philosophical frameworks and prompted scholars to seek alternative approaches to understanding knowledge and reality.

Another crucial social factor was its geographical location within Central Europe. Situated at the crossroads between Western and Eastern intellectual traditions, scholars associated with the School were exposed to diverse philosophical currents from both sides. This unique position fostered an intellectual environment characterized by rich pluralism and cross-pollination of ideas. After WWI, the cultural diversity of Central Europe contributed to a vibrant dialogue among scholars, resulting in an eclectic blend of influences within their epistemological framework. For instance, Tadeusz Kotarbiński drew inspiration from Western analytic philosophy and Eastern phenomenology, incorporating elements from each tradition into their knowledge acquisition and justification conceptualizations.

Relevance and Limitations of Poznań School's Historical Approach

While the historical approach of the Poznań School of Methodology provides valuable insights into the development of scientific knowledge, it has certain limitations. A more interdisciplinary perspective incorporating sociological, psychological, and philosophical aspects would enhance understanding the complexities of generating scientific ideas. Additionally, recognizing the role of institutions and collaborations alongside individual contributions would provide a more comprehensive view of how scientific knowledge evolves. Despite these limitations, historical analysis remains relevant in illuminating past developments and guiding future scientific endeavors.

This School emphasized the importance of historical context and its influence on scientific knowledge. Examining scientific ideas' development over time seeks to uncover how ideas emerged, evolved, and shaped our understanding of reality. This approach is relevant as it highlights the dynamic nature of science and acknowledges that scientific knowledge is not fixed but rather subject to change.

One limitation of the Poznań School's historical approach is its heavy reliance on historical analysis alone. While historical context undoubtedly provides valuable insights into how scientific ideas have developed, it does not necessarily provide a complete understanding of why specific ideas gained prominence over others. The complex interplay between social, cultural, political, and economic factors that shape scientific knowledge cannot be adequately captured through a purely historical lens. An interdisciplinary approach incorporating sociological, psychological, and philosophical perspectives would be necessary to comprehend these dynamics fully.

Another limitation lies in the potential for bias when interpreting history. Historical analysis is inherently subjective since historians must select which events or figures to include or exclude from their narratives based on their judgments and interpretations. This subjectivity can introduce biases that may distort our understanding of past events or misrepresent the contributions made by specific individuals or groups.

Contemporary Criticisms and Alternative Perspectives on Poznań School of Methodology

Contemporary criticisms and alternative perspectives on the Poznań School have emerged in recent years, challenging the traditional understanding of its methodology. I see a challenge in reducing historical epistemology to formal logical rules or methods and argue that this reductionist approach oversimplifies complex cognitive processes involved in knowledge production throughout history. Instead, I propose adopting a more holistic perspective that considers social dynamics, power structures, and contextual factors influencing epistemic practices.

Another alternative perspective suggests reframing historical epistemology as an interdisciplinary field rather than solely relying on philosophical analysis. This approach incorporates insights from other disciplines, such as sociology, anthropology, and psychology, to enrich our understanding of shaping knowledge over time. By incorporating diverse disciplinary perspectives into historical epistemology research, scholars can explore new avenues for investigating the complexities inherent in human cognition.

Moreover, overemphasizing textual analysis within the School's methodologies is another limitation. Focusing solely on written documents may overlook valuable sources of knowledge embedded in oral traditions or material culture. By expanding research methods beyond textual analysis to include oral histories or archaeological evidence, researchers can gain a more nuanced understanding of epistemic practices and their historical development.

Poznań School of Methodology and its Unique Marxist-Inspired Epistemology

The Poznań School of Methodology was critical of the traditional Marxist emphasis on economic determinism and argued that other factors, such as culture, ideology, and human agency, also influenced historical development. The School also developed a new method of historical analysis that emphasized the importance of studying concrete historical processes rather than abstract concepts. The School's work is still relevant today, providing a valuable framework for understanding the complex forces that shape historical development. Its emphasis on the importance of human agency and concrete historical processes is a reminder that history is not simply the product of economic forces but is shaped by the actions of individuals and groups.

Marxist Influence on Poznań School's Epistemology Methodology

The Poznań School, which should be renowned for its groundbreaking epistemological theories, owes a significant debt to Marxist influence. This distinctive intellectual movement challenged prevailing notions of knowledge and truth. Drawing inspiration from Marxist philosophy, the scholars sought to dismantle traditional epistemological frameworks and develop an alternative perspective rooted in social praxis. These intellectuals crafted a perplexing and thought-provoking approach to understanding reality by intertwining Marxist theory with their unique insights.

Central to the Marxist-inspired epistemology of the Poznań School is a recognition of the inherently social nature of knowledge production. Rejecting individualistic conceptions of truth-seeking prevalent in mainstream philosophy, they emphasized that knowledge is intricately tied to human interaction and collective experience. In this view, reality is not an objective entity waiting to be discovered by detached observers; instead, it emerges through dialectical processes shaped by historical and social contexts.

The Marxist influence on the Poznań School's epistemology becomes even more evident when examining their critique of positivism. Positivism, with its focus on empirical observation and scientific methods as the only valid sources of knowledge acquisition, was seen by these scholars as inadequate for comprehending complex social phenomena. They argued that positivism's reductionist tendencies failed to capture the intricate web of relationships that constitute human existence. Instead, they advocated for a holistic approach grounded in Marxism's emphasis on class struggle and historical materialism.

Another perplexing aspect of the School's epistemology lies in its rejection of objectivity as an attainable goal. Departing from traditional epistemologies that strive for impartiality and neutrality in knowledge production, these scholars embraced subjectivity as inherent within all forms of cognition. They argued that every act of knowing is inevitably influenced by one's social position, biases, and interests. Acknowledging this subjectivity aims to uncover hidden power structures and challenge dominant narratives perpetuating inequality and oppression. Thus, epistemology has broader implications for understanding reality beyond academia. It underscores the inseparability of theory and praxis, urging scholars to engage actively in transformative action rather than remaining detached observers. By intertwining theory with practice, these intellectuals sought to dismantle oppressive systems and pave the way for social change. Marxist influence permeates every facet of the School's unique epistemological framework. Bursting with revolutionary energy and challenging established traditions, their approach emphasizes the social nature of knowledge production, critiques positivism's limitations, embraces subjectivity as central to cognition, and advocates for a praxis-oriented engagement with reality. By blending Marxist theories with their distinct insights, the scholars of the Poznań School have crafted an intellectually stimulating perspective that continues to perplex and inspire scholars today.

The Role of Dialectics in Poznań School of Methodology

The Poznań School, with its unique Marxist-inspired epistemology, emphasizes the role of dialectics in understanding social phenomena. Dialectics, as a method of inquiry, aims to reveal the contradictions and tensions within a given social system. The School examined these contradictions to uncover the underlying dynamics that shape societal structures and relationships by recognizing that society is a complex web of opposing forces constantly interacting and influencing one another (presently labeled as entanglement). These opposing forces can manifest as class struggle, power differentials, or conflicting ideologies. Through the lens of dialectics, researchers analyze these contradictions to gain insights into how social systems function.

Dialectical thinking allows researchers to perceive society as a dynamic process rather than a static entity. They recognize that societal phenomena are not fixed but constantly evolving through internal conflicts and external influences. This perspective challenges traditional positivist approaches that view social phenomena as stable objects with predefined characteristics. In addition to analyzing societal contradictions, dialectical thinking also enables the identification of potential points of transformative action. Scholars can identify opportunities for intervention or change within existing systems by understanding how contradictory forces interact and influence each other. This approach aligns with Marxism's goal of transforming oppressive structures through class struggle.

The role of dialectics in the Poznan School's epistemology extends beyond theoretical analysis. It also informs the methodology employed by researchers within this School of thought. Dialectical materialism, a key component of Marxist philosophy, emphasizes the importance of grounding theoretical understanding in empirical evidence. Therefore, researchers engage in rigorous empirical research to support their dialectical analyses and ensure their work remains grounded in reality.

Analysis of Poznań School's of Methodology Critique of Positivism

The Poznań School, with its distinctive Marxist-inspired epistemology, offers a thought-provoking critique of positivism. The analysis the Schools' representatives provided challenges the foundations of traditional positivist methodologies by asserting that they fail to capture the complex and contradictory nature of social reality. The scholars argue that positivism's reliance on empirical observation and quantification oversimplifies social phenomena and disregards the inherent dialectical processes.

Positivism essentially reduces the epistemology of social reality into isolated variables that can be measured and manipulated statistically. This reductionist approach neglects the intricate interconnections between different elements within a social system. By fragmenting reality into separate entities, positivism fails to acknowledge how these entities interact and influence one another in dynamic ways. In contrast, the Marxist-inspired epistemology proposed by the Poznań School emphasizes the importance of understanding society as a totality where various elements are interconnected in a web of relations.

Another dimension of this critique revolves around the limitations imposed by positivism's emphasis on objectivity and neutrality. While positivism claims to be objective, it is inevitably influenced by underlying power structures and dominant ideologies. Scientific inquiry cannot be divorced from its socio-political context and must recognize its inherent biases. Instead of striving for illusory objectivity, they advocate for critical reflexivity, which acknowledges researchers' positionality and actively engages with power dynamics.

Another central tenet underpinning this critique is rejecting positivism's linear notion of progress. The proponents of Marxist-inspired epistemology assert that societal development is not unidirectional but instead marked by contradictions and conflicts. Positivist methodologies overlook these contradictions by assuming a linear trajectory towards progress based on simplistic cause-and-effect relationships. In contrast, the Poznań School argues for a dialectical understanding of social reality that recognizes the interplay between opposing forces, tensions, and contradictions and, thus, the conventional methodologies employed in social sciences.

By highlighting the reductionist tendencies, the inherent biases of objectivity, and the oversimplification of societal development within positivism, the School's Marxist-inspired epistemology offers an alternative framework for understanding complex social phenomena. The emphasis on interconnectedness, critical reflexivity, and dialectical thinking provokes scholars to reconsider their research approaches and encourages a more nuanced understanding of society. This critique serves as a valuable contribution to contemporary debates in social sciences by expanding our perspectives on how knowledge is constructed and how it can better capture the intricate realities of our world.

3.3 *Beyond Marxist Economic Determinism*

Evaluation of Historical Materialism of the Poznań School of Methodology

The examination of historical materialism within the framework of Poznań School of Methodology presents a captivating exploration into the intricate tapestry of Marxist-inspired epistemology. The scholars affiliated with this intellectual movement have embarked upon an audacious endeavor to redefine and reinvigorate the tenets of historical materialism, breathing new life into a theory that has long been subject to fervent debate and critique. By delving deep into the annals of history, these erudite thinkers seek to unravel its complex web of dialectical relationships, unveiling hidden truths and uncovering hitherto unexplored dimensions.

At the heart of my evaluation lies an ardent commitment to understanding history as a dynamic process shaped by material conditions. In stark contrast to traditional interpretations emphasizing class struggle as the primary driving force behind historical change, these scholars proposed a more nuanced approach that embraces multiple determinants. They argued that economic factors alone could not fully account for the intricacies and complexities witnessed throughout history; instead, cultural, social, and political forces must also be considered to grasp the full scope of human development. They have ventured beyond conventional Marxist analysis by incorporating interdisciplinary methodologies in their quest for knowledge. Drawing from sociology, anthropology, and linguistics, they illuminate previously uncharted realms within historical materialism. By analyzing linguistic structures embedded in historical texts or examining cultural practices intertwined with economic systems, they unravel hidden narratives that shed light on broader societal transformations.

The School's scholars tackle one of the most contentious debates surrounding Marxism head-on: its teleological nature. They challenge deterministic interpretations by positing that historical development is not predetermined but contingent upon an array of factors operating within specific contexts. This rejection of teleology injects a degree of unpredictability into their analysis—history becomes less linear and more enigmatic—a notion that perplexes and captivates those eager to comprehend humanity's past.

The scholars also demonstrated remarkable self-awareness and reflexivity in pursuing historical truth. They acknowledge the limitations inherent in any theoretical framework, including their Marxist-inspired epistemology. By openly engaging with alternative perspectives and critically examining their assumptions, they foster an atmosphere of intellectual rigor that is refreshing and indispensable in the academic realm. The evaluation of historical materialism within Poznań School encapsulates a profound engagement with Marxist-inspired epistemology that challenges conventional wisdom and expands the boundaries of knowledge. Through a multidimensional approach incorporating diverse disciplines and rejecting determinism, these erudite thinkers offer a fresh lens through which to view history's intricate tapestry. Their commitment to intellectual reflexivity ensures a dynamic and ever-evolving discourse that enriches our understanding of humanity's past while paving the way for future scholarship.

Comparison with Other Marxist-Inspired Epistemological Approaches

It is crucial to contextualize this perspective by examining its comparisons with other Marxist-inspired epistemological approaches. One notable point of departure between the Poznań School and other Marxist-inspired epistemological approaches lies in their respective ontological foundations. While some approaches prioritize a more deterministic view of reality, emphasizing objective structures and material conditions as primary forces shaping knowledge production, the Poznań School offers a refreshing departure from this deterministic stance. Instead, it recognizes the dialectical interplay between subjective agency and objective structures. This recognition allows for a more intricate understanding of how social actors actively shape knowledge production while being influenced by existing power relations.

Additionally, the Poznań School reveals its distinctiveness by emphasizing interdisciplinary collaboration when considering methodological differences among various Marxist-inspired epistemologies. Unlike approaches that might adopt more rigid disciplinary boundaries or rely solely on traditional modes of research inquiry, the Poznań School advocated for an integrative approach that draws upon multiple disciplines, such as sociology, philosophy, linguistics, and psychology. This interdisciplinary orientation enables comprehensive analyses encompassing diverse aspects of human experience while acknowledging their entanglement within broader socio-political contexts.

Lastly, it is worth noting that the Poznań School's Marxist-inspired epistemology stands out due to its emphasis on praxis. While many other approaches might prioritize theoretical analysis or abstraction, the School greatly emphasized the practical application of knowledge. This approach seeks to bridge the gap between theory and action by emphasizing praxis, aiming for transformative social change grounded in critical understanding.

3.4 Non-Marxist Historical Materialism

The Poznań School contributed significantly to advancing non-Marxist historical materialism as a valuable approach for studying complex social phenomena through concepts like intentional structures and interdisciplinary research methods. Non-Marxist Historical Materialism and the Poznań School of Methodology have emerged as prominent approaches in historical analysis, offering alternative perspectives to traditional Marxist frameworks. The School was pivotal in developing this non-Marxist strand, contributing innovative theories and methodologies that challenged traditional Marxist interpretations.

Rejecting reductionism, they argued that a complete analysis should encompass observable facts, individual experiences, and mental representations. One influential concept introduced by the Poznań School was "the intentional structure." According to this notion, individuals have intentions or purposes guiding their actions within a broader social context. These intentions can be shaped by cultural norms or ideologies prevalent at a given period. By examining these intentional structures alongside economic factors, historians gain insight into how ideas interact with material conditions to shape historical outcomes. The concept of "open systems" was also central to their methodology, positing that historical processes should not be viewed as closed or predetermined but rather as complex and interconnected webs of interactions. This departure from deterministic thinking allowed scholars within the Poznań School to explore historical events with greater flexibility and openness. Another crucial concept introduced by the Poznań School was "multiple determination." This notion challenged reductionist explanations by highlighting how historical events were shaped by numerous interrelated factors rather than being determined by a single cause or force. The School argued that understanding these interactions required an interdisciplinary approach incorporating insights from various fields such as sociology, psychology, and anthropology. This holistic perspective enabled researchers to unravel complex causal chains and shed light on previously overlooked aspects of historical processes.

One of the fundamental principles advocated by the Poznań School was the rejection of teleology in historical analysis. Unlike traditional Marxist approaches that emphasized predetermined outcomes based on economic determinism, the Poznań School argued for a more contingent view of history. They contended that social phenomena were not driven solely by economic forces but were shaped by multiple factors such as culture, ideology, and individual agency. By acknowledging this multifaceted nature of historical development, scholars within the Poznań School could delve deeper into understanding how different elements interacted and influenced each other within specific contexts.

The innovative ideas and methodologies introduced by the Poznań School have had a lasting impact on historical materialism. Their emphasis on open systems, rejection of teleology, multiple determination, and empirical research has broadened the horizons of historical analysis. By transcending traditional Marxist frameworks and embracing a more nuanced understanding of social phenomena, scholars within the Poznań School have enriched our comprehension of history as a complex interplay between various factors. Another significant contribution by the Poznań School was their emphasis on interdisciplinary research methods. They advocated incorporating insights from various disciplines, such as sociology, psychology, and linguistics, into historical analysis. This interdisciplinary approach enriched the understanding of historical phenomena by illuminating different facets of human behavior and societal dynamics. These contributions continue to inspire researchers in their quest for a deeper understanding of human societies and their development over time.

The Poznań School's methodology further enriched our understanding by introducing innovative analytical tools to capture the intricate nature of historical processes. Its emphasis on interdisciplinary collaboration allows historians to transcend disciplinary boundaries and integrate insights from adjacent fields such as sociology, anthropology, and economics. This multidimensional approach broadens the scope of analysis beyond traditional historiography by incorporating diverse perspectives that bring fresh angles into focus.

The implications of non-Marxist Historical Materialism also extend beyond mere reinterpretations; they challenge deterministic interpretations prevalent in conventional historiography. Instead of viewing history as an inexorable march towards predetermined outcomes driven solely by economic forces or class struggles—as espoused by Marxist theories—this framework recognizes the agency and contingency inherent in human actions throughout history. It acknowledges that individuals possess autonomy to shape their destinies within structural constraints—a departure from deterministic narratives that tend to oversimplify complex historical processes.

By embracing this alternative perspective rooted in non-Marxist Historical Materialism and the Poznań School's methodology, scholars gain access to a toolbox of analytical techniques to uncover hidden layers of historical phenomena. Incorporating material factors expands the horizons of historical inquiry, allowing for a more comprehensive understanding of past events and their enduring consequences. Moreover, by challenging deterministic interpretations, this approach encourages historians to engage with the complexities and uncertainties inherent in human history.

The implications of non-Marxist Historical Materialism and the Poznań School's methodology have transformative potential for our understanding of historical phenomena. Scholars can unveil previously unnoticed patterns by emphasizing material conditions, adopting interdisciplinary approaches, challenging deterministic narratives, and shedding light on obscured aspects of history. This alternative lens allows us to grasp the multifaceted nature of historical processes and enriches our comprehension beyond traditional historiography's limits.

4 Conclusions

On Philosophy and Archaeology Philosophy provides a framework for questioning assumptions, exploring epistemological issues, and reflecting on the nature of knowledge production in archaeology. Recognizing the relationship between philosophy, science, and archaeology is essential for bridging gaps between these disciplines. The ambiguity and open-endedness surrounding ancient artifacts give rise to differing interpretations, necessitating a deeper exploration of philosophical underpinnings. Philosophical debates in interpretations of ancient artifacts highlight the complexities involved in understanding past societies through material remains, emphasizing the need for critical thinking and multiple perspectives. Understanding the impact of philosophical perspectives on archaeological methodologies helps us to critically evaluate different approaches to excavation, analysis, interpretation, and dissemination. Whether it is grappling with questions about intentionality, the role of context in interpretation, or ethical considerations in archaeological practices, these debates challenge us to reflect on the nature of knowledge itself.

The philosophy of archaeology of the Poznań School provides a comprehensive framework for understanding various aspects of archaeological practice. This School offers valuable insights into how archaeologists can approach their work more effectively by addressing epistemological foundations, interpretation processes, objectivity-subjectivity dynamics, historical contexts' influence on analysis outcomes, and challenges faced within their methodology.

On Historical Epistemology Delving into historical epistemology enables us to appreciate the dynamic nature of knowledge. By examining the evolution of knowledge, understanding the role of historical context, recognizing cultural biases, exploring scientific methodologies, and critically evaluating traditional frameworks, we can better understand how our understanding of truth has evolved. This interdisciplinary field invites us to question and refine our epistemological foundations while fostering a more nuanced and comprehensive approach to knowledge acquisition.

Challenges to empirical methods in historical epistemology have emerged. The reliance on archival sources and textual analysis may limit our understanding of lived experiences and alternative ways of knowing. Incorporating interdisciplinary approaches from anthropology, sociology, and philosophy can help address these limitations. The implications for contemporary knowledge production are significant. Historical epistemology reminds us that our current understanding is not fixed or absolute but somewhat shaped by social, cultural, and political factors. It encourages us to examine our assumptions and biases when constructing knowledge critically.

Recognizing the influence of historical context allows us to appreciate how societal factors shape our understanding and interpretation of knowledge. Cultural biases play a significant role in shaping epistemology as well. Different cultures have different ways of knowing and valuing knowledge, which can lead to diverse perspectives on truth and reality. Understanding these biases is crucial for fostering cross-cultural dialogue and promoting a more inclusive approach to knowledge production.

Critiques of traditional epistemological frameworks challenge us to question established ways of knowing and encourage us to consider alternative perspectives and approaches that may offer new insights or address existing limitations.

The Poznań School has provided a critical perspective that challenges traditional positivist approaches by adopting Marxist principles. The School's critique of positivism highlights its limitations in explaining complex social phenomena. By emphasizing objectivity and empirical observation, positivism fails to account for subjective experiences and historical context. The Poznań School's critique offers an alternative social relations and historical development framework.

On Historical Materialism, Historical Epistemology, and Methods Evaluation of historical materialism reveals its significance in understanding societal change. Historical materialism provides a framework for analyzing how economic factors shape social structures and ideologies. This approach allows for a comprehensive analysis of society's development over time. Comparison with other Marxist-inspired epistemological approaches further highlights the uniqueness of the School's perspective. While sharing standard foundations with other Marxist theories, such as class struggle and historical materialism, the School offers distinct insights through its emphasis on dialectics and critique of positivism. The role of dialectics in the Poznań School cannot be understated. Dialectical thinking allows for a dynamic understanding of reality, recognizing contradictions and conflicts as essential elements of social processes. This approach enables a deeper analysis of societal structures and power relations.

While the Poznań School's historical approach offers valuable insights into understanding scientific knowledge production, it also has limitations. Its heavy reliance on historical analysis may overlook other important factors contributing to scientific progress.

The Poznań School's Marxist-inspired epistemology offers valuable insights into understanding society from a critical perspective. Its emphasis on dialectics, critique of positivism, evaluation of historical materialism, and comparison with other Marxist-inspired approaches contribute to a rich body of knowledge within philosophy and social sciences. Furthermore, social factors have played a significant role in shaping the epistemological framework of the Poznań School. The School's emphasis on historical analysis can be seen as a response to broader societal shifts towards contextual understanding and cultural relativism. The epistemological foundations of archaeology emphasize the importance of empirical evidence and logical reasoning in constructing knowledge about the past. This approach highlights the need for rigorous methodology and critical thinking in archaeological research. The role of interpretation in archaeological practice is another crucial aspect addressed by the School. However, its scholars acknowledge that interpretation can be subjective and influenced by personal biases, highlighting the need for reflexivity and transparency in archaeological interpretations. They advocate for a balanced approach that acknowledges both objective facts and subjective interpretations.

The influence of historical contexts on archaeological analysis is another significant aspect discussed by the Poznań School. They emphasize that historical contexts shape our understanding of material culture and influence our interpretations. Therefore, archaeologists must consider historical contexts when analyzing artifacts or sites. Despite its archaeological contributions, the Poznań School's methodology also faces challenges and limitations. These include issues related to bias in data collection, difficulties in interpreting fragmentary evidence, and limitations imposed by available resources or technology.

On Non-Marxist Historical Materialism The study of non-Marxist historical materialism offers a valuable alternative perspective to Marx's interpretation of historical materialism. By examining the fundamental principles of this theory, it becomes evident that it provides a more nuanced understanding of historical development, considering factors beyond economic determinism. The criticisms of Marx's interpretation highlight the limitations and oversimplifications inherent in his approach, emphasizing the need for a broader framework that incorporates cultural and ideological influences. Within non-Marxist historical materialism, various alternative perspectives have emerged, each offering unique insights into the complexities of historical processes. These perspectives challenge traditional Marxist notions and provide a more comprehensive understanding of social change and development.

The role of culture and ideology in shaping history is another crucial aspect explored within this framework. Recognizing that ideas and beliefs significantly shape societal structures allows a more holistic analysis of historical events. However, it is essential to evaluate the applicability and limitations of this theory. While non-Marxist historical materialism offers valuable insights, it may not fully capture all aspects of historical development. Its focus on material conditions may overlook other influential factors, such as individual agency or geopolitical dynamics.

Studying non-Marxist historical materialism offers valuable insights into understanding historical phenomena. The origins and development of non-Marxist historical materialism have shown how this approach emerged as a response to the limitations of traditional Marxist theory. By incorporating diverse factors such as culture, ideology, and individual agency, non-Marxist historical materialism provides a more comprehensive understanding of historical processes.

The key concepts and principles of the Poznań School further enhance our understanding by emphasizing the importance of interdisciplinary approaches and methodological pluralism. This allows for a more nuanced analysis that considers various perspectives and sources of evidence.

Two Final Thoughts It might sound trivial to some, but it is worth reminding, especially to those who entrenched themselves in knowing-it-all epistemological orthodoxies and judge others from the heights of their academic pedestals, that knowledge is neither absolute nor unquestionable but must be contextualized in a particular time and space. Any adequate conceptual and theoretical framework developed in studying the past must incorporate reflection upon archaeology as a professional discipline in the present. We should not seek to reduce the past to the mechanical application of naive positivism dressed up as scientific procedure (in which methodology is confused with theory); equally, we should not believe that criteria of testability and falsification should be abandoned in favor of speculations about unrecorded intentions in which anyone's opinion is as good as anyone else's. Although seeing alternative approaches to the past is refreshing, we cannot abandon tests of adequacy or more satisfying approaches, which may also mean more explanatory than others. Therefore, in addition to pursuing new theories and methodologies, we want to examine the claims of various archaeological schools against a broader historical and geographical perspective of archaeological work. Archaeologists should not seek to reduce the past to the mechanical application of naive positivism dressed up as scientific procedure (in which methodology is confused with theory); equally, they should not abandon criteria of testability and falsification in favor of speculations about unrecorded intentions in which anyone's opinion is as good as anyone else's.

It seems that any adequate conceptual and theoretical framework developed in studying the past must incorporate reflection upon archaeology as a professional discipline in the present. We have only been able to suggest that a satisfactory explanation for the various trends in archaeology must incorporate a sociological analysis of the way discipline is structured, the knowledge is produced, and the purpose for which it is put.

References

The following publications inspired my thoughts presented in this essay:

Bachelard, G. 1938. *La formation de l'esprit scientifique*. Paris: Vrin.

Borbone, G. 2011. The legacy of Leszek Nowak. *Epistemologia* 34:233–258.

Brzechczyn, K., ed. 2022. *Non-Marxian historical materialism: Reconstructions and comparisons*. Leiden/Boston: Brill.

Chapman, R., and A. Wylie. 2016. *Evidential reasoning in archaeology*. London: Bloomsbury.

Daston, L. 1994. Historical epistemology. In *Questions of evidence*, ed. J. Chandler, A. I. Davidson, and H. Harootunian, 282–289. Chicago: University of Chicago Press.

Feest, U., and Th. Sturm. 2011. What (good) is historical epistemology? Editors' introduction. *Erkenntnis* 75:285–302.

Flannery, K. V. 1982. The Golden Marshalltown: A parable for the archeology of the 1980s. *American Anthropologist* 84 (2): 265–278.

Foucault, M. 1969. *L'archéologie du savoir*. Paris: Gallimard.

Foucault, M. 1972. *The archaeology of knowledge. And the discourse on language*. New York: Pantheon Books.

Fuller, S., and J. H. Collier. 2004. *Philosophy, rhetoric, and the end of knowledge. A new beginning for science and technology studies*. 2nd ed. London: Routledge.

Gingras, Y. 2010. Naming without necessity: On the genealogy and uses of the label 'historical epistemology'. *Revue de Synthese* 131:439–454.

Hodder, I. 1987. *Reading the past. Current approaches to interpretation in archaeology*. Cambridge: Cambridge University Press.

Humphreys, P., ed. 2016. *The Oxford handbook of philosophy of science*. Oxford: Oxford University Press.

Hyder, D. 2003. Foucault, Cavaillès, and Husserl on the historical epistemology of the sciences. *Perspectives on Science* 11:107–129.

Kmita, J. 1988. *Problems in historical epistemology*. Dordrecht: Reidel.

Kmita, J. 1991. *Essays on the theory of scientific cognition*. Warszawa/Dordrecht/Boston/London: PWN, Kluwer Academic.

Kmita, J. 1996. Towards cultural relativism "with a small 'R'". In *Epistemology of history. Poznań studies in the philosophy of the sciences and the humanities*, ed. A. Zeidler-Janiszewska, vol. 47, 541–615. Amsterdam: Brill.

Krieger, W. H. 2006. *Can there be a philosophy of archaeology?* Lanham: Lexington Books.

Kuhn, T. S. 1996. *The structure of scientific revolutions*. 3rd ed. Chicago: University of Chicago Press.

Kuokkanen, M. 1988. The Poznań School methodology of idealization and concretization from the point of view of a revised Structuralist theory conception. *Erkenntnis* 28 (1):97–115.

Lecourt, D. 1975. *Marxism and epistemology: Bachelard, Canguilhem, and Foucault*. London: New Left Books.

Lozny, L., ed. 2011. *Comparative archaeologies. A sociological view of the science of the past*. New York: Springer.

McQuillan, C. 2010. Philosophical archaeology in Kant, Foucault, and Agamben. *Parrhesia* 10:39–49.

Minta-Tworzowska, D. 2022. Teoretyczne i metodologiczne podstawy archeologicznego dyskursu w Polsce. Zarys problematyki. *Przegląd Archeologiczny* 70:5–40.

Nowak, L. 1983. *Property and power: Towards a non-Marxian historical materialism*. Dordrecht: D. Reidel Publishing.

Pałubicka, A., and S. Tabaczyński. 1986. Społeczeństwo i kultura jako przedmiot badań archeologicznych. In *Teoria i praktyka badan archeologicznych*, ed. W. Hensel and S. Tabaczyński, 57–183. Warszawa/Wrocław: Zakład Narodowy im. Ossolińskich, PAN, Instytut Histrorii Kultury Materialnej.

Pearce, D., and J. Woleński, eds. 1988. *Logischer Rationalismus. Philosophische Schriften der Lemberg-Warschauer Schule*. Frankfurt am Main: Athenäum.

Peña-Guzmán, D. M. 2020. French historical epistemology: Discourse, concepts, and the norms of rationality. *Studies in History and Philosophy of Science. Part A* 79:68–76.

Psillos, S., and M. Curd, eds. 2014. *The Routledge companion to philosophy of science*. 2nd ed. London: Routledge.

Rabinow, P., W. M. Sullivan, and H. L. Dreyfus. 1983. *Michel Foucault: Beyond structuralism and hermeneutics*. 2nd ed. Chicago: University of Chicago Press.

Rheinberger, H.-J. 2005. Reassessing the historical epistemology of Georges Canguilhem. In *Continental philosophy of science*, ed. C. Gutting, 187–197. Oxford: Blackwell.

Salmon, M. H. 1982. *Philosophy and archaeology*. New York: Academic.

Topolski, J. 1983. *Teoria wiedzy historycznej*. Poznań: Wydawnictwo Poznańskie.

Topolski, J. 1984. *Metodologia historii*. Warszawa: PWN.

Topolski, J. 1996. *Jak się pisze i rozumie historię. Tajemnice narracji historycznej*. Warszawa: Wydawnictwo Rytm.

Trigger, B. G. 1978. *Time and traditions: Essays in archaeological interpretation*. New York: Columbia University Press.

Wybraniec-Skardowska, U., and Á. Garrido, eds. 2018. *The Lvov-Warsaw school. Past and present*. Cham: Springer – Birkhauser.

Wylie, A. 2002. *Thinking from things*. Berkeley: University of California Press.

Wylie, A. 2017. From the ground up: Philosophy and archaeology. John Dewey lecture delivered at the ninety-first annual eeting of the Pacific Division of the American Philosophical Association, Seattle, on April 14, 2017. *Proceedings of the American Philosophical Association* 91:118–136.

Late Mesolithic and Early Neolithic Hunter–Gatherer Cultural Processes in Cis-Baikal, Eastern Siberia: A Case Study in Darwinian Evolutionary Archaeology

Andrzej W. Weber

Abstract Late Mesolithic (LM) and Early Neolithic (EN) hunter–gatherers (HG) of the Cis-Baikal region, Eastern Siberia, display substantial spatio-temporal variation in adaptive strategies highlighted by a few cultural transitions. These transitions are examined focusing on the role of the following factors: (1) Changes in the distribution of the boreal forest; (2) Technological innovations; (3) Intensification of fishing; and (4) Their combined impacts on subsistence and social structure. The expansion and retreat of boreal forest was important because it directly affected the distribution and abundance of large and medium terrestrial game, the core of subsistence for all LM and EN HGs in the region. All other things being equal, expanding forests meant less game while shrinking forest meant more game and more living space for HGs. The bow was crucial not only due to its technological superiority over the spear and atlatl as a game hunting weapon but also because its flexibility allowed it to work equally well for groups of any size and in any environment—forest or steppe. The much higher return rates associated with bow hunting freed enough labour to be allocated to other activities such as the intensification of fishing. Fishing and fisheries were important because they worked as a differentiating force: the more intensive the fishing, the more limited its spatial distribution and the greater the micro-regional differences between cultural patterns. Intensive fishing and game hunting pulled the adaptive strategies in two opposing directions: fishing towards differences and hunting towards similarities between groups and micro-regions. Social relations were important because they fine-tuned the social fabric to optimize the general strategy. Together, these factors account for all cultural transitions and variation documented for the LM and EN HGs of Cis-Baikal: Transition 1—Formation of the LM cultural pattern with incipient formal cemeteries *~8630 cal. BP*; Transition 2—Replacement of the LM system on the Angara and in Southwest Baikal, but not in the Little Sea or on the Upper Lena, by the EN Kitoi pattern with

A. W. Weber (✉)
Department of Anthropology, University of Alberta, Edmonton, Canada
e-mail: aweber@ualberta.ca

A. Marciniak et al. (eds.), *The Poznań School of Archaeology*, Outstanding Contributions to Archaeology, https://doi.org/10.1007/978-3-032-07848-3_12

its very large cemeteries *~7560 cal. BP*; Transition 3—Collapse of the Kitoi by *~6660 cal. BP* followed by the formation of the Middle Neolithic pattern with no cemeteries.

Keywords Mesolithic · Neolithic · Cis-Baikal region · Adaptation strategy · Fishing · Technology · Innovation · Subsistence

1 Introduction

This study is an exercise in application of archaeological theory to a concrete body of data in order to explain the history of a specific prehistoric cultural pattern. In stark contrast, all other chapters included in this volume are exercises in archaeological theory in order to engage with debates on archaeological theory or, in some cases, with debates on the theory of humanities or social sciences more broadly. In other words, all these pieces lack the practical archaeological aspect, something that this study engages with from start to end. This examination is a shorter and somewhat revised version of a synthetic assessment (Weber 2020) of the history of Middle Holocene (~8700–3400 cal. BP) hunter–gatherers (HG) in the Cis-Baikal region, Eastern Siberia (Table 1, Fig. 1).[1] The longer paper and two other equally substantial studies (Bronk Ramsey et al. 2021; Weber et al. 2021) constitute together the latest and the most complete summary of the long-term research project led by the author over the last ~30 years.

The account presented here is grounded in Darwinian evolutionary archaeology, but is not its theoretical explication. It does not delve into theoretical structure and

Table 1 A.P. Okladnikov and current models of Middle Holocene culture history in Cis-Baikal, Siberia. (After Okladnikov 1950; Weber et al. 2021)

	Okladnikov model	Current model	HPD cal. BP[a]
Period	Mortuary traditions or groups		
Late Mesolithic	Khin	Khin Group	*8630–7560*
Early Neolithic	**Isakovo**[b]	Kitoi, Khin Group	*7560–6660*
Middle Neolithic	**Serovo**	None	*6660–6050*[c]
Late Neolithic	**Kitoi**	Isakovo, Serovo	*6050–4970*
Early Bronze Age	Glazkovo	Glazkovo	*4970–3470*

[a]All dates are modelled and, by convention, presented in italics
[b]Bold font denotes units in the incorrect chronological position
[c]Defined indirectly by the upper and lower boundaries calculated for the Kitoi and Isakovo–Serovo datasets, respectively

[1] Cis-Baikal is the region located to the west of Lake Baikal without precisely defined its west and north boundaries. For the purpose of the Baikal Archaeology Project, Cis-Baikal is defined as the area between the north and south ends of Lake Baikal, including the Ol'khon island, and Ust'-Ilimsk on the Angara.

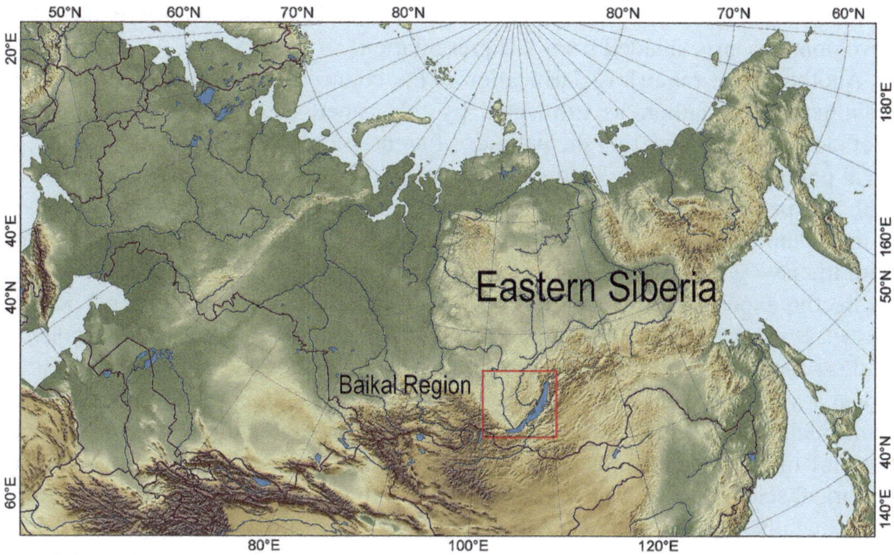

Fig. 1 Eastern Siberia and the Baikal region

underpinnings of Darwinian evolutionary theory, it is not a review of the history of evolutionary thought in archaeology, nor of its ascendance over other schools of evolutionary thought within the discipline. There already are a few publications that deal with these matters sufficiently and more competently (e.g. Bettinger 1991; Teltser 1994; O'Brien and Lyman 2000; Bettinger et al. 2015). Instead, apart from a brief discussion of the Darwinian perspective presented in the "Background and Approach" section, the goal here is to address certain key research questions as effectively as possible with the empirical evidence currently at hand to provide a fresh and insightful explanation of the processes affecting the Middle Holocene HG groups in Cis-Baikal.

For a few reasons, the paper explicitly focuses on the processes leading to the formation and eventual dissolution of the Early Neolithic (EN) Kitoi cultural pattern.[2] First, in the original culture history developed by A.P. Okladnikov (1950) he placed the Kitoi culture incorrectly at the end of the Neolithic sequence whereas its correct place is at its start (Table 1). This error, unsurprisingly, had confusing and profoundly frustrating consequences for Cis-Baikal Middle Holocene archaeology through to the end of the twentieth century at both empirical and explanatory levels which reverberate to this day. Second, the Kitoi cultural pattern lends itself particularly cleanly to Darwinian evolutionary treatment in a rather simple, compact, and well-defined package of principles and mechanisms. And third, the space

[2] In Siberian archaeology, Neolithic is defined by the appearance of technological innovations such as the bow and arrow, clay pots, and ground stone tools rather than domestication of plants and animals as generally accepted in much of the Old World.

restrictions of this edited volume require that Middle Neolithic to Early Bronze Age developments are excluded from this presentation.

Another point requiring clarification is the connection between the theoretical leanings of the author, as explicated by this paper, and the Poznań School of Methodology, its impact on the practice of archaeology, in Poland and beyond, the focal point of this edited volume. The connections are both personal and intellectual. The personal ones go back to my years as undergraduate and graduate student at the Adam Mickiewicz University, Poznań, in the 1970s and 1980s, and mingling with the new generation of archeologists. The intellectual links regard, directly or indirectly, exposure to and influence by such mentors as Professors Jerzy Kmita, Leszek Nowak, Anna Pałubicka, Jan Żak etc., some of them more ardent exponents of this school of thought than others, but all invariably emphasising the role of theory in any empirical science, archaeology included. It is important to note that one of the main points promoted by the Poznań School of Methodology was the replacement of the dominant at the time, at least in Poland, the positivist approaches in empirical sciences (i.e., also in archaeology) with the anti-positivist paradigm and, consequently, replacement of the inductive reasoning with the deductive one.

This intellectual engagement with the Poznań School of Methodology resulted in a few publications and my doctoral thesis. In one of these papers (Weber 1988, 1990[3]), I developed a model of interactions between a Mesolithic population (hunter-gatherers) on one side and a Neolithic one (farmers) on the other with the goal of tracking and understanding better the process of neolithization, an "in vogue" topic in European archaeology back then. To the best of my knowledge, this work attracted very little or, more precisely, no attention among the Polish or European archaeological community and, so it seems, for a good reason. The approach to build a simplified model, as all models are, was explicitly theoretical and rather formalized as well as entirely detached from any empirical record. Apparently, such abstract models invoked little interest in the archaeological readership as it was still dominated (mainly in Central Europe) by the positivist culture-historical model of practicing archaeology. Although the piece promised that more realistic models and empirical testing would follow—the promise never materialized for the simple reason that I left Poland in 1987 and later changed my research focus entirely. This study deliberately attempts not to repeat the same mistake and from start to end engages with theory and empirical material on equal terms.

One other intellectual link, probably more critical, involves explicitly the Darwinian evolutionary theory which, like Marxism, also fell into the focus of the thinkers loosely associated with the Poznań School of Methodology (e.g. Maciej Henneberg, Krzysztof Łastowski, Janusz Piontek, Jan Strzałko), although not from its very start in the 1960s, as Marxism, but a little later, coinciding with my undergraduate and graduate studies. Collaborating closely for several years with this group, I found the Darwinian stream of thinking of particular appeal and potential to archaeology. I returned to it with full interest and vigour after immigrating to

[3] The second paper is the English language version of the first.

Canada, establishing myself professionally within quite a different academic environment which also offered very different research opportunities compared to what was possible back in Poland, and, most importantly, after embarking on a fascinating research adventure in Siberia investigating the prehistory of Middle Holocene HG in Cis-Baikal with a family of modern scientific methods. The need for an overarching theoretical approach to make sense of the wealth of entirely new and diverse empirical data became obvious, necessary, and pressing. That both Marxism, including its interpretation promoted by the Poznań School of Methodology, and the modern Darwinian evolutionary theory explicitly subscribe to the materialist view of reality (c.f., the next section) created a very advantageous platform from which to develop and employ such an approach.

2 Background and Approach

The HG archaeological record is rich in evidence for comprehensive and relatively abrupt changes in adaptive strategies effected either by major climatic shifts or competition with food producers. Explanation of HG cultural change when these two factors are not applicable, however, is rather challenging. The inadequacies of these dominant explanations are particularly apparent when applied to the Holocene HGs of Eurasian boreal settings. The environmental paradigm, i.e., climatic fluctuations of major magnitude such as those documented for the Pleistocene and the Pleistocene–Holocene transition, is insufficient when applied on its own because the Holocene climate was relatively quiescent. The economic paradigm of direct competition with farmers or pastoralists is also only marginally useful because boreal environments have never been friendly to this type of food procurement and food producers expanded into the southern fringes of the boreal forest only late in the Holocene.

The inapplicability of these two explanatory approaches to such settings has created an important gap in the current understanding of Holocene HGs in boreal environments, by far the largest biogeographical zone in Eurasia, though continuously occupied over long periods only during the last 10,000 years after the retreat of the glaciers. Furthermore, these circumstances have led to a lingering impression that boreal HGs remained relatively static and marginalized over long periods. Consequently, what drove the processes of culture change is arguably the most fundamental question regarding Holocene HG archaeology of the Eurasian north.

This matter has been an overarching objective of the research conducted in the Baikal region of Eastern Siberia by the Baikal Archaeology Project (BAP) since the mid-1990s. The Baikal HG archaeological sequence is unique in that its Middle Holocene portion (Table 1) is dominated by materials from cemeteries (Table 2), a category of archaeological finds rare in most boreal settings. Furthermore, the sequence is unaffected directly by food producers (i.e., pastoralists) who did not arrive in the area until its end.

Table 2 Geographic, chronological, and cultural distribution of archaeologically documented Middle Holocene cemeteries, graves, and burials in Cis-Baikal. In cases where more than one mortuary tradition was represented at a given location, each was counted as a separate cemetery

		Microregion				
Period and mortuary tradition or group	Category	Angara	SW Baikal	Upper Lena	Little Sea	Total
Late Mesolithic–Early Neolithic	Cemeteries	6		6	10	22
Khin	Graves	8		12	38	58
	Individuals[a]	8		14	44	66
Early Neolithic	Cemeteries	13	1			14
Kitoi	Graves	147	99			246
	Individuals	215	159			374
Middle Neolithic		Lack of documented formal cemeteries				
Late Neolithic	Cemeteries	23		1		24
Isakovo	Graves	94		1		95
	Individuals	124		1		125
Late Neolithic	Cemeteries	4		5	10	19
Serovo	Graves	19		30	42	91
	Individuals	20		51	70	141
Early Bronze Age	Cemeteries	47	1	12	16	76
Glazkovo	Graves	193	12	65	200	470
	Individuals	211	14	77	230	532
Total	Cemeteries	93	2	24	36	155
	Graves	461	111	108	280	960
	Individuals	578	173	143	344	1238

[a]Separate burials or skeletons

As mentioned earlier, the insights generated by BAP are conceptualized employing a theoretical perspective grounded in modern Darwinian evolutionary theory (e.g. Lewontin 1970; Mayr 1976; Mayr and Provine 1980) and such related bodies of work as evolutionary ecology (Pianka 1981), game theory (Maynard Smith 1982), and foraging theory (Krebs and Davies 1981, 1991; Stephens and Krebs 1986). These approaches have gained increased prominence in archaeology and anthropology since the early 1980s, primarily through explications of human behaviour via micro-economics (Sugden 1986), human behavioural ecology (Winterhalder and Smith 1981, 1992; Bettinger 1991; Smith and Winterhalder 1992; Kelly 1995, 2013), culture transmission, change, and ethnicity (Boyd and Richerson 1985), and the like (Kennett and Winterhalder 2006). The applicability of the Darwinian evolutionary theory to archaeology being straightforward, it will suffice to highlight a few key points building on the conceptual framework succinctly formulated by Ernst Myer (Mayr 1959, 1982, 1983, 1997).

1. Evolution is perhaps most simply defined as differential persistence of variation of any kind (biological or behavioral, genotypic or phenotypic) within a population of individuals through time and space.
2. All evolutionary changes are cumulative (i.e., gradual).

3. Evolution requires only three mechanisms:
   (a) mechanisms that generate variation,
   (b) mechanisms that transmit variation between generations and individuals, and
   (c) mechanisms that select for some variants over others.
4. Individuals are important because they differ in ways subject to forces affecting the generation, transmission, and selection of variation (i.e., individuals are the source of variation).
5. Populations are important because they register change as the result of these forces and provide the environment in which individuals operate (i.e., populations evolve).
6. In higher animals, and especially in humans, behaviour is a major source of variation subject to evolutionary forces.
7. Further, to a much greater degree than in other species, human behavior is culturally transmitted, producing individual and population level change that is faster than and, in some, ways different from evolutionary change in species whose behavior is mainly transmitted genetically.
8. One effect of this difference is human reliance on decision-making: generalized ability to evaluate costs and benefits in ways that optimize returns, reducing exposure to selection, which is the basis for optimal foraging theory and explains why humans have been so extraordinarily successful in adapting to widely varying environments (i.e., polar to desert).
9. A second important difference is human reliance on prestige-biased cultural transmission, which causes otherwise independent behaviors (i.e., of prestigious or successful individuals) to be transmitted as packages, producing distinct styles that become the locus of selection (i.e., rather than the constituent behaviors). This is why the behavior of humans is so much more "stylized" and ethnically differentiated than that of other species.
10. Finally, as a consequence of all the above, in the case of humans especially, evolutionary relationships are complicated, occurring simultaneously at multiple levels, requiring that attention be paid to both variation and lack of variation wherever they occur.

This evolutionary "decalogue" has a number of additional implications for the practice of all empirical sciences engaged in examination of the living world, including archaeology. First, there is room here for the hardline Darwinian materialist emphasis on variation as well as more essentialist-leaning schools recognizing the possibility of types with emergent properties (Mayr 1959, 1997). Human reliance on cultural transmission, specifically prestige biased transmission, permits, but certainly does not assure, the development of types and styles (packages of behavior) which also become loci of selection. Second, on the same premise, one needs to be alert to variation wherever it occurs, attending to differences between unit boundaries (e.g. archaeological cultures, periods etc.) as well as, perhaps more importantly, to tracking changes within them.

The point that all evolutionary changes are cumulative, or gradual, is of further relevance to archaeology. Although the tempo of evolutionary changes may differ, sometimes giving an impression, enhanced by the substantial incompleteness of the archaeological record, of a relatively rapid change, any population level change is always gradual. Populations do not have the capacity to change their characteristics "wholesale", no matter how small the number of characteristics involved and how small the population: the change must first take place at the level of the individual to make a difference at the population level. This also underscores the importance of quantitative methods in archaeology without which it is impossible to measure cumulative changes.

In a most succinct formulation, evolutionary research is a two-step process: documentation of variation and explanation of the differential distribution of variation through time and space. Explanations of cumulative changes can employ any kind of evolutionary mechanisms as long as they invoke no idealist, vitalistic or teleological factors. Rejecting these agents of change, evolutionary archaeology also rejects the notion of human exceptionalism but in a more limited way. Humans like all other species are subject to forces of evolution that generate, transmit, and differentially select variation. That humans depend so overwhelmingly on cultural transmission, however, will set them apart on that basis from other species. This difference, however, does not permit explanations that do not specifically identify the forces that generate, transmit, and differentially select for behavioral variation.

The evolutionary program has proven particularly useful in understanding HG subsistence activities, while human behavioural ecology, and to a lesser extent culture transmission theory, formalized the dynamic interaction between environment and technology, permitting testable predictions about resource choices and aiding in understanding patterns of long-term culture change via the modeling of adaptive strategies (e.g. Winterhalder and Smith 1981, 1992; Smith and Winterhalder 1992; Bettinger 1991; Kelly 2013; Bettinger et al. 2015). A recent synthesis of California prehistory provides the most comprehensive application of this approach to past HG culture change (Bettinger 2015).

The argument for Middle Holocene HG evolutionary history in the Cis-Baikal region advanced here is threefold:

- Gradual, perhaps even imperceptible on a generational scale, climate and environmental trends can effect cumulative changes in prehistoric HG adaptive strategies leading, in turn, to tipping points at which rapid system restructuring occurs;
- The overall impact of these gradual climate and environmental changes on HG adaptive strategies depends to a large degree on concurrent technological, economic, and social innovations; and
- Even rather simple environmental settings with limited food options can support rather diverse HG adaptations and complicated evolutionary trajectories.

3 Recent Advances in Cis-Baikal Middle Holocene Hunter–Gatherer Archaeology

BAP was founded around the time when the culture history model developed by A.P. Okladnikov (1950, 1955) collapsed under the weight of the radiocarbon evidence (Konopatskii 1982; Mamonova and Sulerzhitskii 1989; Weber 1995; Weber et al. 2002; Weber et al. 2006). Three out of the four Middle Holocene mortuary traditions with good archaeological visibility (i.e., Isakovo, Serovo, Kitoi, and Glazkovo[4]) occupied incorrect chronological positions in the model (Table 1). Phrased differently, the entire Neolithic sequence was wrong. Furthermore, the Okladnikov sequence was continuous while the new model shows a long gap between the Early Neolithic (EN) and Late Neolithic (LN) mortuary traditions. Since the Okladnikov system guided archaeological research in Cis-Baikal for half a century its collapse created much confusion, well-reflected in publications dating as late as the 2000s (e.g. Aseev 2003).

This development, however, also created a "clean slate" for BAP research. In order to document afresh the spatio-temporal variation in Cis-Baikal Middle Holocene HG strategies, the focus from the very beginning was on the examination of human skeletal materials using a range of bioarchaeological methods and employing the individual life history approach (Zvelebil and Weber 2013). Examination of ~700 individuals revealed a large number of new insights with regard to diet, subsistence, genetic structure, population size and distribution, number and size of cemeteries, health and activity patterns, mobility and migrations, mortuary protocols, socio-political differentiation, and inter- and intra-microregional variation (c.f. Weber 2020 and the references therein).

The most intriguing aspect of this variation is that the EN HG system appears to show more structural and spatial diversity than subsequent systems. Recently, this new knowledge has acquired fresh meaning due to the following developments:

First, we addressed the problem of direct radiocarbon dating of human skeletal remains by developing equations to correct the effect of old carbon, which transferred to the skeletons of Baikal HGs via the consumption of aquatic foods (Schulting et al. 2022 and the references therein). Applying this method first to ~310 and then to 560 directly dated individuals, we revised the Middle Holocene culture history of the region, tied it more closely to climate history, examined chronological structure—in chronometric and relative terms—of many cemeteries including a few larger ones, and identified previously unknown dietary trends exhibited by Cis-Baikal HGs (Weber et al. 2021 and the references therein; Bronk Ramsey et al. 2021).

Second, assessment of faunal remains from several camp-sites hinted at a gradual transition from hunting large game (red deer) to hunting medium game (roe deer) and fishing, a trend consistent with the results from stable isotope analyses on human bones (Losey and Nomokonova 2017b; Weber et al. 2021 and the references therein).

[4] Only a few graves were attributed to the Late Mesolithic period.

And third, chronological examination of several cemeteries uncovered variation in what was initially assumed to be a continuous pattern of cemetery usage: some indeed seem to have been used continuously, but others only sporadically, and some show long periods of disuse (Bronk Ramsey et al. 2021). Evidence has likewise been found for variation in the timing and duration of cemetery use.

The above describes Cis-Baikal Middle Holocene HGs as far from static, monolithic, or marginal to the shaping of their own destiny. These groups display degrees of spatio-temporal variation in many aspects of their adaptive strategies rarely seen not only in the boreal world but in other environmental settings too. Importantly, these results allow us to see changes in HG behaviour not only at period boundaries but also within periods. The new insights show that Cis-Baikal HGs experienced periods of continuity and discontinuity as well as stability and change. Some changes were gradual, some rapid, but together resulting in several important transitions:

Transition 1: Formation of the Late Mesolithic (LM) cultural pattern with incipient cemeteries and undifferentiated mortuary protocols around *8630*[5] *calibrated years before present (cal. BP)*;
Transition 2: Replacement of the LM system on the Angara and in Southwest Baikal by the EN Kitoi pattern with its large cemeteries around *7560 cal. BP*;
Transition 3: Collapse of the Kitoi by *6660 cal. BP* followed by establishment of the MN pattern with no cemeteries;
Transition 4: Re-appearance of cemeteries around *6060 cal. BP*, or somewhat earlier, and the formation of the LN pattern;
Transition 5: Establishment of the EBA system around *4970 cal. BP*; and lastly
Transition 6: The end of the EBA socio-economic pattern by *3470 cal. BP*.

Fresh assessment of this variation and the patterns of temporal change is the primary objective of this paper, focusing on the first three transitions and paying attention to the following questions:

- Why did all these cultural transitions happen when they did, and not earlier or later?
- What kind of barriers—environmental or cultural—affected the spatial distribution of the different HG patterns across Cis-Baikal?
- What was the significance of the MN break in the use of formal cemeteries?

The proposed evolutionary approach is especially well suited to this exercise because evolutionary forces act at different levels: from artifact to pan-regional techno-complex, from individual to population, and from decades to millennia. New methods and findings make assessment of these different scales possible, while the evolutionary framework offers the most promising avenue to understanding the cumulative patterns of culture change.

[5] By convention, the dates presented in italics are modelled, all other dates are not.

4 Environment

Since several good recent accounts of the modern environment and climate in Cis-Baikal, as relevant to HG archaeology, already exist (e.g. Weber et al. 2002; Weber 2003; McKenzie 2006; White 2006; Weber and Bettinger 2010; Nomokonova 2011; Fraser-Shapiro 2012; Losey and Nomokonova 2017a) only the most important points will be mentioned here.

Cis-Baikal (the area of about 200,000–250,000 km^2 immediately west of Lake Baikal—between its northwest coast, including the Ol'khon Island, and Ust'-Ilimsk on the Angara; Fig. 1) features a markedly continental climate with Bailey's effective temperatures consistently around 11 °C (Bailey 1960; Breken 1966). Topography, geology, hydrography, precipitation, vegetation, and terrestrial and aquatic fauna are all variable across the region. These, in combination with archaeological criteria, define four main microregions: (1) The Angara River Valley; (2) The Upper Lena River Valley; (3) The Little Sea (or Ol'khon) area; and (4) Southwest Baikal (Fig. 2). Of these, the Little Sea stands out due to its aridity, with annual precipitation of only 160–190 mm, roughly less than a half of what the other microregions receive (Kozhov 1963; Breken 1968).

With the exception of the four archaeological microregions, which all feature at least some open landscapes (steppe and forest-steppe), the region is covered by thick taiga. Historically, only the Angara and the Upper Lena have been connected by a stretch of open landscape along the Kuda and Manzurka rivers. The distribution and abundance of terrestrial game varies with vegetation (Shvetsov et al. 1984), and more precisely with the balance between boreal forest on the one side and open landscape (forest-steppe and steppe) on the other. Red and roe deer favour open and ecotonal landscapes and thus are available in all microregions, while moose, boar, and musk deer prefer more densely forested parts. Red deer and roe deer form large herds and are migratory whereas the boreal forest moose and musk-deer are not and are solitary. The boar either forms small sounders or is solitary. The aridity of the Little Sea would have made its red and roe deer smaller in body size and probably less abundant than in places with more precipitation. Hare and marmot are small game potentially also available for hunting but their diminutive size, like the musk-deer, would not make them high rank game. Ethnographically confirmed plant foods include pine nuts, inner bark (pine, birch or willow), young shoots and buds, roots, bulbs (onion, garlic, Siberian lily, and Martagon lily), berries, mushrooms, moss, lichen, etc., most available seasonally (Okladnikov 1950, 1955).

Although fish are available in all four microregions, the fisheries differ much in abundance, diversity, accessibility, and reliability (Table 3). Historically, the Angara fishery has been very rich and diverse, much more so than its large western tributaries (Irkut, Kitoi, and Belaia) and an order of magnitude more abundant than the Upper Lena (Kozhov 1950). The Baikal fishery has been diverse overall with several distinct habitats each preferred by a different assortment of fishes (Kozhov 1963, 1972). The Little Sea fishery is patchy in distribution, with a number of shallow and quiet coves in the south end, most separated from one another by stretches

Fig. 2 Cis-Baikal, micro-regions, and main archaeological sites mentioned in the paper

of inaccessible cliffs. In contrast, the Kultuk Bay fishery in Southwest Baikal, with an open coastline and direct access to the shallows running uninterrupted for about 15–20 km, is linear and thus similar to the riverine fisheries although not nearly as stretched out. It is safe to assume that inter-annual variability in fish abundance would be higher in the small coves and rivers, due to their lower species diversity, compared to the more diverse fisheries in large rivers and expansive shallows. The Baikal seal is another aquatic resource available from the lake, but also of seasonally limited HG procurement to late winter and early spring.

Table 3 Cis-Baikal fisheries based on data published by M.M. Kozhov (1950), summarized first by Weber and Bettinger (2010) and modified further for this table. Data assembled by Kozhov are useful because they predate construction of the three hydroelectric power dams on the Angara River (Irkutsk, Bratsk, and Ilimsk), which effectively destroyed its fishery

River section	Section length [km]	Fish [kg/year]	Fish [kg/km/year]	References
Angara				
Angara Section 1 (Baikal–Irkut)	71	200,000	2817	Kozhov 1950, 46
Angara Section 2 (Irkut–Bratsk)	680	500,000	735	Kozhov 1950, 49
Angara Section 3 (Bratsk–Ilim)	240	700,000	2917	Kozhov 1950, 51
Angara, Baikal to Belaia	148	400,000	2703	
Lena				
Lena (source to Vitim)	1300	297,100	229	Kozhov 1950, 84–85
Lena (~Anga to Zhigalovo)	160	18,300	114	

In sum, the Middle Holocene HGs in Cis-Baikal would not have had that many food choices. Large and medium game (moose, red deer, roe deer, and boar) would have been the primary choice, small game (musk deer, hare, and marmot) would have been less important, and fish would be a viable option too. Due to their narrow seasonal availability, scarcity, or low caloric content, plant foods and Baikal seal are expected to have been minor dietary elements.

5 Main Factors of Middle Holocene Hunter–Gatherer Culture Change in Cis-Baikal

The evolutionary approach to explaining Middle Holocene hunter–gatherer culture change in Cis-Baikal requires that environmental and cultural factors (e.g., technology, subsistence, demography, and social organization) are examined from the perspective of the costs, benefits, and risks involved in the actions taken by these people. Of particular interest are the interactions between factors that produce gradual changes in HG behaviour, the accumulation of which can lead, in turn, to rapid shifts in adaptive strategies. A good place to begin is to assess the effects of changes in the distribution of the boreal forest, technological innovations, and intensification of fishing, as well as their combined impacts on subsistence and social structure.

5.1 Distribution of the Boreal Forest

The changes in the distribution of the boreal forest during the Holocene are singularly important because they control the distribution of terrestrial game, the most critical resource for all inland North Eurasian HGs. Large, medium, and to some extent small game provide not only food but also hides and furs for clothes and shelter, as well as bone, antler, and sinew to make a range of essential tools, weapons, utensils, and ornaments. That Cis-Baikal offers only a few species of large and medium game underscores the importance of this resource. In this regard, the Baikal seal, with its short limbs and seasonally restricted accessibility, would not be a practical substitution for the resources provided by red and roe deer and moose. Seal meat could be stored, however, and the fur would likely have been quite useful.

Following the dynamic climatic oscillations associated with the Pleistocene–Holocene transition, Siberia experienced a long trend towards warmer temperatures and increased precipitation. On a subcontinental scale the boreal forest expanded from a relatively narrow zone at the end of the Late Pleistocene, replacing tundra to the north and steppe to the south (Khotinskii 1984a, b). In places with complex topography, such as Eastern Siberia, the southward expansion of the boreal forest also took on longitudinal and altitudinal dimensions. In Cis-Baikal, additional variation to forest distribution was related to the size and volume of Lake Baikal acting as a barrier to both the southeast Asian monsoon and North Atlantic weather and climate impacts.

The expansion of coniferous forests in Cis-Baikal, already well on its way around 9500 cal. BP, reached its maximum extent around 7000–6500 cal. BP. It was associated with thicker and longer-lasting snow cover, resulting in generally deteriorating conditions for HG game hunting (White and Bush 2010; Bezrukova et al. 2013, 2014; Tarasov et al. 2015; Kobe et al. 2020). What was initially, very likely, a vast expanse of open landscape connecting all different areas of Cis-Baikal, turned at the time of maximum forest distribution into small patches isolated from one another by dense taiga. From 7000–6500 cal. BP the climate stabilized and then, slowly, the trend began to reverse. Gradually drier and cooler conditions across the region meant thinner snow cover and shrinking forests. Such an environment was potentially able to support larger deer populations which, particularly in places with forest-steppe and steppe landscape, were also much easier to procure. This trend ended ~3000 cal. BP when the modern climatic regime was established. Patches of open landscape expanded again, though never to their Preboreal distribution. This scenario is a product of work on environmental proxies with large catchment (e.g., Lake Baikal cores) and variation in local conditions and timing is expected although not yet well documented (White and Bush 2010; Bezrukova et al. 2013; Tarasov et al. 2015; Kobe et al. 2020).

In the Cis-Baikal region, the four archaeological microregions would likely have been the areas into which the boreal forest expanded last and from which it retreated first. However, good empirical data on this matter are lacking. The greater ecological diversity of these ecotones would have been particularly attractive to HG groups although, with the forests closing in, the forest–steppe ecotone would have been shrinking too. All these points emphasize the importance of changes in vegetation distribution for understanding of the history of Cis-Baikal Middle Holocene HG groups. It is this shifting balance between open and forest landscapes that likely should be considered the stage for all cultural events to unfold.

The history of the boreal forest in Cis-Baikal is also important because, following the principle of ideal free distribution (IFD) (Fretwell 1972), it is expected that the dispersal of HG groups across the landscape would generally reflect the distribution of available food resources. It is also expected that HG settlement patterns and movement across the landscape would tend to maximize procurement of the most important food source. For example, although the Hadza of eastern Africa depend on plant foods for ~65% of their subsistence, they still adjust their camp size, movement, and population size and density relative to hunting returns (Woodburn 1982). In the Cis-Baikal setting, it is red and roe deer that should be considered the highest ranked resources, albeit also high risk, as most high-ranked resources are.

Comparison between the environmental and cultural histories, particularly regarding the timing of the identified cultural transitions, shows that they are not aligned (Table 4), which is not particularly surprising given the very gradual nature of the environmental trends compared to the rather rapid nature of culture change. This lack of association implies that the environmental explanation on its own, as expected, is not readily applicable here and makes the questions listed earlier even more pressing.

Table 4 Summary of climate and environmental history for northern Eurasia (Walker et al. 2009, 2012) and the Baikal region (Tarasov et al. 2015; Kobe et al. 2020), and Cis-Baikal Early and Middle Holocene hunter–gatherer culture history (Weber et al. 2021 and the references therein). All ages in cal. years BP

Eurasian climate	Climate and environment in the Baikal region	Archaeological periods and transitions
Preboreal–Boreal 11,700–8200	Cool, dry, rising temperature, expansion of the boreal forest from ~9500	EM ~10,000–8630 Transition 1 ~8630
Atlantic 8200–5700	Long trend towards warmer temperatures and more precipitation, expansion of coniferous forest, thicker and longer-lasting snow cover	LM 8630–7560 Transition 2 ~*7560*
	Continued forest expansion reaching maximum at 7000–6500	EN *7560–6660* Transition 3 ~*6660*
	Stabilization and reversal of the earlier trend after 7000–6500	MN *6660–6060* Transition 4 ~*6060* LN *6060–4970*
Subboreal 5700–2500	Continued reversal: cooler, drier, shrinking forests	LN *6060–4970* cont. Transition 5 ~*4970*
	Stabilization towards modern conditions around 3000	EBA *4970–3470* Transition 6 ~*3470*

5.2 Technological Innovations

Of all technological innovations characterizing the culture historical sequence examined in this paper, the introduction of the bow-and-arrow at the beginning of the EN surely is the most important, its potential impacts reaching far beyond hunting alone. While the atlatl is better than the spear, the bow-and-arrow is superior to both (Table 5) (e.g., Bettinger 2013). Because of their low accuracy and short range, spear and atlatl are effective from a close distance and on large targets (i.e., individual animals or compact herds). They require the cooperation of many hunters dispatching their weapons simultaneously not to miss the target to minimize risks of failure. Conversely, the bow is more accurate, has a longer range (153%), and covers a larger area (235%). It is even more silent than the atlatl, can be used from a few body positions, and is easier to master due to much simpler biomechanics. A bow hunter can dispatch many arrows in quick succession quietly with minimal body movement, thus lowering the risk of startling the prey (Okladnikov 1950). Due to these advantages, the bow does not require cooperation and is effective also on small targets and in the woods. Bettinger (2013) estimates that one hunter with a self-bow (made of wood only) is as good as 2–3 atlatl hunters. Moreover, the bow can also be employed as a bow trap, with many such devices potentially installed at any given location, an additional advantage over the spear or atlatl. Overall, the bow offers much higher return rates per capita and lower risks of hunting failure.

The social and demographic consequences of the bow-and-arrow are important too. While the atlatl requires large groups, the bow makes small groups viable too on the account of its better returns per hunter and risk reduction, thus also reducing the need to move as frequently as before. Which of these two group size options would prevail after adoption of the bow, or whether both would prove to be equally

Table 5 Comparison between hunting with atlatl and bow-and-arrow. (After: Okladnikov 1950; Bettinger 2013)

Variable	Atlatl	Bow-and-arrow
Biomechanics & learning	Complex & difficult to master	Simple & easier to master
Accuracy	Lower	Higher
Target distance	Shorter	Longer (by 153%)
Target area	Smaller	Larger (by 235%)
Repetition rate	Lower	Higher
Risk of startling prey	Higher	Lower
Unattended use	Not suitable for traps	Suitable for traps
Target size	Effective on large targets	Effective on large to small targets
Hunting environment	Effective in open area	Effective also in woods
Returns per capita	Lower	Higher
Overall risk of failure	Higher	Lower
Effect on group size	Favours large groups	Favours large and small groups
Effect on group mobility	Favours higher mobility	Favours lower mobility

practical, would depend on other circumstances such as the requirements of other forms of food procurement. The important point is that while the spear and atlatl favour large groups only, the bow-and-arrow makes room for units from small to large.

Lastly, better returns from hunting with the bow-and-arrow also create opportunities for: (1) Population growth by a combination of such demographic factors as higher birth and infant survival rates, higher fecundity and fertility, and lower morbidity etc., all relative to the times prior to the adoption of the bow; and (2) Changes in population distribution by crowding defined as the combined effects of individuals and families forming larger groups and the tendency of such groups to live relatively nearby, thus leading to higher variability in microregional and regional population densities.

While the role of fishing gear is introduced in the next section and assessed in more detail below, the socio-economic impacts of the other Neolithic innovations (e.g., ceramic pots, and new forms and kinds of tools) are expected to be only minor and, thus, are discussed only as necessary. Introduction of copper and bronze objects (e.g., knives, rings, needles, and fishhooks) defines the EBA but is not relevant to the cultural transitions discussed here.

5.3 *Intensification of Fishing*

The critical difference between game and fish resources in most HG settings, including Cis-Baikal, is that fish lends itself to both non-intensive and intensive procurement, but large game only to non-intensive—otherwise resulting quickly in overhunting (Winterhalder and Goland 1993). Of all the resources in Cis-Baikal, only the fisheries are amendable to intensification, though differences in abundance, distribution, access, and reliability mean that not all are equally suited to this practice. The Angara and the shallows of Kultuk Bay in the southwest corner of Lake Baikal offer the best opportunities for intensification, while the Upper Lena is the least suitable, and the coves of the Little Sea only with difficulty, mostly due to their patchy and unpredictable nature. Ice cover is another important aspect of accessibility and only the very upper section of the Angara fishery remains unfrozen during long winters while the ice on the others and the lower sections of the Angara can easily be 1 m thick (e.g., Kozhov 1963).

The numerous differences between game and fish resources are outlined in Table 6. Relative to game hunting, fishing— particularly its intensive from—is heavily front-loaded, requiring a major cooperative investment of labour in the manufacture and maintenance of fishing gear, in procurement itself, and in post-harvest handling for consumption and storage (Lindström 1996; Tushingham and Bettinger 2013). Post-harvest handling is particularly relevant where large amounts of fish can be procured with relatively simple, less intensive, methods. Additionally, small fish can be a very cost-effective resource when mass harvested whereas small game (e.g., hare) cannot (Lindström 1996).

Table 6 Comparison between game hunting with bow and fishing

Variable	Game hunting with bow	Fishing
Territory	Larger	Smaller
Group mobility	Higher	Lower
Returns per capita	Higher	Lower
Risk of failure	Higher	Lower
Labour investment	Back-loaded	Front-loaded
Risk of resource depletion	High	Low
Subject to intensification	No	Yes
Organization	Individual hunting	Individual (non-intensive) and cooperative (intensive) fishing

In Cis-Baikal, fish lends itself to storage by freezing in winter and drying and smoking in summer, the latter requiring considerable cooperation in order to be done quickly to avoid spoilage. The easily distributable food packages would also help solve a problem inherent to sharing game meat: no need to haggle over who gets what or how much because each unit is relatively the same. Lastly, intensive fishing requires much higher levels of cooperation than game hunting, especially with the bow. While bow hunting would probably work equally well for groups of any size, intensive and, to some extent, non-intensive fishing would need the cooperation of large social units to work properly.

As in many other HG settings (e.g., Fitzhugh 2003; Kennett 2005; Prentiss et al. 2007, 2014; Tushingham et al. 2016), there is no reason to believe that Middle Holocene Cis-Baikal HGs were on an inevitable course to develop intensive fishing sooner or later simply because there was fish just about everywhere around the region. Therefore, When, Where, and Why the intensification of fishing took place around Cis-Baikal, are all good questions to ask. Moreover, Bettinger (2015, 30, 38) notes that HG intensification is as much a social and political problem as it is a subsistence and technological problem. A shift to more intensive fishing can be facilitated by a change in the mode of hunting to one that is more efficient and frees up the labour required by the heavily front-loaded intensive fishing. In sum, an understanding of fishing is as critical to the understanding of HG evolution in Cis-Baikal as is the history of boreal forest distribution and the socio-economic effects of hunting with the bow.

6 Discussion

The following argument focuses on the interactions between the main factors identified above: changes in the distribution of the boreal forest and approaches to game hunting and fishing, as well as their combined impacts on HG adaptive strategies, together leading to gradual changes within each cultural pattern and the transitions between them. Other factors are brought into the discussion as necessary. Since the materials from camp-sites in this region do not correlate well with mortuary groupings, the empirical foundations for this assessment are provided mainly by the recent archaeological and bioarchaeological studies of materials from cemeteries.

The narrative is presented within the traditional culture historical units and deliberately emphasizes regional patterns. Microregional or local particularities are invoked when necessary, not to divert attention from meaningful generalizations and mechanisms, but to suggest future research directions. The chronological ranges for each analytical unit (mortuary tradition or archaeological period) have been updated based on the recent analysis of 560 radiocarbon dates (Weber et al. 2021). On a regional scale, most of the boundaries are considered relatively firm with the exception of the start of the LN—till defined by a rather small number of dates—which is likely to shift slightly earlier. More revisions are expected on a microregional scale, where a few samples of radiocarbon dates are still relatively small. New fieldwork, as well as the dating of materials excavated in the past but not yet analyzed, will expand the radiocarbon dataset further. Continued research may bring microregional differences in the timing of relevant mortuary traditions into sharper focus, but is not expected to move their boundaries enough to make much difference for this analysis.

6.1 Mesolithic: Early (~10,000–8630 cal. BP) *and Late* (8630–7560 cal. BP[6])

The Mesolithic can further be divided into Early Mesolithic (EM) and LM, differentiated by the appearance of individual graves, sometimes forming very small cemeteries, in the latter. The Mesolithic Khin mortuary group was represented in the Okladnikov sequence by only two graves from the Angara valley and subsequent fieldwork identified two more: one on the Angara and one on the Upper Lena. Recently, a dozen or so graves from several locations across Cis-Baikal have been assigned to the latter half of the Mesolithic based on a combination of typological and radiocarbon criteria, thus defining the LM (Bazaliiskii 2010; Weber et al. 2021). In cultural terms, the EM is essentially the LM without the mortuary component. It is possible that other differences in cultural characteristics exist but these are currently difficult to identify.

The LM mortuary record appears highly variable with regard to body position, orientation, use of red ochre or rocks, and grave goods (Bazaliiskii 2010; Goriunova et al. 2021; Weber et al. 2021). Thus, the term mortuary tradition is not fully applicable and a more neutral concept of *mortuary group* seems more appropriate. In this paper, to keep the matter simple, the label of *Khin Group* is used in reference to all

[6] All period boundaries are from Weber et al. 2012.

graves of this kind from the entire region (Tables 1 and 2). Grave goods are generally small in number but relatively diverse and show similarities with the EM (prismatic blades and points on prismatic blades, known from camp-sites) and the EN (fishing tackle including fishhooks—some composite—and leisters, known mostly from cemeteries). Ornaments are rare and include beads, red deer canine pendants, and boar tusk pendants—all also known from numerous EN, LN, and EBA graves. Bazaliiskii (2010) underscores the absence of pottery and bifacially formed arrowheads, but many other categories relatively common in EN, LN, and EBA graves are lacking too: nephrite tools and ornaments, composite tools, bone points, etc. (Weber et al. 2021).

The stable isotope data in this group imply at least some consumption of aquatic resources across Cis-Baikal and, on the Angara specifically, the start of a dietary trend towards increased reliance on the local fishery (Weber et al. 2021 and the references therein). For the Little Sea and the Upper Lena, radiocarbon dates show sometimes sporadic and sometimes clustered use of cemeteries continuing well into the EN without significant changes in mortuary characteristics (Bronk Ramsey et al. 2021; Goriunova et al. 2021).

As mentioned, at the beginning of the Holocene Cis-Baikal was essentially a vast and continuous expanse of steppe and forest-steppe with ample large game roaming around. The available hunting technology, the atlatl and the spear, required that HG groups were relatively large in order to have enough hunters to cooperate: the only way to compensate for the inherent deficiencies of this technology, to provide enough food, and to satisfy needs for clothes, shelter, etc. Consequently, these groups allocated much of their time to searching for and pursuing large game. The EM groups were very mobile and their transient camps were scattered in a pattern dictated as much by hunting success as by a preference for specific locations. Discard rates of archaeological materials at these locations were low and their archaeological visibility today is equally low. These groups also subsisted on plant foods and fish, the latter harvested employing a range of techniques for individual capture. Hunting with spear and atlatl and high residential mobility would have accommodated only non-intensive fishing, which could be practiced on many rivers and along lake shallows whenever the opportunity arose. Controlled and sustained access to such fisheries, including by legitimizing it via formal cemeteries, was neither necessary nor practical due to the high level of mobility.

This strategy continued through the EM and, in general approach, was not much different from that employed in the Late Pleistocene or at the Pleistocene–Holocene transition. It would likely have continued for much longer if not for the expanding forests, which had two perceptible consequences: (1) An increasingly patchy environment; and (2) Population crowding within the gradually shrinking open landscape and along the forest–steppe ecotones. Together, this resulted in increased inter-group competition for game resources and created socio-economic problems with no solution (technological or otherwise) readily available.

Table 7 Formation of the Late Mesolithic cultural pattern—Transition 1

Variable	Description
Environment	Warming and wetter climate; forest expansion; increasing patchiness
Population	Small, stable in size; crowding within open landscape and along ecotones
Group size	Small number of relatively large groups
Group mobility	Decreasing
New technology	No important innovations
Hunting	Group with atlatl and spear; large game
Fishing	Non-intensive
Social relations	Limited social differentiation; increasing inter-group competition for game resources; emerging leaders
Mortuary activities	Rare and unstructured formal burials; single graves, emergent small cemeteries

The appearance of the first single graves and possibly also of very small cemeteries around *8630 cal. BP* perhaps mark the earliest attempts to address these troubles, thus marking also the beginning of the LM—**Transition 1** (Table 7). Crowding and competition would have created opportunities for charismatic leaders to emerge and to organize hunting and individual fishing (using the same techniques as before) into more efficient operations. There were, perhaps, even attempts to control access to some resources, leading to episodes of increased socio-political differentiation. Since the long-term wellbeing of such groups would have relied on personal leadership qualities rather than on institutionalized mechanisms, it is unlikely that such arrangements would have lasted beyond the lifetime of a leader or a few generations at most. It could be these leaders and their families that were interred in the LM graves. The fleeting nature of such arrangements and the relative rarity of leader deaths were not conducive to the development of a formalized mortuary protocol which, therefore, varied greatly from place to place across Cis-Baikal.

Around ~8000 cal. BP, that is, at the beginning of the warming and wetter trend associated with the onset of the Atlantic period, forest expansion would have likely accelerated. Such expansion would have led to even more crowding and perhaps more competition for game, now under growing pressure from hunting. The HGs of Cis-Baikal lacked a long-term solution to the mounting problem of too many people in places with probably still enough food but no technology, time, or hands to harvest it efficiently. The intensification of fishing was still not viable due to the constraints created by inefficient game hunting with atlatl and spear as well as high mobility, and thus a lack of labour surpluses. Due to this population pressure, the overall size of the EM and LM population was unlikely to grow much. The short-lasting efforts of leaders to organize their people into economically more efficient units were also unlikely to lift this pressure for long enough to make a difference. Socially, all Mesolithic groups display essentially the same limited intra- and inter-group social differentiation.

6.2 Early Neolithic (~7560–6660 cal. BP)

Formation of the Kitoi Pattern The EN started with the formation of the Kitoi cultural pattern known primarily from mortuary sites on the Angara and Southwest Baikal (Table 2) (Weber 2020 and the references therein).

It featured many technological innovations including the bow-and-arrow and ceramic pots traditionally used to define the Neolithic in Siberian archaeology. Nets, new types of leisters and harpoons, fish lures, composite fishhooks, green nephrite tools (adzes, knives, and chisels), and a range of composite tools (e.g., spearheads, daggers, and knives) were also either novel or much more morphologically variable than during the LM. The shanks of composite fishhooks, one of the most diagnostic Kitoi objects and a common grave good, were morphologically rather stable and varied mainly in size. The pottery, with simple profile and round or pointed bottom, was rather uniform. The two types of decoration, net and cord impressions, directly imply the knowledge of weaving and cordage.

The Kitoi mortuary protocol is typically defined by grave-pits lacking rocks, the presence of red ochre, extended body position, a northern orientation, graves with mostly single but not infrequently also with multiple interments (occasionally in the head-to-toe position), and grave goods of the kind mentioned above. Many of these mortuary traits repeat with only minor deviations from grave to grave and from cemetery to cemetery. There are aspects, however, that differ between cemeteries such as perimortem removal of the head, post-mortem disturbances, and bear or fire rituals. Kitoi cemeteries in the Angara valley also vary in size from one or two graves to a hundred or more. Nonetheless, the material culture was basically the same and the highly variable distribution of grave goods, in both kind and number, suggests substantial levels of social differentiation and inequality.

Bioanthropological studies indicate a population operating under a combination of skeletal "wear and tear" and physiological stress, as evidenced by the prevalence of joint osteoarthritis and tooth enamel hypoplasia (both higher than in LN and EBA groups), stature and child skeletal development rates (both lower than in LN and EBA groups), as well as infant mortality rates during breastfeeding (likely higher than in the LN). Moreover, the differential distribution of joint osteoarthritis suggests a division of labour along gender lines, a characteristic absent in LN and EBA groups. Radiocarbon dates and stable isotope data reveal that most, but not all, Kitoi groups experienced a gradual and relatively short—lasting only a few centuries—trend towards the increased contribution of local aquatic foods to their diet and most used their large cemeteries continuously. At Shamanka II on Southwest Baikal, however, a break of a few hundred years was found between the long Phase 1 and short Phase 2, though both showed similar dietary trends.

All this suggests that the evolutionary landscape for Cis-Baikal HGs changed dramatically and rapidly with the advent of the bow-and-arrow. Where the bow came from and why it arrived on the Cis-Baikal stage around 7560 cal. BP is a different matter and inconsequential to the argument advanced here. Clearly, the quick adoption of the bow—at least on the Angara and Southwest Baikal—is a proof that,

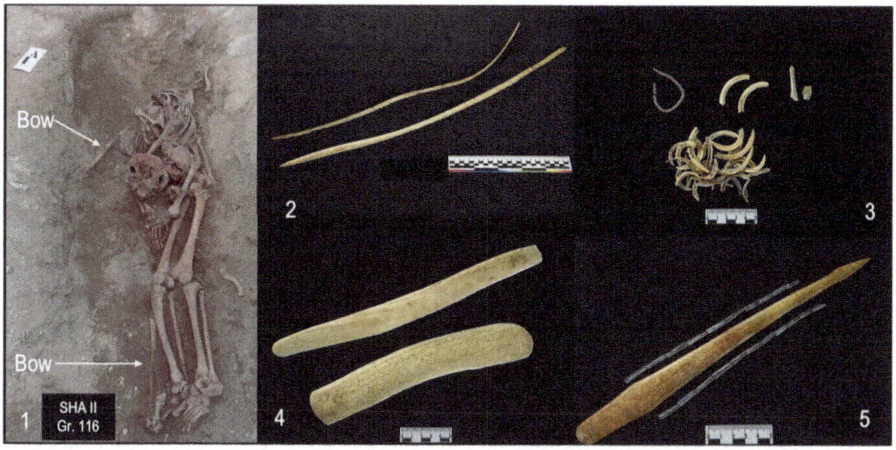

Fig. 3 Shamanka II, Grave 116 with two bows (photos by P. Kurzybov and A.W. Weber): 1–burial level; 2–two bow stiffeners; 3–beads, marmot incisors, ring fragments, and small lithics; 4–antler implements; 5–composite tool (dagger or spearhead) with prismatic blades of rock crystal

from the perspective of LM groups facing mounting problems, the timing of this innovation could not have been better. It provided exactly what was needed most at this particular time: an improved technology to harvest not only large game but also medium and small with substantially better per-capita return rates and a lower risk of failure, thus releasing enough labour to be potentially channeled towards other activities.

The archaeological evidence shows that the Kitoi people used a composite bow about 0.80–1.2 m long with bone or antler stiffeners such as those found in at least 16 graves at Shamanka II (e.g., Gr. 46, 51, and 116; BAP unpublished data[7]; Fig. 3) and a few graves at the Glazkovo and Lokomotiv cemeteries in Irkutsk (e.g. Okladnikov 1974, 34, 44, Plate 4: 2). The relatively short Kitoi composite bow, which could be used from a few different positions (i.e., not only standing but also crouching or kneeling) would have been quite effective in the forest and along ecotones where game needed to be approached closely and still powerful enough due to the stiffeners.

The radiocarbon dates for Shamanka II imply that bow stiffeners were employed already at the early stages of the Kitoi cultural pattern. This suggests either that the bow arrived in Cis-Baikal with the stiffeners already a part of the technology or that they were developed locally shortly after the bow arrived. Since Bettinger's assessment of bow firepower used data collected for the less powerful self-bow, it is not unreasonable to expect that already from the beginning of the EN, one Kitoi composite bow hunter was much better than 2–3 atlatl hunters.

[7] A monograph of the Shamanka II cemetery is in its final stages of preparation for publication. It should appear in print in 2023 or 2024 at the latest.

Excepting the fishing gear discussed below, none of the other Kitoi technological innovations would have had an impact on subsistence or social organization similar to the bow-and-arrow. The introduction of ceramic vessels may have been important for the management of household chores as their use for cooking would have allowed people to work on other tasks at the same time (Bondetti et al. 2020) and, perhaps, would assist in preparation of weaning foods. However, since EN ceramic vessels were rather small and are generally rare at cemeteries and camp-sites, their socio-economic impact was probably minor. Green nephrite tools were probably part of the toolkit for working wood, bone, and antler to make bows, arrows, and the wide assortment of bone and antler composite tools and utensils so well-known from Kitoi graves. Nephrite tools would also have been very practical for building weirs across rivers and other hunting and fishing contraptions such as basket traps, platforms, lifting devices, and drying racks, all well-documented ethnographically across Siberia (Okladnikov 1950, 1955).

The intensification of fishing was the obvious allocation of the emerging labour surpluses, particularly in places with rich, predictable, accessible, and reliable fisheries located within or nearby the patches of remaining, but continuously shrinking, open landscape. The upper section of the braided, fast flowing, and never-freezing Angara and the open shallows of Kultuk Bay on Southwest Baikal, were the obvious places for this process to set out first. The Angara fishery would also have been particularly amenable to scheduling seasonal complementarity between hunting and fishing.

Although Bettinger (2015, 29) notes that the distinction between non-intensive and intensive hunting–gathering is quantitative, thus subtle, and difficult to measure archaeologically, the archaeological evidence is unequivocal. Many of the methods described by Lindström (1996). in her three main groups of fishing techniques, were likely part of Kitoi fishing: (1) Attended techniques for individual fish capture (single fishhook lines, leisters, and harpoons); (2) Attended techniques for mass capture (lift nets, seines, and dragnets); and (3) Unattended techniques for mass capture (trot lines, gill nets, weirs, and basket traps). The last two groups support intensive fishing, while the first—non-intensive. Nettle, hemp, willow bark, and wild flax were all broadly available providing fibers for cordage and net making (Okladnikov 1950, 1955). Furthermore, chronological analysis of grave goods and isotopic evidence from the Shamanka II cemetery shows that all new fishing techniques were introduced at the beginning of the EN and then used without any major improvements or additions while reliance on fish in the diet gradually increased (Scharlotta et al. 2016). Evidently, these EN groups allocated more and more time to fishing.

These developments quickly led to the formation of the Kitoi pattern—**Transition 2** (Table 8)—simultaneously on the Angara and in Southwest Baikal (Weber et al. 2021). The recent discovery of a Kitoi cemetery in Moty-Novaia Shamanka on the lower Irkut River, about 40 km west from its confluence with the Angara (Bazaliiskii et al. 2016), suggests that the fisheries of the lower sections of the Angara's left tributaries were also suitable for intensive fishing. Unfortunately, the cemetery has been entirely destroyed by a modern housing project and details of its size and use are unknown.

Table 8 Formation of the Early Neolithic Kitoi cultural pattern—Transition 2

Variable	Description
Environment	Warming and wetter climate; forest expansion; decreasing patchiness
Population	Growth in size; crowding within open landscape and along ecotones
Group size	Few very large, and some medium to small groups
Group mobility	Low
New technology	Bow-and-arrow; new methods of fish mass capture; ceramic pots
Hunting	Individual with bow; large, medium, and small game; pressure on large game
Fishing	All forms, from non-intensive through to intensive
Social relations	Substantial social differentiation; increasing inter- and intra-group competition for access to resources; strong leaders; craft specialization
Mortuary activities	Structured formal burials; large and small cemeteries

Introduction of the bow, quickly followed by the intensification of fishing, also had important demographic and social consequences reflected in the Kitoi pattern. First, improved return rates from bow hunting lifted the pressure that had long been keeping the Mesolithic population in check, thus allowing Kitoi groups to grow and crowd further. Groups that were large enough engaged in intensive fishing. Chances are that the overall Kitoi population grew rapidly at the beginning of the EN and then stabilized. Second, both bow hunting and non-intensive fishing can be practiced by bands of any size but, as mentioned, intensive fishing requires large groups. This combination creates room for an array of socio-economic arrangements.

While an alliance of large cooperating units—perhaps on an intermittent basis-appears to have established itself as the dominant Kitoi socio-economic model, the structure was flexible enough to accommodate much smaller social units, able to eke out a living thanks to the advantages offered by the bow as well as the variety of fishing techniques now available. The variable size of Kitoi cemeteries was perhaps thus linked to the equally variable size of the social units using them. Well-defined home ranges were likely based along the Irkut, Kitoi, and Belaia rivers and adjoining sections of the Angara, each with a cemetery located at its respective confluence—perhaps the symbolic centre of each unit and its home range. Likely, it is these cooperating units that are behind the dietary trend towards an increased reliance on fish discovered at Shamanka II and Lokomotiv, whereas smaller units are behind the groups of burials there (Weber et al. 2021) and Kitoi cemeteries further downstream on the Angara (Galashikha and Ust'-Belaia) which show no dietary change over time.

By favoring cooperation, this new evolutionary HG landscape also created opportunities for leaders to coordinate such team efforts as well as room for specialization and competition within and between groups. All new toolmaking, including bow-and-arrow and fishing gear, would likely have been subject to at least some craft specialization and, perhaps, the division of labour along gender lines, although this is difficult to see in the archaeological record. Competition would have focused not only on the accumulation of personal wealth, prestige, political influence, and

Table 9 Formation of the Middle Neolithic cultural pattern—Transition 3

Variable	Description
Environment	Still warming and wetter climate; forest expansion; decreasing patchiness
Population	Dispersal of the Kitoi population into the forest
Group size	Many small groups
Group mobility	High
New technology	No important innovations
Hunting	Individual with bow; large, medium, and small game
Fishing	Non-intensive
Social relations	Low social differentiation; low intra- and inter-group competition for access to resources
Mortuary activities	No or rare formal burials

access to resources, but likely would have also involved a search for capable marriage partners and the recruitment of skilled craftsmen and talented individuals to lead cooperative activities. This, in sum, would have resulted in much intra- and inter-group social differentiation, with small groups being potentially less differentiated than large ones.

Together, these points suggest that life at the beginning of the EN was essentially quite good—certainly much better than any time before—but the situation likely changed quickly, particularly under the persistent pressure of the forest closing in on the open landscape. Returns from game hunting also probably levelled off relatively quickly, later perhaps even decreased while the risk of hunting failure increased once again, both due to the combined effects of a large HG population, bow efficiency resulting in overhunting, and habitat loss for game. All this put Kitoi groups under renewed population pressure and halted further growth. Crowding, competition, and the need for cooperation likely increased, potentially leading to even more social differentiation. From this point on, life was no longer so good, as demonstrated by the ample evidence for physical, physiological, and developmental stress experienced by these people as mentioned above. Sources of this stress are easy to imagine: extended search and pursuit times while game-hunting meant bringing home heavy loads of meat from further and further afield or food shortages if returning empty-handed, as well as all the heavy lifting and pulling required to handle the attended and unattended devices used for mass fish capture.

The Kitoi pattern continued until it collapsed no later than 6660 cal. BP, roughly at the time when the boreal forest reached its maximum expansion around 7000–6500 cal. BP (Table 9). The substantial radiocarbon evidence suggests a relatively quick breakdown, although not as quick as its formation (Weber et al. 2021). More cooperation to increase returns from fishing was an insufficient solution to the mounting range of problems as fish was not a viable substitute for the resources that only game could provide. The system lacked the capacity to counteract the negative impacts of growing competition, population crowding, environmental changes, and diminishing returns from game hunting. At this point, ironically perhaps, the bow also facilitated a solution that the atlatl and spear could not: dispersal of the Kitoi groups into the forest.

The same radiocarbon evidence suggests variable tempo and timing of the breakdown, implying a rather complicated process (c.f., Bronk Ramsey et al. 2021; Weber et al. 2021 for chronological details). First, Shamanka II on Southwest Baikal shows two phases of use. Such discontinuity is lacking on the Angara as a whole and specifically at Lokomotiv, the largest cemetery with enough radiocarbon dates (n = 80) to document such break, if present. Second, Phase 1 at Shamanka II starts at the same time as the Kitoi on the Angara but it ends earlier. Third, Phase 2 begins after a long break, but ends even more quickly than Phase 1. And fourth, the dietary trend documented for Phase 2 repeats exactly the main dietary trend from Phase 1 (Weber et al. 2021). Over all, then, the Kitoi pattern began to vanish with the end of Shamanka II Phase 1, followed by the end of Lokomotiv, then of the smaller groups associated with the other Kitoi cemeteries on the Angara, and disappeared for good at the end of Shamanka II Phase 2.

While it can be argued that some of these chronological offsets are within the statistical confidence intervals of Bayesian analysis (Weber et al. 2021), they may yet be real. Already, evidence shows many similar temporal offsets (Bronk Ramsey et al. 2021) across Middle Holocene Cis-Baikal. Furthermore, there is no reason to expect that the end of the Kitoi should be chronologically the same everywhere. If these temporal differences are genuine, the argument advanced here helps make sense of them. The Kitoi pattern required bow technology, an open landscape with sufficient game, and fisheries suitable for intensification—the latter two in close proximity to one another—otherwise the Kitoi pattern could neither form nor function properly. The technology being the same everywhere, differences in the timing of the Kitoi dissolution must then be related to the other two factors.

In the Angara valley, good fisheries coincided spatially with open landscape along much of the first 200–300 km of the river and along the lower sections of its tributaries (Irkut, Kitoi, Belaia, and Kuda). Such a setting likely provided some socio-economic flexibility to the Kitoi pattern, making room for groups of different sizes to function side-by-side: the small ones more flexible, or resilient, than larger ones. But the ecological setting of Kultuk Bay was more restricted. Here, the fishery was limited to a relatively short stretch of lake coast, frozen for about 4–5 months every winter, and the open landscape extended westward from the lake (away from the fishery), into the middle Irkut valley where fisheries were inadequate for intensification. The result was that in Kultuk Bay, the Kitoi pattern could only develop and function within a much smaller area than on the Angara, making the arrangement less stable overall.

The expansion of the boreal forest could also have contributed to the earlier collapse of the Kitoi pattern in Kultuk Bay. The area immediately west of the bay and along the middle Irkut lies between two high mountain ranges: Eastern Sayan to the north and Khamar-Daban to the south, both with much higher precipitation (two to threefold) than the rest of Cis-Baikal (Breken 1968). This would have translated into a much faster advance of the forest into the valley and the concomitant reduction of the open landscape and its game resources.

Given these limitations, it seems reasonable to suggest that Phase 2 at Shamanka II represents an attempt to reintroduce the Kitoi strategy into the area at a time when it was still functioning on the Angara, although already with much difficulty. The abandonment of the Kultuk Bay area at the end of Phase 1 would have allowed game resources to rebound somewhat, depending on how long it was before the Kitoi groups returned, but it would not have stopped the forest expansion. That the diet of Phase 1 was more diverse than that of Phase 2, where all individuals neatly fit into a single trend (Weber et al. 2021), suggests that Phase 2 likely relied on a narrower range of fishing techniques. Since Phase 2 repeats the trend of the SE Cluster row burials from Phase 1, it is possible that Phase 2 represents the direct descendants of these Phase 1 individuals returning to Shamanka II and employing the same strategy for fishing intensification.

After dispersal, the diet of the Phase 1 people is expected to have "reset" to what it was at the beginning of Phase 1 (a lower consumption of aquatic foods), which should also characterize the first returners of Phase 2. The direct biocultural continuity between Phase 1 and Phase 2 people at Shamanka II is supported by the spatial patterns of cemetery use. Rather than forming a new group, Phase 2 graves were added to the existing spatial units — sectors, clusters, and rows — with some burials added to graves built during Phase 1 (Weber et al. 2021). Moreover, the entire mortuary protocol of Phase 2 emulates that of Phase 1. Overall, the evidence points towards the rather short-lived Phase 2 as an attempt to re-establish the previous Kitoi pattern in Kultuk Bay that failed due to the less abundant game resources, relative to what was available at the beginning of Phase 1. The valley west of Kultuk Bay was the natural direction for Phase 2 groups to disperse.

Early Neolithic Patterns on the Upper Lena and in the Little Sea

On the Upper Lena and in the Little Sea, the EN mortuary record is confined to either individual graves or small (fewer than 10 graves) cemeteries, totalling a few dozen burials only (Table 2). It is probably significant that only two graves display a few mortuary characteristics attributable to the "*classic*" Kitoi: one at Nikol'skii Mys on the Upper Lena (Bazaliiskii personal communication) and one at Shamanskii Mys on the Ol'khon Island in the Little Sea (Konopatskii 1982), both showing some differences too. Among the rest of the graves assigned to the EN arrowheads are conspicuously rare (Vetrov et al. 1995; Goriunova et al. 2021) and grave goods from the better documented burials on the Upper Lena (e.g., Obkhoi; BAP unpublished data) and in the Little Sea reveal continuity from the LM, with the rare addition of net and cord impressed clay pots (Goriunova et al. 2021). It seems that attempts to "transplant" the Kitoi strategy were made, as suggested by these two Kitoi graves as well as the rare presence in the Little Sea of a few graves with typical Kitoi composite fishhook shanks (Goriunova et al. 2021). They were unsuccessful, however, evidently because the Kitoi pattern was only viable where labour surpluses generated by bow hunting could be efficiently diverted towards the intensification of fishing. Outside of the Kitoi centers, several factors prevented such developments.

Due to its northern location, the forest on the Upper Lena would have closed in well before the Kitoi groups ran into problems in their homeland on the Angara and in Southwest Baikal. Thus, very likely, the Upper Lena was never a viable option for the Kitoi people as a fall-back or expansion area. The same factor apparently limited the northward distribution of the Kitoi pattern beyond the first 200–300 km of the Angara River, despite its rich fishery (Table 3). Evidence for use of the bow on the Upper Lena is weak, and while it would have made hunting in the forest worthwhile, it would not have affected search time, which would have been much higher than in more open landscapes. Fishing probably employed mainly single capture techniques as the more intensive approaches were not compatible with high mobility or the poor local fisheries that were, moreover, subject to long winter freezing.

In contrast, the fisheries of the Little Sea are thought to have been rich and Middle Holocene fishing is well documented by recent excavations at Ityrkhei Cove in Kurkut Bay (Losey et al. 2008, 2012). Moreover, while the Little Sea's considerable aridity is expected to have slowed down the advancement of forest into the area, it likely also restricted the size and abundance of game relative to the other microregions. Whether the open landscape disappeared entirely and when the forest reached its maximum expansion is still unclear. Regardless, our current understanding implicates the combined effects of several elements as the most parsimonious reason for the absence of the Kitoi pattern in the Little Sea.

Shallow cove fisheries in the Little Sea are patchy in distribution and low in species diversity, as demonstrated by the materials from Ityrhkei where 93% of identified fish bones belonged to only 3 species: perch (65%) and dace/roach (28%) (Losey et al. 2008). This lack of diversity, coupled with the piscivorous perch's high position on the trophic ladder, would have made the abundance of these fisheries more seasonally and annually variable than that of the linear and diverse fisheries of the Angara and Kultuk Bay. The complex topography of the coastline would have resulted in an equally patchy distribution of HG groups, with travel between their homes requiring major inland detours. In contrast, the distribution of HG groups on the Angara and in Kultuk Bay was mostly linear, and travel and communication would not have been a problem, an advantage in organizing cooperative fishing. A practical solution to this problem in the Little Sea would have been effective watercraft technology, which was not as necessary on the Angara and in Kultuk Bay. Thus, that fishing in the Little Sea did not advance to the same intensification level might also be related to available watercraft technology. Lastly, the lack of rivers with substantial spawning runs in the Little Sea, as well as the annual freezing of the shallow coves from November to May, further lowered the utility of its fisheries. All these factors together would likely have made the Little Sea fisheries less amenable to intensification than they may appear at first glance.

Overall, the Kitoi strategy never collapsed on the Upper Lena and in the Little Sea, but only because it never formed there in the first place. The evidence suggests that in both areas the EN pattern, including mortuary practices (i.e., the Khin Group), continued from the LM times without much change. Hunting still relied mostly on the atlatl and spear, while fishing perhaps used some of the advanced techniques but never on a scale similar to that of the Angara and Kultuk Bay. Without much game and with fisheries inadequate for intensification, the impact of the bow on social and economic life was not as profound.

6.3 *Formation of the Middle Neolithic Pattern (~6660–6060 cal. BP)*

The MN period began with the breakdown of the Kitoi culture—**Transition 3**, which took place around the time of maximum forest expansion throughout Cis-Baikal (Table 9).

Our knowledge of the MN is limited by a few factors: (1) That the period has been defined only recently (Weber et al. 2002), substantially postdating most archaeological fieldwork in the region; (2) The absence of bioarchaeological data due to a complete lack of MN cemeteries; (3) The inadequate stratigraphic resolution of most camp-sites around Cis-Baikal; and (4) The fact that past attempts to correlate cultural strata at camp-sites with mortuary traditions failed to recognize that the MN lacked cemeteries and continued to employ Okladnikov's flawed culture history sequence.

A complete depopulation of Kitoi centers, as well as of the other microregions, is highly unlikely. That people were still living in the region, and on occasion still engaging with EN burial sites, is suggested by a small number of interments (e.g., at Shamanka II in Southwest Baikal and Verkholensk on the Upper Lena) (Weber et al. 2021; White et al. 2021), as well as by archaeological strata at camp-sites (e.g., Sagan-Zaba II in the Little Sea) (Nomokonova et al. 2013, 2015) with radiocarbon dates within MN boundaries. Based on the data at hand, the most likely scenario seems to have been a major socio-economic reshuffling, resulting in a rather substantial decrease in HG population density, directly related to the dispersal of much of the Kitoi population into the forest. New materials from well-stratified settlements, which are challenging to find, are needed to shed fresh light on this matter.

As the MN progressed, however, the region began to see a gradual reversal of the long climatic and environmental trends that characterized the preceding periods. The first third of the MN perhaps still saw some forest expansion but the trend halted as the climate stabilized towards the middle of the period and the final third saw forests retreat. These changes to forest cover in the first and final thirds of the MN were so gradual they would have been barely perceptible on a generational scale. MN groups were likely small, making ends meet in the forest and any remaining patches of open landscape with whatever resources available, and with the bow, proving its superiority over the atlatl and spear in an environment dominated by forest. Subsistence for these groups was based on some combination of terrestrial game, aquatic resources, and plant foods, as suggested by the fauna from camp-sites and the limited stable isotope results. Emphasizing residential mobility, they traveled a great deal, allocating much of their time to search for game dispersed across the taiga. Under such conditions, the intensification of fishing made little sense and formal cemeteries, having no purpose to serve, did not form. Also, many of the Kitoi technological novelties were rare, or wholly absent, in the MN package.

The transition to the MN pattern on the Angara and in Southwest Baikal was fast, needing perhaps only a few generations to complete. Elsewhere, however, the transition was probably more gradual. Differences between EN and MN environmental conditions on the Upper Lena and in the Little Sea were not substantial, requiring only minor adjustments to existing strategies or perhaps none at all. In these micro-regions the EN pattern, itself still fundamentally similar to that of the LM, may have continued well into MN times.

6.4 *Late Neolithic* (6060–4970 cal. BP) *and Early Bronze Age* (4970–3470 cal. BP)

The matters related to the history (formation, development, and termination) of the Late Neolithic and Early Bronze age cultural patterns and relevant transitions are beyond the scope of this paper. They are discussed in more detail in the original study (Weber 2020) demonstrating the unique character of each cultural pattern and transition even though some of the most important factors appear on the surface to be similar to those affecting the earlier developments. Yet, in each case the relevant configuration of environmental, social, and technological factors is somewhat different resulting eventually in the formation of quite different cultural patterns.

7 Summary

The Middle Holocene HG groups of Cis-Baikal obviously had no control over the long-term climatic and environmental changes affecting the region. Of these, the most important in terms of impacting the entire HG adaptive strategy, were changes in the distribution of boreal forest and the corresponding changes in the distribution of open landscape and its game resources. Although these HG groups might have had equally little say over which technological innovations were introduced into the region and when, they were in full command of whether or not to adopt them and how to adjust them to the local environmental and cultural settings or, alternatively, how to modify the existing socio-economic systems to make them work.

The bow-and-arrow was clearly the driving force behind the entire Kitoi pattern. Local HG groups had been "primed" for this technology for some time, as evidenced by LM developments, which imply that already back then people were trying to cope with expanding forests, population crowding, and diminishing returns from hunting, but were unable to solve these problems with available means. If the bow had arrived earlier, the Kitoi pattern likely would have formed earlier and may have also lasted longer. If it had arrived somewhat later, the Kitoi pattern would very likely have been short-lived and, perhaps, would not have had enough time to develop to the same degree. If the bow had appeared much later, when the forests had reached their maximum expansion or were already in retreat, the Kitoi probably would never have developed at all.

While the bow-and-arrow "formed" the Kitoi pattern, its spatial distribution across Cis-Baikal was controlled by the environment: the combined effects of the availability of open landscape with its large and medium game as well as access to fisheries that were suitable for intensification—the latter a practical use of the labour surpluses generated by bow hunting and a necessary response to the growing risks of hunting failure and diminishing returns. This is why the Kitoi pattern was confined to the upper section of the Angara and Southwest Baikal, the rest of the Angara and the Upper Lena probably already overgrown by forest, and the fishery of the latter never good enough. In the Little Sea, game was probably too thin and the fisheries did not meet all the conditions of intensification.

The bow also "wrote" the final chapter in the history of the Kitoi pattern, serving as a crucial factor in its dismantling. The bow allowed Kitoi groups to disband and disperse, the only viable solution under the changing environmental conditions, leading directly to the formation of the MN cultural pattern. The new pattern endured across Cis-Baikal for a long time and would likely have continued for much longer, if it were not for the change in climate which gradually reversed the forest trend from expansion to stabilization and, eventually, retreat.

Overall, the expansion and retreat of boreal forest was important because it directly affected the distribution and abundance of large and medium terrestrial game, the core of HG subsistence in Middle Holocene Cis-Baikal. The bow was important because its increased return rates provided a solution to the problems created by the expanding forest. Moreover, adoption of the bow as the dominant hunting technique was the most likely, and perhaps only, means of reorganizing food procurement in Cis-Baikal enough to set the intensification of fishing in motion (Bettinger 2015, 44). These two processes were the main factors in the formation of the Kitoi pattern. While the bow subsequently created a series of new problems for these people, at the same time it offered a solution: dispersal into the expanding forest, which would not have been equally successful with the atlatl or spear. The bow made life on the taiga viable until the retreating forests opened enough landscape for people to start leaving the woods. And again, it was the bow that made this a relatively smooth transition. Obviously, the bow was crucial not only because of its technological superiority in game hunting but also because of its flexibility: working equally well for HG groups of any size and in any environment.

Fishing and fisheries were important because they worked as a differentiating force: the more intensive the fishing, the more limited its spatial distribution and the greater the microregional differences between cultural patterns. Thus, intensive fishing and game hunting pulled the adaptive strategies in two opposing directions: fishing towards differences and hunting towards similarities between groups and microregions.

Lastly, social relations were important because they provided something that none of the other factors could: finetuning the social fabric that helped optimize operation of the general strategy. It is clear that the social organization changed not only change a few times from the Late Mesolithic to the Middle Neolithic but it also varied spatially, particularly during the Early Neolithic times.

It is possible that the EN Kitoi people, facing mounting economic difficulties, attempted to reform their social structure but any such efforts were unsuccessful under the overwhelming pressure of encroaching forests and diminishing returns from hunting. It is also possible that no new social solutions were explored because there was not enough time and the pressures were too strong. Phase 2 at Shamanka II, for example, suggests a return to the same pattern that forced the abandonment of the area not long before.

8 Conclusion

This examination provides a general account of the evolution of Late Mesolithic to Early Neolithic HGs in Cis-Baikal, emphasising internal factors and developments on a regional scale. The remaining part of these processes, i.e., from the Middle Neolithic to Early Bronze Age, are described in the full version of this paper (Weber 2020). This second part of the Middle Holocene evolutionary sequence is interesting because it shows how essentially the same factors (changes in the distribution of the boreal forest, food resources, technology, balancing game hunting with fishing, social organization etc.) resulted at each major junction in a different configuration of cultural patterns and their distribution across the entire Cis-Baikal. In other words, the rather subtle differences between these factors individually, each time resulted together in a different HG cultural landscape.

This account highlights also many microregional and local differences and details, though their explanation is beyond the scope of this assessment and, in many instances, the empirical data are still insufficient to address these matters. For example, from the analysis of the currently available set of 560 radiocarbon dates (Weber et al. 2021), it is already evident that the regional scale masks differences between microregions in the timing and duration of cultural developments. Logically, any microregional offsets in the timing of a given phenomenon would make its duration longer on a regional scale. Whether or not these differences are an artifact of several still relatively small samples is not yet clear (Bronk Ramsey et al. 2021; Weber et al. 2021).

Certainly, there are other equally pressing inquiries. First is the history of boreal forest distribution within each archaeological microregion as not enough is currently known about this matter. Next is the spatio-temporal variation in the genetic make-up of these groups for, again, little has been done so far and the data currently available offer ambiguous insights. The potential for this rapidly growing approach to provide a range of novel information is very strong. About 400 individuals with sufficient dental remains representing most microregional HG group from Late

Mesolithic to Early Bronze Age have been submitted for genomic research using the most recent generation of methodological advances. This work is still in progress but results are expected soon. Lastly, many new details of social relations can be revealed through a systematic assessment of the large body of mortuary data.

Acknowledgments The research reported in this paper has been part of the Baikal Archaeology Project and Baikal–Hokkaido Archaeology Project supported by grants from the Social Sciences and Humanities Research Council of Canada: Major Collaborative Research Initiative Nos. 410-2000-1000, 412-2005-1004, and 412-2011-1001; and Partnership Grant No. 895-2018-1004 as well as matching funding provided by the University of Alberta and other participating partner organizations. Additional funding was provided by the Government of the Russian Federation, grant No. 075-15-2019-866 ("Baikal Siberia in the Stone Age: At the crossroads of the worlds"). Many thanks to R.L. Bettinger for his comments on the Background and Approach section and to E. Jessup for help with the maps. I am particularly grateful to Professor Dr. W. Rączkowski for inviting me to contribute to this volume and for his long and most dependable friendship since our graduate student years at the Adam Mickiewicz University, Poznań, Poland.

References

Aseev, I. V. 2003. *Iugo-Vostochnaia Sibiri' v epokhu kamnia i metalla*. Novosibirsk: Izdatel'stvo Instituta Arkheologii i Etnografii, Sibirskoe otdelenie, Rossiiskaia akademiia nauk.

Bailey, H. P. 1960. A method of determining the warmth and temperateness of climate. *Geografiska Annaler* 42 (1):1–16.

Bazaliiskii, V. I. 2010. Mesolithic and Neolithic mortuary complexes in the Baikal region of Siberia. In *Prehistoric hunter–gatherers of the Baikal region, Siberia: Bioarchaeological studies of past lifeways*, ed. A. W. Weber, M. A. Katzenberg, and T. Schurr, 51–86. Philadelphia: University of Pennsylvania Press.

Bazaliiskii, V. I., S. A. Peskov, A. A. Shchetnikov, and A. A. Tiutrin. 2016. Ranneneoliticheskii mogil'nik Moty-Novaia Shamanka v doline r. Irkut. *Izvestiia Irkutskogo gosudarstvennogo universiteta. Seriia "Geoarkheologiia, Etnologiia, Antropologiia"* 18:40–72.

Bettinger, R. L. 1991. *Hunter–gatherers: Archaeological and evolutionary theory*. New York: Plenum Press.

Bettinger, R. L. 2013. Effects of the bow on social organization in Western North America. *Evolutionary Anthropology* 22:118–123.

Bettinger, R. L. 2015. *Orderly anarchy: Sociopolitical evolution in aboriginal California*. Oakland: University of California Press.

Bettinger, R. L., R. Garvey, and S. Tushingham. 2015. *Hunter-gatherers: Archaeological and evolutionary theory*. New York: Springer.

Bezrukova, E. V., S. Hildebrandt, P. P. Letunova, E. V. Ivanov, L. A. Orlova, S. Müller, and P. E. Tarasov. 2013. Vegetation dynamics around Lake Baikal since the middle Holocene reconstructed from the pollen and botanical composition analyses of peat sediments: Implications for paleoclimatic and archeological research. *Quaternary International* 290–291:35–45.

Bezrukova, E. V., A. V. Belov, P. P. Letunova, and N. V. Kulagina. 2014. The response of the environment of the Angara-Lena Plateau to global climate change in the Holocene. *Russian Geology and Geophysics* 55:463–471.

Bondetti, M., A. Lucquin, N. A. Savel'ev, A. W. Weber, O. Craig, and P. Jordan. 2020. Resource-processing, early pottery and the emergence of Kitoi culture in Cis-Baikal: Insights from lipid residue analysis of an Early Neolithic ceramic assemblage from the Gorelyi Les habitation site, Eastern Siberia. *Archaeological Research in Asia* 29:100324.

Breken, N. S. 1966. (Ed.), *Spravochnik po klimatu SSSR. Vypusk 22. Irkutskaia oblast' i zapadnaia chast' Buriatskoi ASSR. Chast' II. Temperatura vozdukha i pochvy*. Leningrad: Gidrometeorologicheskoe izdatel'stvo.

Breken, N. S. 1968. (Ed.), *Spravochnik po klimatu SSSR. Vypusk 22. Irkutskaia oblast' i zapadnaia chast' Buriatskoi ASSR. Chast' IV. Vlazhnost' vozdukha, atmosfernye osadki i snezhnyi pokrov*. Leningrad: Gidrometeorologicheskoe izdatel'stvo.

Bronk Ramsey, C., R. J. Schulting, V. I. Bazaliiskii, O. I. Goriunova, and A. W. Weber. 2021. Spatio-temporal patterns of cemetery use among Middle Holocene hunter-gatherers of Cis-Baikal, Eastern Siberia. *Archaeological Research in Asia* 25:100253.

Fitzhugh, B. 2003. *The evolution of complex hunter-gatherers*. Boston: Springer.

Fraser-Shapiro, I. 2012. *Studying hunter–gatherer mobility using isotopic and trace elemental analysis*. Unpublished Ph.D. thesis, University of Alberta, Edmonton.

Fretwell, S. D. 1972. *Populations in a seasonal environment*. Princeton: Princeton University Press.

Goriunova, O. I., A. G. Novikov, G. V. Turkin, and A. W. Weber. 2021. Middle Holocene hunter–gatherer mortuary practices in the Little Sea micro-region on Lake Baikal, Part I: Early Neolithic. *Archaeological Research in Asia* 26:100224.

Kelly, R. L. 1995. *The foraging spectrum: Diversity in hunter-gatherer lifeways*. Washington, DC: Smithsonian Institution Press.

Kelly, R. L. 2013. *The lifeways of hunter–gatherers: The foraging spectrum*. Cambridge: Cambridge University Press.

Kennett, D. J. 2005. *The island Chumash: Behavioral ecology of a maritime society*. University of California Press.

Kennett, D.J., Winterhalder, B. (Eds.), 2006. Behavioral Ecology and the Transition to Agriculture. University of California Press, Berkeley

Khotinskii, N. A. 1984a. Holocene climatic changes. In *Late quaternary environments of the Soviet Union*, ed. A. A. Velichko, 305–312. Minneapolis: University of Minnesota Press.

Khotinskii, N. A. 1984b. Holocene vegetation history. In *Late quaternary environments of the Soviet Union*, ed. A. A. Velichko, 179–200. Minneapolis: University of Minnesota Press.

Kobe, F., E. V. Bezrukova, C. Leipe, A. A. Shchetnikov, T. Goslar, M. Wagner, and P. E. Tarasov. 2020. Holocene vegetation and climate history in Baikal Siberia reconstructed from pollen records and its implications for archaeology. *Archaeological Research in Asia* 23:100209.

Konopatskii, A. K. 1982. *Drevnie kul'tury Baikala (o. Ol'khon)*. Novosibirsk: Nauka.

Kozhov, M. M. 1950. *Presnye vody Vostochnoi Sibiri (bassein Baikala, Angary, Vitima, verkhnego techenia Leny i Nizhnei Tunguski)*. Irkutsk: Irkutskoe oblastnoe gosudarstvennoe izdatel'stvo.

Kozhov, M. M. 1963. *Lake Baikal and its life*. The Hague: W. Junk.

Kozhov, M. M. 1972. *Ocherki po baikalovedeniiu*. Irkutsk: Vostochno-Sibirskoe knizhnoe izdatel'stvo.

Krebs, J. R., and N. B. Davies. 1981. *An introduction to behavioural ecology*. Oxford: Blackwell.

Krebs, J. R., and N. B. Davies. 1991. (Eds.), *Behavioural ecology: An evolutionary approach*. 3rd ed. Oxford: Blackwell.

Lewontin, R. C. 1970. The units of selection. *ARES* 1:1–18.

Lindström, S. 1996. Great Basin fisherfolk: Optimal diet breadth modeling the Truckee River aboriginal subsistence fishery. In *Prehistoric hunter–gatherer fishing strategies*, ed. M. G. Plew, 114–179. Boise: Boise State University.

Losey, R. J., and T. Nomokonova. 2017a. Background, methods, and questions posed. In *Holocene Zooarchaeology of Cis-Baikal. Archaeology in China and East Asia*, vol. 6, ed. R. J. Losey and T. Nomokonova, 5–14. Mainz: Nünnerich–Asmus Verlag & Media GmbH.

Losey, R. J., and T. Nomokonova. 2017b. Discussion and conclusions. In *Holocene Zooarchaeology of Cis-Baikal. Archaeology in China and East Asia*, ed. R. J. Losey and T. Nomokonova, vol. 6, 121–128. Mainz: Nünnerich–Asmus Verlag & Media GmbH.

Losey, R. J., T. Nomokonova, and O. I. Goriunova. 2008. Fishing ancient Lake Baikal, Siberia: Inferences from the reconstruction of harvested perch (Perca fluviatilis) size. *Journal of Archaeological Science* 35:577–590.

Losey, R. J., T. Nomokonova, and D. White. 2012. Fish and fishing in Holocene Cis-Baikal, Siberia: A review. *Journal of Island and Coastal Archaeology* 7:126–145.
Mamonova, N. N., and L. D. Sulerzhitskii. 1989. Opyt datirovaniya po 14C pogrebenii Pribaikal'ia epokhi golotsena. *Sovetskaia arkheologiia* 1:19–32.
Maynard Smith, J. 1982. *Evolution and the theory of games*. Cambridge: Cambridge University Press.
Mayr, E. 1959. Typological versus population thinking. In *Evolution and anthropology: A centennial appraisal*, ed. B. Meggers, 409–412. Washington, DC: Washington Anthropological Society.
Mayr, E. 1976. *Evolution and the diversity of life*. Cambridge, MA: Harvard University Press.
Mayr, E. 1982. *The growth of biological thought: Diversity, evolution, and inheritance*. Cambridge, MA: Harvard University Press.
Mayr, E. 1983. How to carry out the adaptationist program? *The American Naturalist* 121 (3):324–334.
Mayr, E. 1997. *What evolution is*. New York: Basic Books.
Mayr, E., and W. B. Provine. 1980. *The evolutionary synthesis: Perspectives on the unification of biology*. Cambridge, MA: Harvard University Press.
McKenzie, H.G. 2006. *Mortuary variability among middle Holocene hunter–gatherers in the Lake Baikal region of Siberia, Russia*. Unpublished Ph.D. thesis, University of Alberta, Edmonton.
Nomokonova, T. 2011. *Holocene sealing and pastoralism at Sagan-Zaba Cove, Siberia*. Unpublished Ph.D. thesis, University of Alberta, Edmonton.
Nomokonova, T., Losey, R. J., Goriunova, O. I., Weber, A. W., 2013. A freshwater oldcarbon offset in Lake Baikal, Siberia, and problems with the radiocarbon dating of archaeological sediments: evidence from the Sagan-Zaba II site. Quat. Int. 290–291
Nomokonova, T., Losey, R. J., Goriunova, O. I., Novikov, A. G., Weber, A. W., 2015. A 9,000 year history of seal hunting on Lake Baikal, Siberia: The zooarchaeology of Sagan-Zaba II. PLoS One 10 (5), e0125314
O'Brien, M. J., and R. L. Lyman. 2000. *Applying evolutionary archaeology: A systematic approach*. New York: Kluwer Academic.
Okladnikov, A. P. 1950. *Neolit i bronzovyi vek Pribaikal'ia (chast' I i II)*. Moscow: Izdatel'stvo Akademii nauk SSSR.
Okladnikov, A. P. 1955. *Neolit i bronzovyi vek Pribaikal'ia (chast' III)*. Moscow: Izdatel'stvo Akademii nauk SSSR.
Okladnikov, A. P. 1974. *Neoliticheskie pamiatniki Angary (ot Shchukino do Bureti)*. Novosibirsk: Nauka.
Pianka, E. R. 1981. *Evolutionary ecology*. New York: Harper & Row.
Prentiss, A. M., N. Lyons, L. E. Harris, M. R. P. Burns, and T. M. Godin. 2007. The emergence of status inequality in intermediate scale societies: A demographic and socio-economic history of the Keatley Creek Site, British Columbia. *Journal of Anthropological Archaeology* 26:299–327.
Prentiss, A. M., H. S. Cail, and L. M. Smith. 2014. At the Malthusian ceiling: Subsistence and inequality at Bridge River, British Columbia. *Journal of Anthropological Archaeology* 33:34–48.
Scharlotta, I., V. I. Bazaliiskii, A. R. Lieverse, and A. W. Weber. 2016. Social consequences of increased reliance on fishing in middle Holocene Cis-Baikal: Relating fishing gear, axes, and social status at the Shamanka II cemetery, Lake Baikal, Siberia. *Quaternary International* 419 (C):99–132.
Schulting, R. J., C. Bronk Ramsey, I. Scharlotta, M. Richards, and A. W. Weber. 2022. Freshwater reservoir effects in cis-Baikal: An overview. *Archaeological Research in Asia* 29:100324.
Shvetsov, I. G., M. N. Smirnov, and G. I. Monakhov. 1984. *Mlekopitaiushchie basseina ozera Baikal*. Novosibirsk: Nauka.
Smith, E. A., and B. Winterhalder. 1992. (Eds.), *Evolutionary ecology and human behavior*. New York: Aldine de Gruyter.

Stephens, D. W., and J. R. Krebs. 1986. *Foraging theory*. Princeton: Princeton University Press.

Sugden, R. 1986. *The economics of rights, co-operation and welfare*. Oxford: Basil Blackwell.

Tarasov, P. E., E. V. Bezrukova, S. Müller, S. S. Kostrova, and D. White. 2015. Climate and vegetation history. In *Holocene zooarchaeology of Cis-Baikal. Archaeology in China and East Asia*, vol. 6, ed. R. J. Losey and T. Nomokonova, 15–26. Mainz: Nünnerich–Asmus Verlag & Media GmbH.

Teltser, P. A. 1994. (Ed.), *Evolutionary archaeology: Methodological issues*. Tucson: University of Arizona Press.

Tushingham, S., and R. L. Bettinger. 2013. Why foragers choose acorns before salmon: Storage, mobility, and risk in aboriginal California. *Journal of Anthropological Archaeology* 32:527–537.

Tushingham, S., J. P. Eidsness, T. Fulkerson, J. Hopt, C. Christiansen, A. Arpaia, and J. Chang. 2016. Late Holocene coastal intensification, mass harvest fishing, and the historical ecology of marine estuaries: The view from the Manila Site (CA-HUM-321), Humboldt Bay, Northwestern Alta California. *California Archaeology* 8 (1):1–35.

Vetrov, V. M., N. E. Berdnikova, V. V. Alukhov, and A. V. Frolov. 1995. Makrushinskii mogil'nik. Renneneoliticheskii kompleks. In *Baikal'skaia Sibir' v drevnosti*, ed. V. M. Vetrov, 112–132. Irkutsk: Irkutskii gosudarstvennyi universitet.

Walker, M., S. Johnsen, S. O. Rasmussen, T. Popp, J. P. Steffensen, P. Gibbard, W. Hoek, J. Lowe, J. Andrews, S. Björck, L. C. Cwynar, K. Hughen, P. Kershaw, B. Kromer, T. Litt, D. J. Lowe, T. Nakagawa, R. Newnham, and J. Schwander. 2009. Formal definition and dating of the GSSP (Global Stratotype Section and Point) for the base of the Holocene using the Greenland NGRIP ice core, and selected auxiliary records. *Journal of Quaternary Science* 24 (1):3–17.

Walker, M. J. C., M. Berkelhammer, S. Björck, L. C. Cwynar, D. A. Fisher, A. J. Long, J. J. Lowe, R. M. Newnham, S. O. Rasmussen, and H. Weiss. 2012. Formal subdivision of the Holocene series/epoch: A discussion paper by a Working Group of INTIMATE (Integration of ice-core, marine and terrestrial records) and the Subcommission on Quaternary Stratigraphy (International Commission on Stratigraphy). *Journal of Quaternary Science* 27 (7):649–659.

Weber, A. 1988. Idealizacyjny model procesu neolityzacji. In *Szkice z antropologii ogólnej*, ed. J. Piontek, 43–62. Poznań: Wydawnictwo Naukowe UAM.

Weber, A. 1990. Theory of Neolithisation: An introductory model. *Hemispheres* V:193–214.

Weber, A. W. 1995. The Neolithic and Early Bronze Age of the Lake Baikal Region, Siberia: A review of recent research. *Journal of World Prehistory* 9:99–165.

Weber, A. W. 2003. Bio-geographic profile of the Lake Baikal region, Siberia. In *Prehistoric foragers of the Cis-Baikal, Siberia. Proceedings of the first conference of the Baikal Archaeology Project*, ed. A. W. Weber and H. G. McKenzie, 51–66. Edmonton: Canadian Circumpolar Institute Press.

Weber, A. W. 2020. Middle Holocene hunter–gatherers of Cis-Baikal, Eastern Siberia: Combined impacts of the boreal forest, bow-and-arrow, and fishing. *Archaeological Research in Asia* 24:100222.

Weber, A. W., and R. L. Bettinger. 2010. Middle Holocene hunter–gatherers of Cis-Baikal, Siberia: An overview for the new century. *Journal of Anthropological Archaeology* 29:491–506.

Weber, A. W., D. W. Link, and M. A. Katzenberg. 2002. Hunter–gatherer culture change and continuity in the Middle Holocene Cis-Baikal, Siberia. *Journal of Anthropological Archaeology* 21:230–299.

Weber, A. W., R. P. Beukens, V. I. Bazaliiskii, O. I. Goriunova, and N. A. Savel'ev. 2006. Radiocarbon dates from Neolithic and Bronze Age hunter–gatherer cemeteries in the Cis-Baikal region of Siberia. *Radiocarbon* 48 (1):1–40.

Weber, A. W., C. Bronk Ramsey, R. J. Schulting, V. I. Bazaliiskii, and O. I. Goriunova. 2021. Middle Holocene hunter–gatherers of Cis-Baikal, Eastern Siberia: Chronology and dietary trends. *Archaeological Research in Asia* 25:100234.

White, D. 2006. *Holocene climate and culture change in the Lake Baikal region, Siberia*. Unpublished Ph.D. thesis, University of Alberta, Edmonton.

White, D., and A. B. G. Bush. 2010. Holocene climate, environmental variability and Neolithic biocultural discontinuity in the Lake Baikal region. In *Prehistoric hunter–gatherers of the Baikal region, Siberia: Bioarchaeological studies of past lifeways*, ed. A. W. Weber, M. A. Katzenberg, and T. Schurr, 1–26. Philadelphia: University of Pennsylvania Press.

White, J. A., R. J. Shulting, P. Hommel, V. Moiseyev, V. Khartanovich, C. Bronk Ramsey, and A. W. Weber. 2021. Turning eastward: New radiocarbon and stable isotopic data for Middle Holocene hunter-gatherers from Fofanovo, Trans-Baikal, Siberia. *Archaeological Research in Asia* 28:100323.

Winterhalder, B., and C. Goland. 1993. On population, foraging efficiency, and plant domestication. *Current Anthropology* 34:710–715.

Winterhalder, B., and A. E. Smith. 1981. (Eds.), *Hunter–gatherer foraging strategies: Ethnographic and archaeological analyses*. Chicago: University of Chicago Press.

Winterhalder, B., and E. A. Smith. 1992. (Eds.), *Hunter–gatherer foraging strategies: Ethnographic and archaeological analyses*. Chicago: University of Chicago Press.

Woodburn, J. 1982. Egalitarian societies. *Man* 17:431–451.

Zvelebil, M., and A. W. Weber. 2013. Human bioarchaeology: Group identity and individual life histories – Introduction. *Journal of Anthropological Archaeology* 32 (6):275–279.

GPSR Compliance

The European Union's (EU) General Product Safety Regulation (GPSR) is a set of rules that requires consumer products to be safe and our obligations to ensure this.

If you have any concerns about our products, you can contact us on ProductSafety@springernature.com

In case Publisher is established outside the EU, the EU authorized representative is:

Springer Nature Customer Service Center GmbH
Europaplatz 3
69115 Heidelberg, Germany

Batch number: 09457520

Printed by Printforce, the Netherlands